T0291627

LONDON MATHEMATICAL SOCIETY LECTURE NOTE SERIES

Managing Editor: Professor M. Reid, Mathematics Institute,
University of Warwick, Coventry CV4 7AL, United Kingdom

The titles below are available from booksellers, or from Cambridge University Press at
http://www.cambridge.org/mathematics

London Mathematical Society Lecture Note Series: 426

Lectures on the Theory of Water Waves

Edited by

THOMAS J. BRIDGES
University of Surrey

MARK D. GROVES
Universität des Saarlandes, Saarbrücken, Germany

DAVID P. NICHOLLS
University of Illinois, Chicago

Shaftesbury Road, Cambridge CB2 8EA, United Kingdom

One Liberty Plaza, 20th Floor, New York, NY 10006, USA

477 Williamstown Road, Port Melbourne, VIC 3207, Australia

314–321, 3rd Floor, Plot 3, Splendor Forum, Jasola District Centre, New Delhi – 110025, India

103 Penang Road, #05–06/07, Visioncrest Commercial, Singapore 238467

Cambridge University Press is part of Cambridge University Press & Assessment, a department of the University of Cambridge.

We share the University's mission to contribute to society through the pursuit of education, learning and research at the highest international levels of excellence.

www.cambridge.org
Information on this title: www.cambridge.org/9781107565562

© Cambridge University Press & Assessment 2016

First published 2016

A catalogue record for this publication is available from the British Library

Library of Congress Cataloging-in-Publication data
Names: Bridges, Thomas J., editor. | Groves, Mark D., editor. |
Nicholls, David P., editor.
Title: Lectures on the theory of water waves / edited by Thomas J. Bridges (University of Surrey), Mark D. Groves (Universität des Saarlandes, Saarbrücken, Germany) and David P. Nicholls (University of Illinois, Chicago).
Other titles: London Mathematical Society lecture note series; 426.
Description: New York, NY: Cambridge University Press, 2016. | © 2016 |
Series: London Mathematical Society lecture note series; 426 |
Includes bibliographical references
Identifiers: LCCN 2015040740 | ISBN 9781107565562 (pbk.) |
ISBN 1107565561 (pbk.)
Subjects: LCSH: Wave-motion, Theory of. | Water waves.
Classification: LCC QC174.26.W28 L43 2016 | DDC 532/.593–dc23
LC record available at http://lccn.loc.gov/2015040740

ISBN   978-1-107-56556-2   Paperback

Contents

Contributors

David P. Nicholls
Department of Mathematics, Statistics, and Computer Science, University of Illinois at Chicago, Chicago, IL 60607, U.S.A.

Benjamin F. Akers
Air Force Institute of Technology, 2950 Hobson Way, WPAFB, OH 45433, U.S.A.

Athanassios S. Fokas
Department of Applied Mathematics and Theoretical Physics (DAMTP), University of Cambridge, Cambridge, CB3 0WA, UK

Konstantinos Kalimeris
Research Center of Mathematics, Academy of Athens, Athens 11527, Greece & Johann Radon Institute for Computational and Applied Mathematics (RICAM), Austrian Academy of Sciences, Altenbergerstrasse 69, 4040 Linz, Austria

Mark D. Groves
Fachrichtung 6.1 - Mathematik, Universität des Saarlandes, Postfach 151150, 66041 Saarbrücken, Germany & Department of Mathematical Sciences, Loughborough University, Loughborough, LE11 3TU, UK

Guido Schneider
Institut für Analysis, Dynamik und Modellierung, Universität Stuttgart, Pfaffenwaldring 57, 70569 Stuttgart, Germany

David M. Ambrose
Department of Mathematics, Drexel University, Philadelphia, PA 19104, U.S.A.

Sijue Wu
Department of Mathematics, University of Michigan, Ann Arbor, MI 48109, U.S.A.

André Nachbin
Instituto Nacional de Matemática Pura e Aplicada, Est. D. Castorina 110, Jardim Botânico, Rio de Janeiro, RJ 22460-320, Brazil

Onno Bokhove
School of Mathematics, University of Leeds, Leeds, LS2 9JT, UK

Anna Kalogirou
School of Mathematics, University of Leeds, Leeds, LS2 9JT, UK

Thomas J. Bridges
Department of Mathematics, University of Surrey, Guildford, Surrey, GU2 7XH, UK

Preface

Waves on the surface of the ocean are a dramatic and beautiful phenomena that impact every aspect of life on the planet. At small length scales, ripples driven by surface tension on the surface of these "water waves" affect remote sensing of surface and underwater obstacles. At intermediate scales, waves on the surface and the interface between internal layers of water of differing densities affect shipping, coastal morphology, and near-shore navigation. At larger lengths, tsunamis and hurricane-generated waves can cause devastation on a global scale. Additionally, water waves play a crucial role at all length scales in the exchange of momentum and thermal energy between the ocean and atmosphere that, in turn, affect the global weather system and climate.

From a mathematical viewpoint, the water wave equations pose severe challenges for rigorous analysis, modelling, and numerical simulation. The governing equations are widely accepted and there has been substantial research into their validity. However, a rigorous theory of their solutions is extremely complex due not only to the fact that the water wave problem is a classical free boundary problem, where the domain shape is unknown, but also because the boundary conditions are strongly nonlinear. The level of difficulty is such that the theory has merely begun to answer the fundamental questions that must be addressed before our understanding can be considered "adequate." For instance, it is well known that the very existence of solutions to the equations that describe fluid motion, even in the absence of free boundaries, is one of the most difficult unanswered questions in mathematics (indeed it is one of the Clay Mathematics Institute Millennium Prize Problems).

In July and August of 2014 a four-week programme on the theory of water waves was convened at the Isaac Newton Institute for Mathematical Sciences in Cambridge,[a] with over 50 leading researchers in the theory of water waves from fifteen countries across the globe. The aim was to share recent research ideas and identify strategies for future directions. The key

---

[a] https://www.newton.ac.uk/event/tww/

themes of the conference were (a) the initial value problem (well-posedness, singularities, numerical methods, simulations), (b) existence of classes of waves (travelling waves, standing waves, multi-periodic waves, solitary waves, patterns, waves with vorticity and viscosity), (c) stability of waves (analytical and numerical methods, wave interactions, rogue waves, energetic stability), and (d) dynamical systems approaches (variational principles, modulation, centre-manifolds, spatial dynamics, bottom topography).

Interspersed among the various discussions, seminars, workshops, and other events, were lectures of an introductory and tutorial nature. The purpose of this volume is to compile a summary of those introductory lectures.

In week one the key theme was numerics. There were introductory talks by Benjamin Akers and David Nicholls on the theory and numerics of High-Order Perturbation of Surfaces methods, and a series of introductory talks on the Unified Transform Method and its application to water waves by Athanassios S. Fokas. All three of these speakers have provided chapters for this volume summarizing their lectures. In week two a London Mathematical Society Spitalfields event was held with four talks on rigorous analysis of nonlinear waves by Mark Groves, Guido Schneider, Steve Shkoller and Eugen Varvaruca, with the first two presenters providing chapters for this volume including an introduction to their talks. In week four a summer school on the theory of water waves was held, where early career researchers from the UK and overseas were invited to attend, with funding provided for the attending participants by the Newton Institute and the Office of Naval Research Global. The summer school had a cross section of talks including rigorous analysis of the initial value problem, numerical analysis of water waves, variational principles, shallow water hydrodynamics, the role of deterministic and random bottom topography, and modulation of nonlinear waves. The lecturers were David Ambrose, Onno Bokhove, Thomas Bridges, André Nachbin, and Sijue Wu, and they have all written up their lectures which are included as chapters herein.

The editors are grateful to the Newton Institute and the Director, John Toland, for supporting and facilitating the water waves programme. Special thanks are due to Christine West for excellent support leading up to and during the programme, and to Almarie Williams for her support during the planning and running of the Summer School.

Overall, the Newton programme covered a very wide range of material and provided a catalyst for future research. We hope this volume will contribute to the momentum that emerged from the programme and stimulate further interest in the fascinating combination of mathematics and water waves.

1

High-Order Perturbation of Surfaces Short Course: Boundary Value Problems

David P. Nicholls

Abstract

In this lecture we introduce two classical High-Order Perturbation of Surfaces (HOPS) computational schemes in the simplified context of elliptic boundary value problems inspired by models in water waves. For the problem of computing Dirichlet–Neumann Operators (DNOs) for Laplace's equation, we outline Bruno & Reitich's method of Field Expansions (FE) and then describe Milder and Craig & Sulem's method of Operator Expansions (OE). We further show how these algorithms can be extended to three dimensions and finite depth, and describe how Padé approximation can be used as a method of numerical analytic continuation to realize enhanced performance and applicability through a series of numerical experiments.

1.1 Introduction

Calculus in general, and Partial Differential Equations (PDEs) in particular have long been recognized as the most powerful and successful mathematical modeling tool for engineering and science, and the study of surface water waves is no exception. With the advent of the modern computer in the 1950s, the possibility of numerical simulation of PDEs at last became a practical reality. The last 50–60 years has seen an explosion in the development and implementation of algorithms for this purpose, which are rapid, robust, and highly accurate. Among the myriad choices are:

1. Finite Difference methods (e.g., [1–4]),
2. Finite Element methods (Continuous and Discontinuous) (e.g., [5–8]),
3. High-Order Spectral (Element) methods (e.g., [9–14]),
4. Boundary Integral/Element methods (e.g., [15, 16]).

The class of High-Order Perturbation of Surfaces (HOPS) methods we describe here are a High-Order Spectral method, which is particularly well suited for PDEs posed on *piecewise homogeneous domains*. Such "layered media" problems abound in the sciences, e.g., in

- free-surface fluid mechanics (e.g., the water wave problem),
- acoustic waves in piecewise constant density media,
- electromagnetic waves interacting with grating structures,
- elastic waves in sediment layers.

For such problems these HOPS methods can be

- *highly accurate* (error decaying *exponentially* as the number of degrees of freedom increases),
- *rapid* (an order of magnitude fewer unknowns as compared with volumetric formulations),
- *robust* (delivering accurate results for rather rough/large interface shapes).

However, these HOPS schemes are *not* competitive for problems with inhomogeneous domains and/or "extreme" geometries.

In this lecture we discuss two classical HOPS methods for the solution of such interfacial problems: Bruno & Reitich's Field Expansions (FE) method [17–24], and Milder and Craig & Sulem's Operator Expansions (OE) method [25–31]. In a future lecture we discuss a stabilized version of the FE method (the Transformed Field Expansions – TFE–method) due to the author and Reitich [32–34]. In addition to specifying the details of these two algorithms (FE and OE) for a particular problem that arises in the study of water waves, we also want to illustrate the accuracy, efficiency, speed, and ease of implementation of HOPS schemes.

The rest of the lecture is organized as follows. In § 1.2 we recall the classical water wave problem and how the Dirichlet–Neumann Operator (DNO) arises as a fundamental object of study. In § 1.3 and § 1.4 we give the details of the Field Expansions and Operator Expansions methods, respectively, as applied to the problem of simulating the DNO. In § 1.5 we present results of numerical simulations realized with a simple MATLAB implementation of these recursions. In § 1.6 we discuss generalization of these algorithms to three dimensions and finite depth. We close with a presentation of the Padé approximation approach in § 1.7 to analytic continuation for these problems, and the extremely beneficial effect this methodology can have on these HOPS methods.

1.2 Water Waves and the Dirichlet–Neumann Operator

To fix on a problem we consider a classical water wave problem [35] which models the evolution of the free surface of a deep, two-dimensional, ideal fluid under the influence of gravity. The widely accepted model [35] is

$$\Delta\varphi = 0 \qquad\qquad y < \eta(x,t),$$
$$\partial_y\varphi \to 0 \qquad\qquad y \to -\infty,$$
$$\partial_t\eta + \partial_x\eta(\partial_x\varphi) = \partial_y\varphi, \qquad\qquad y = \eta(x,t),$$
$$\partial_t\varphi + (1/2)\nabla\varphi \cdot \nabla\varphi + \tilde{g}\eta = 0 \qquad y = \eta(x,t).$$

In these $\varphi(x,y,t)$ is the velocity potential ($\vec{u} = \nabla\varphi$), $\eta(x,t)$ is the air–water interface, and $\tilde{g}$ is the gravitational constant.

At the center of this problem is the solution of the elliptic Boundary Value Problem (BVP)

$$\Delta v = 0 \qquad y < g(x),$$
$$\partial_y v \to 0 \qquad y \to -\infty,$$
$$v = \xi \qquad y = g(x).$$

In particular, upon solving this problem, the *DNO*

$$G(g)[\xi] := \left[\partial_y v - (\partial_x g)\partial_x v\right]_{y=g(x)},$$

allows one to recast the water wave problem as [29, 36]

$$\partial_t\eta = G(\eta)\xi,$$
$$\partial_t\xi = -\tilde{g}\eta - A(\eta)B(\eta,\xi),$$

where

$$A = \left[2\left(1 + (\partial_x\eta)^2\right)\right]^{-1},$$
$$B = (\partial_x\xi)^2 - (G(\eta)\xi)^2 - 2(\partial_x\eta)(\partial_x\xi)(G(\eta)\xi).$$

For many problems of practical interest it suffices to consider the classical periodic boundary conditions, e.g.,

$$v(x+L,y) = v(x,y), \quad g(x+L) = g(x), \quad L = 2\pi,$$

which permits us to express functions in terms of their Fourier Series

$$g(x) = \sum_{p=-\infty}^{\infty} \hat{g}_p e^{ipx}, \quad \hat{g}_p = \frac{1}{2\pi}\int_0^{2\pi} g(s)e^{-ips}\,ds.$$

Thus, from here we focus on the BVP

$$\Delta v = 0 \qquad\qquad y < g(x), \qquad\qquad (1.2.1a)$$

$$\partial_y v \to 0 \qquad\qquad y \to -\infty, \qquad\qquad (1.2.1b)$$

$$v = \xi \qquad\qquad y = g(x), \qquad\qquad (1.2.1c)$$

$$v(x + 2\pi, y) = v(x, y), \qquad\qquad (1.2.1d)$$

and the DNO it generates.

1.3 The Method of Field Expansions

Our first HOPS approach for approximating DNOs solves the BVP, (1.2.1), directly. Its origins can be found in the work of Rayleigh [37] and Rice [38]. The first *high-order* implementation is due to Bruno & Reitich [18–20] and was originally denoted the "method of Variation of Boundaries." To prevent confusion with subsequent methods it was later renamed the method of Field Expansions (FE). The "key" to the method is the realization that interior to the domain (i.e., $y < -|g|_\infty$) the solution of Laplace's equation by separation of variables is

$$v(x, y) = \sum_{p=-\infty}^{\infty} a_p e^{|p|y} e^{ipx}. \qquad\qquad (1.3.1)$$

This HOPS approach uses the fact that, for a sufficiently smooth boundary perturbation $g(x) = \varepsilon f(x)$, the field, $v = v(x, y; \varepsilon)$, depends *analytically* upon ε.

Assume that the interface is shaped by $g(x) = \varepsilon f(x)$, where $f \sim \mathcal{O}(1)$ and, initially, $\varepsilon \ll 1$. We will be able to show *a posteriori* that v depends *analytically* upon ε so that

$$v = v(x, y; \varepsilon) = \sum_{n=0}^{\infty} v_n(x, y) \varepsilon^n.$$

Inserting this expansion into the governing equations, (1.2.1), and equating at orders $\mathcal{O}(\varepsilon^n)$ yields

$$\Delta v_n = 0 \qquad\qquad y < 0, \qquad\qquad (1.3.2a)$$

$$\partial_y v_n \to 0 \qquad\qquad y \to -\infty, \qquad\qquad (1.3.2b)$$

$$v_n = Q_n \qquad\qquad y = 0, \qquad\qquad (1.3.2c)$$

$$v_n(x + 2\pi, y) = v_n(x, y). \qquad\qquad (1.3.2d)$$

The crucial term is the boundary inhomogeneity

$$Q_n(x) = \delta_{n,0} \xi(x) - \sum_{m=0}^{n-1} F_{n-m}(x)\, \partial_y^{n-m} v_m(x, 0),$$

where $F_m(x) := \frac{f^m(x)}{m!}$ and $\delta_{n,m}$ is the Kronecker delta. This form comes from the expansion

$$v(x, \varepsilon f; \varepsilon) = \sum_{n=0}^{\infty} v_n(x, \varepsilon f)\varepsilon^n = \sum_{n=0}^{\infty} \varepsilon^n \sum_{m=0}^{n} F_{n-m}(x)\, \partial_y^{n-m} v_m(x, 0).$$

Bounded, periodic solutions of Laplace's equation can be expressed as

$$v_n(x, y) = \sum_{p=-\infty}^{\infty} a_{n,p} e^{|p|y} e^{ipx}. \tag{1.3.3}$$

Inserting this form into the surface boundary condition, (1.3.2c), delivers

$$\sum_{p=-\infty}^{\infty} a_{n,p} e^{ipx} = \sum_{p=-\infty}^{\infty} \hat{Q}_{n,p} e^{ipx},$$

where, since

$$e^{|p|\varepsilon f} = \sum_{m=0}^{\infty} \varepsilon^m F_m |p|^m,$$

we have

$$Q_n(x) = \delta_{n,0} \sum_{p=-\infty}^{\infty} \hat{\xi}_p e^{ipx} - \sum_{m=0}^{n-1} F_{n-m}(x) \sum_{p=-\infty}^{\infty} |p|^{n-m} a_{m,p} e^{ipx}.$$

Summarizing, we have the *FE Recursions*

$$a_{n,p} = \delta_{n,0}\hat{\xi}_p - \sum_{m=0}^{n-1} \sum_{q=-\infty}^{\infty} \hat{F}_{n-m,p-q} |q|^{n-m} a_{m,q}. \tag{1.3.4}$$

The FE recursions deliver the solution everywhere *well inside* the problem domain. However, two questions immediately arise: Is the expansion

$$v(x, y) = \sum_{p=-\infty}^{\infty} a_p e^{|p|y} e^{ipx}$$

valid at the *boundary*? Is this expansion valid *near* the boundary? For rigorous answers to these questions we refer to Bruno & Reitich's first contribution [17], the work of the author and Reitich [32–34], and the third lecture in this series.

Assuming for the moment that there is some validity at the boundary, recall that we wish to compute the Neumann data

$$\nu(x) = \left[\partial_y v - (\partial_x g)\partial_x v \right]_{y=g(x)}.$$

Expanding in ε

$$\sum_{n=0}^{\infty} \nu_n(x)\varepsilon^n = \sum_{n=0}^{\infty} \left[\partial_y v_n(x, \varepsilon f) - \varepsilon(\partial_x f)\partial_x v_n(x, \varepsilon f) \right] \varepsilon^n,$$

and equating at order $\mathcal{O}(\varepsilon^n)$ gives

$$v_n(x) = \sum_{m=0}^{n} F_{n-m} \partial_y^{n+1-m} v_m(x,0) - \sum_{m=0}^{n-1} (\partial_x f) F_{n-1-m} \partial_x \partial_y^{n-1-m} v_m(x,0).$$

At each wavenumber we have

$$\hat{v}_{n,p} = \sum_{m=0}^{n} \sum_{q=-\infty}^{\infty} \hat{F}_{n-m,p-q} |q|^{n+1-m} a_{m,q}$$

$$- \sum_{m=0}^{n-1} \sum_{q=-\infty}^{\infty} \hat{F}'_{n-1-m,p-q} (iq) |q|^{n-1-m} a_{m,q}, \qquad (1.3.5)$$

where $F'_m(x) := (\partial_x f) F_m(x)$. Together, formulas (1.3.4) and (1.3.5) can be implemented in a high-level computing language to deliver a fast and accurate method for simulating the action of the DNO, $G : \xi \to v$.

1.4 The Method of Operator Expansions

The second HOPS approach we investigate considers the DNO alone without explicit reference to the underlying field equations. For this reason the method has been termed the method of Operator Expansions (OE). The first *high-order* implementation for electromagnetics (the Helmholtz equation) is due to Milder [25, 26] and Milder & Sharp [27, 28]. The first *high-order* implementation for water waves (the Laplace equation) is due to Craig & Sulem [29]. Once again, we use, in a fundamental way, the representation, (1.3.1),

$$v(x,y) = \sum_{p=-\infty}^{\infty} a_p e^{|p|y} e^{ipx}.$$

This HOPS method uses the fact that, for a boundary perturbation $g(x) = \varepsilon f(x)$, the DNO, $G = G(\varepsilon f)$, depends *analytically* upon ε.

Again, assume that the interface is shaped by $g(x) = \varepsilon f(x)$, where $f \sim \mathcal{O}(1)$ and, initially, $\varepsilon \ll 1$. We now focus on the definition of the DNO, G,

$$G(g)[\xi] = v,$$

and seek the action of G on a *basis function*, $\exp(ipx)$. To achieve this we use a bounded, periodic solution of Laplace's equation

$$v_p(x,y) := e^{|p|y} e^{ipx}. \qquad (1.4.1)$$

Inserting the solution $v_p(x,y)$ into the definition of the DNO gives

$$G(g)[v_p(x,g(x))] = \left[\partial_y v_p - (\partial_x g) \partial_x v_p \right]_{y=g(x)}.$$

We assume that everything is *analytic* in ε and expand

$$\left(\sum_{n=0}^{\infty} \varepsilon^n G_n(f)\right)\left[\sum_{m=0}^{\infty} \varepsilon^m F_m |p|^m e^{ipx}\right] = \sum_{n=0}^{\infty} \varepsilon^n F_n |p|^{n+1} e^{ipx}$$

$$-\varepsilon(\partial_x f)\sum_{n=0}^{\infty} \varepsilon^n F_n (ip) |p|^n e^{ipx}.$$

At $\mathcal{O}\left(\varepsilon^0\right)$ this reads

$$G_0\left[e^{ipx}\right] = |p| e^{ipx},$$

so that we can conclude that

$$G_0[\xi] = G_0\left[\sum_{p=-\infty}^{\infty} \hat{\xi}_p e^{ipx}\right] = \sum_{p=-\infty}^{\infty} \hat{\xi}_p G_0\left[e^{ipx}\right] = \sum_{p=-\infty}^{\infty} |p| \hat{\xi}_p e^{ipx} =: |D|\xi,$$

which defines the order-one Fourier multiplier $|D|$. At order $\mathcal{O}\left(\varepsilon^n\right)$, $n > 0$, we find

$$\sum_{m=0}^{n} G_m(f)\left[F_{n-m} |p|^{n-m} e^{ipx}\right] = F_n |p|^{n+1} e^{ipx} - (\partial_x f)F_{n-1}(ip) |p|^{n-1} e^{ipx},$$

which we can write as

$$G_n(f)\left[e^{ipx}\right] = F_n |p|^{n+1} e^{ipx} - (\partial_x f)F_{n-1}(ip) |p|^{n-1} e^{ipx}$$

$$-\sum_{m=0}^{n-1} G_m(f)\left[F_{n-m} |p|^{n-m} e^{ipx}\right].$$

or, using $\partial_x e^{ipx} = (ip)e^{ipx}$,

$$G_n(f)\left[e^{ipx}\right] = F_n |D|^{n+1} e^{ipx} - (\partial_x f)F_{n-1}\partial_x |D|^{n-1} e^{ipx}$$

$$-\sum_{m=0}^{n-1} G_m(f)\left[F_{n-m} |D|^{n-m} e^{ipx}\right].$$

Since $(ip)^2 = -|p|^2$ we deduce that $|D|^2 = -\partial_x^2$ and we arrive at

$$G_n(f)\left[e^{ipx}\right] = \left(-F_n\partial_x^2 - (\partial_x f)F_{n-1}\partial_x\right) |D|^{n-1} e^{ipx}$$

$$-\sum_{m=0}^{n-1} G_m(f)\left[F_{n-m} |D|^{n-m} e^{ipx}\right].$$

Next, since

$$\partial_x[F_n\partial_x J] = F_n\partial_x^2 J + (\partial_x f)F_{n-1}\partial_x J,$$

we have

$$G_n(f)\left[e^{ipx}\right] = -\partial_x F_n\partial_x |D|^{n-1} e^{ipx} - \sum_{m=0}^{n-1} G_m(f)\left[F_{n-m} |D|^{n-m} e^{ipx}\right].$$

As we have the "action" of G_n on any complex exponential $\exp(ipx)$, we write down the *Slow OE Recursions*

$$G_n(f)[\xi] = -\partial_x F_n \partial_x |D|^{n-1}\xi - \sum_{m=0}^{n-1} G_m(f)\left[F_{n-m}|D|^{n-m}\xi\right], \qquad (1.4.2)$$

for any function

$$\xi(x) = \sum_{p=-\infty}^{\infty} \hat{\xi}_p e^{ipx}.$$

So, what is wrong with this set of recursions, (1.4.2)? To compute G_n one must evaluate G_{n-1}, which requires the application of G_{n-2}, etc. Since the *argument* of G_m *changes* as m changes, these cannot be precomputed and stored. Therefore, a naive implementation will require time proportional to $\mathcal{O}(n!)$. One can improve this by storing G_m as an *operator* (a matrix in finite dimensional space), and thus computing G_n requires time proportional to $\mathcal{O}(nN_x^2)$. Happily we can do even better by using the *self-adjointness* properties of the DNO.

It can be shown that the DNO, G, and all of its Taylor series terms G_n are *self-adjoint*: $G^* = G$ and $G_n^* = G_n$. This can be used to advantage by recalling that $(AB)^* = B^*A^*$, $\partial_x^* = -\partial_x$, and $F_n^* = F_n$. Now, one takes the adjoint of G_n to realize the *Fast OE Recursions*

$$G_n(f)[\xi] = G_n^*(f)[\xi] = -|D|^{n-1}\partial_x F_n \partial_x \xi - \sum_{m=0}^{n-1} |D|^{n-m} F_{n-m} G_m(f)[\xi].$$

$$(1.4.3)$$

As above, formula (1.4.3) can be implemented on a computer to deliver an alternative, fast and accurate method for simulating the action of the DNO, $G : \xi \to \nu$.

1.5 Numerical Tests

Now that we have *two* HOPS schemes for approximating DNOs, we can test them and compare their performance. For this we make use of the following exact solution. Recall the solution we used for the OE formula

$$v_p(x,y) := e^{|p|y} e^{ipx}.$$

If we choose a wavenumber, say $p = r$, and a profile $f(x)$, for a given $\varepsilon > 0$, it is easy to see that the Dirichlet data

$$\xi_r(x;\varepsilon) := v_r(x, \varepsilon f(x)) = e^{|r|\varepsilon f(x)} e^{irx}$$

generates Neumann data

$$\nu_r(x;\varepsilon) := \left[\partial_y v_r - \varepsilon(\partial_x f)\partial_x v_r\right](x,\varepsilon f(x))$$
$$= \left[|r| - \varepsilon(\partial_x f)(ir)\right]e^{|r|\varepsilon f(x)}e^{irx}.$$

Using this we can, with a Fourier spectral method in mind [9, 10], sample the ξ_r at equally spaced points, appeal to either the FE or OE algorithms described above, and compare our outputs to ν_r evaluated at these same gridpoints.

To be more specific, for either HOPS algorithm we choose a number of equally spaced collocation points, N_x, and perturbation orders, N. For the FE algorithm we utilize (1.3.4) to find approximations $a_{n,p}^{N_x}$ for $-N_x/2 \le p \le N_x/2 - 1$ and $0 \le n \le N$ and form

$$v_{FE}^{N_x,N}(x) := \sum_{n=0}^{N} \sum_{p=-N_x/2}^{N_x/2-1} a_{n,p}^{N_x} e^{ipx} \varepsilon^n. \tag{1.5.1}$$

All nonlinearities are approximated on the physical side using pointwise multiplication, while Fourier multipliers are implemented in wavenumber space by invoking a Fast Fourier Transform (FFT), applying the (diagonal) Fourier multiplier operator, and then appealing to the inverse FFT algorithm.

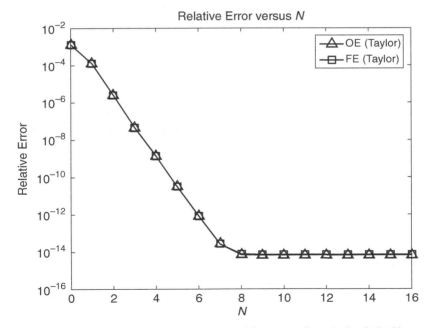

Figure 1.1. Relative Error in FE and OE Algorithms versus Perturbation Order N for Smooth Interface Configuration, (1.5.2), with Taylor Summation.

In the same way, the OE method uses (1.4.3) to provide approximations $v_{n,p}^{N_x}$, which are then used to generate $v_{OE}^{N_x,N}$ just as in (1.5.1).

We consider an example problem with geometric and numerical parameters

$$L = 2\pi, \quad \varepsilon = 0.02, \quad f(x) = \exp(\cos(x)), \quad N_x = 64, \quad N = 16, \quad (1.5.2)$$

and note that f is real analytic and that all of its derivatives are $L = 2\pi$–periodic (so that its Fourier series decays exponentially fast). In Figure 1.1 we display results of our numerical experiments with both the FE and OE algorithms as N is increased from 0 to 16. Here we note the stable and rapid convergence one can realize with these algorithms as the perturbation order N is increased.

1.6 Generalizations

Having described two rather simple and efficient algorithms for the simulation of solutions to Laplace's equation on a semi-infinite domain in two dimensions, one can ask, are these algorithms restricted to this simple case? Happily we can answer in the negative and now describe how to generalize the algorithms to three dimensions (§ 1.6.1) and finite depth (§ 1.6.2). Other generalizations are possible (e.g., to Helmholtz [39–41] and Maxwell [20, 43] equations, and the equations of elasticity [42]) but would take us rather far afield.

1.6.1 Three Dimensions

A generalization of crucial importance is to the more realistic situation of a genuinely three-dimensional fluid. In this case the air–fluid interface, $y = g(x) = g(x_1, x_2)$ is two-dimensional rather than one-dimensional. Such a generalization for Boundary Integral/Element methods requires a new formulation as the fundamental solution changes from

$$\Phi_2(r) = C_2 \ln(r),$$

to

$$\Phi_3(r) = C_3 r^{-1}.$$

One of the most appealing features of our HOPS methods is the *trivial* nature of the changes required moving from two to three dimensions. This can be seen by inspecting the solution of Laplace's equation, c.f. (1.3.1),

$$v(x,y) = \sum_{p_1=-\infty}^{\infty} \sum_{p_2=-\infty}^{\infty} a_p e^{|p|y} e^{ip \cdot x}, \quad p = (p_1, p_2) \in \mathbf{Z}^2.$$

Once again assuming $g(x_1, x_2) = \varepsilon f(x_1, x_2)$, $f \sim \mathcal{O}(1)$, $\varepsilon \ll 1$, we expand

$$v = v(x, y; \varepsilon) = \sum_{n=0}^{\infty} v_n(x, y) \varepsilon^n.$$

We find the same inhomogeneous Laplace problem, (1.3.2), at every perturbation order n, which has solution, c.f. (1.3.3),

$$v_n(x, y) = \sum_{p_1=-\infty}^{\infty} \sum_{p_2=-\infty}^{\infty} a_{n,p} e^{|p|y} e^{ip \cdot x}.$$

Following the development from before we find, c.f. (1.3.4),

$$a_{n,p} = \delta_{n,0} \hat{\xi}_p - \sum_{m=0}^{n-1} \sum_{q_1=-\infty}^{\infty} \sum_{q_2=-\infty}^{\infty} \hat{F}_{n-m,p-q} |q|^{n-m} a_{m,q}.$$

As before, if we seek the Neumann data

$$\nu(x) = \left[\partial_y v - (\partial_x g) \partial_x v \right]_{y=g(x)},$$

and expand

$$\nu(x; \varepsilon) = \sum_{n=0}^{\infty} \nu_n(x) \varepsilon^n,$$

then FE delivers approximations to ν_n from the $a_{n,p}$. More precisely, for $F'_m(x) := (\nabla_x f) F_m(x)$, c.f. (1.3.5),

$$\hat{\nu}_{n,p} = \sum_{m=0}^{n} \sum_{q_1=-\infty}^{\infty} \sum_{q_2=-\infty}^{\infty} \hat{F}_{n-m,p-q} |q|^{n+1-m} a_{m,q}$$

$$- \sum_{m=0}^{n-1} \sum_{q_1=-\infty}^{\infty} \sum_{q_2=-\infty}^{\infty} \hat{F}'_{n-1-m,p-q} \cdot (iq) |q|^{n-1-m} a_{m,q}.$$

Regarding the OE methodology, we once again assume that the interface is shaped by $g(x_1, x_2) = \varepsilon f(x_1, x_2)$, $f \sim \mathcal{O}(1)$, $\varepsilon \ll 1$. As before, we seek the *action* of G on a basis function, $\exp(ip \cdot x)$. To achieve this we use a bounded, periodic solution of Laplace's equation, c.f. (1.4.1),

$$v_p(x, y) := e^{|p|y} e^{ip \cdot x}.$$

We make the expansion

$$G(\varepsilon f) = \sum_{n=0}^{\infty} G_n(f) \varepsilon,$$

and seek forms for the G_n. Using the methods outlined ealier, we can show that

$$G_0[\xi] = |D| \xi = \sum_{p_1=-\infty}^{\infty} \sum_{p_2=-\infty}^{\infty} |p| \hat{\xi}_p e^{ip \cdot x},$$

and, for $n > 0$,

$$G_n(f)[\xi] = -\text{div}_x\left[F_n\nabla_x |D|^{n-1}\xi\right] - \sum_{m=0}^{n-1} G_m(f)\left[F_{n-m}|D|^{n-m}\xi\right],$$

c.f. (1.4.2). Again, these can be accelerated by *adjointness* considerations to

$$G_n(f)[\xi] = -|D|^{n-1}\text{div}_x\left[F_n\nabla_x\xi\right] - \sum_{m=0}^{n-1}|D|^{n-m}F_{n-m}G_m(f)[\xi],$$

c.f. (1.4.3).

1.6.2 Finite Depth

As we will now show, the generalization to *finite depth* is immediate. The problem statement, (1.2.1), is the same, save we replace $\partial_y v \to 0$ with

$$\partial_y v(x, -h) = 0.$$

Once again, the key is the periodic solution of Laplace's equation satisfying this boundary condition, c.f. (1.4.1),

$$v_p(x,y) := \frac{\cosh(|p|\,(y+h))}{\cosh(|p|\,h)} e^{ipx}.$$

Since v_p and $\partial_y v_p$ *evaluated at* $y = 0$ will clearly become important, we introduce the symbol

$$T_{n,p} = T_{n,p}(h) := \begin{cases} 1 & n \text{ even} \\ \tanh(h\,|p|) & n \text{ odd} \end{cases}.$$

We can show that the *FE Recursions* become

$$a_{n,p} = \delta_{n,0}\hat{\xi}_p - \sum_{m=0}^{n-1}\sum_{q=-\infty}^{\infty} \hat{F}_{n-m,p-q}\,|q|^{n-m}\,T_{n-m,q}(h)a_{m,q},$$

c.f. (1.3.4), and that

$$\hat{v}_{n,p} = \sum_{m=0}^{n}\sum_{q=-\infty}^{\infty} \hat{F}_{n-m,p-q}\,|q|^{n+1-m}\,T_{n+1-m}(h)a_{m,q}$$

$$-\sum_{m=0}^{n-1}\sum_{q=-\infty}^{\infty} \hat{F}'_{n-1-m,p-q}(iq)\,|q|^{n-1-m}\,T_{n-1-m}(h)a_{m,q},$$

c.f. (1.3.5).

For the OE recursions we can show that, at order zero,

$$G_0[\xi] = \sum_{p=-\infty}^{\infty} |p|\tanh(h\,|p|)\hat{\xi}_p e^{ipx} = |D|\tanh(h\,|D|)\xi,$$

while at higher orders (after appealing to *self-adjointness*)

$$G_n(f)[\xi] = -|D|^{n-1}T_{n+1,D}\partial_x F_n \partial_x \xi - \sum_{m=0}^{n-1} |D|^{n-m} T_{n-m,D} F_{n-m} G_m(f)[\xi],$$

c.f. (1.4.3).

1.7 Padé Summation

One of the classical problems of numerical analysis is the approximation of an analytic function given a truncation of its Taylor series. Simply evaluating the truncation will be spectrally accurate in the number of terms for points *inside* the disk of convergence of the Taylor series. However, one may be interested in points of analyticity *outside* this disk, and a numerical "analytic continuation" is of great interest. Padé approximation [44] is one of the most popular and successful choices for this procedure, and we refer the interested reader to the insightful calculations of § 8.3 of Bender & Orszag [45] for more details.

To summarize this procedure, consider the analytic function

$$c(\varepsilon) = \sum_{n=0}^{\infty} c_n \varepsilon^n,$$

which we approximate by its truncated Taylor series

$$c^N(\varepsilon) := \sum_{n=0}^{N} c_n \varepsilon^n.$$

If ε_0 is in the disk of convergence then

$$\left| c(\varepsilon_0) - c^N(\varepsilon_0) \right| < K\rho^N.$$

However, if ε_0 is a point of analyticity *outside* the disk of convergence of the Taylor series, c^N will produce *meaningless* results.

The idea behind Padé summation is to approximate c^N by the rational function

$$[L/M](\varepsilon) := \frac{a^L(\varepsilon)}{b^M(\varepsilon)} = \frac{\sum_{l=0}^{L} a_l \varepsilon^l}{\sum_{m=0}^{M} b_m \varepsilon^m}$$

where $L + M = N$ and

$$[L/M](\varepsilon) = c^N(\varepsilon) + \mathcal{O}\left(\varepsilon^{L+M+1}\right).$$

For convenience we choose the *equiorder* Padé approximant $[N/2, N/2](\varepsilon)$. For the purposes of the following discussion we assume that either N is even or, in the case N odd, that the highest order term c_N is ignored so that $L = M = N/2$ is an integer.

To derive equations for the $\{a_l\}$ and $\{b_m\}$ we note that

$$\frac{a^M(\varepsilon)}{b^M(\varepsilon)} = c^{2M}(\varepsilon) + \mathcal{O}\left(\varepsilon^{2M+1}\right)$$

is equivalent to

$$b^M(\varepsilon)c^{2M}(\varepsilon) = a^M(\varepsilon) + \mathcal{O}\left(\varepsilon^{2M+1}\right),$$

or

$$\left(\sum_{m=0}^{M} b_m \varepsilon^m\right)\left(\sum_{n=0}^{2M} c_n \varepsilon^n\right) = \left(\sum_{m=0}^{M} a_m \varepsilon^m\right) + \mathcal{O}\left(\varepsilon^{2M+1}\right).$$

Multiplying the polynomials on the left-hand-side and equating at equal orders in ε^m for $0 \leq m \leq 2M$ reveals two sets of equations

$$\sum_{j=0}^{m} c_{m-j} b_j = a_m \qquad 0 \leq m \leq M, \tag{1.7.1a}$$

$$\sum_{j=0}^{M} c_{m-j} b_j = 0 \qquad M+1 \leq m \leq 2M. \tag{1.7.1b}$$

Given that the c_m are known, if the b_m can be computed then the first equation, (1.7.1a), can be used to find the a_m. Without loss of generality we make the classical specification $b_0 = 1$ so that the second equation, (1.7.1b), becomes

$$\sum_{j=1}^{M} c_{m-j} b_j = -c_m, \quad M+1 \leq m \leq 2M,$$

or $H\tilde{b} = -r$ where

$$H = \begin{pmatrix} c_M & \cdots & c_1 \\ c_{M+1} & \cdots & c_2 \\ \vdots & \ddots & \vdots \\ c_{2M-1} & \cdots & c_M \end{pmatrix}, \quad \tilde{b} = \begin{pmatrix} b_1 \\ b_2 \\ \vdots \\ b_M \end{pmatrix}, \quad r = \begin{pmatrix} c_{M+1} \\ c_{M+2} \\ \vdots \\ c_{2M} \end{pmatrix}.$$

To see the extremely beneficial effects this procedure can have upon HOPS simulations, we revisit the calculations of § 1.5 save that we consider a deformation of size $\varepsilon = 0.2$ (ten times as large). More specifically, we once again consider the problem (1.5.2) (with $\varepsilon = 0.2$) and compare Taylor summation with Padé approxmation. In Figure 1.2 we show the discouraging results delivered by FE and OE in this challenging configuration with Taylor summation. However, one is able to recover two extra digits of accuracy from the FE and OE approximations of the Taylor coefficients provided one appeals to the analytic continuation of Padé approximation.

Dedication. The author would like to dedicate this contribution to his beautiful wife, Kristy. Without her love and support none of this work would

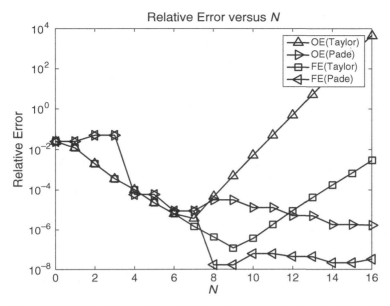

Figure 1.2. Relative Error in FE and OE Algorithms versus Perturbation Order N for Smooth, Large Interface Configuration, (1.5.2) ($\varepsilon = 0.2$) with Taylor and Padé Summation.

have been possible. This was evident most recently during our family's one-month stay in Cambridge to participate in the Isacc Newton Institute programme "Theory of Water Waves" where this lecture was delivered.

Acknowledgment

The author gratefully acknowledges support from the National Science Foundation through grant No. DMS–1115333.

The author also would like to thank the Issac Newton Institute at the University of Cambridge for providing the facilities where this meeting was held, as well as T. Bridges, P. Milewski, and M. Groves for organizing this exciting meeting.

References

[1] Strikwerda, John C. 2004. *Finite difference schemes and partial differential equations*. Second edn. Society for Industrial and Applied Mathematics (SIAM), Philadelphia, PA.

[2] Morton, K. W., and Mayers, D. F. 2005. *Numerical solution of partial differential equations*. Second edn. Cambridge University Press, Cambridge. An introduction.

[3] LeVeque, Randall J. 2007. *Finite difference methods for ordinary and partial differential equations*. Society for Industrial and Applied Mathematics (SIAM), Philadelphia, PA. Steady-state and time-dependent problems.

[4] Thomas, J. W. 1995. *Numerical partial differential equations: finite difference methods*. Texts in Applied Mathematics, vol. 22. Springer-Verlag, New York.

[5] Johnson, Claes. 1987. *Numerical solution of partial differential equations by the finite element method*. Cambridge: Cambridge University Press.

[6] Karniadakis, George Em, and Sherwin, Spencer J. 1999. *Spectral/hp element methods for CFD*. Numerical Mathematics and Scientific Computation. New York: Oxford University Press.

[7] Braess, Dietrich. 2001. *Finite elements*. Second edn. Cambridge: Cambridge University Press. Theory, fast solvers, and applications in solid mechanics, Translated from the 1992 German edition by Larry L. Schumaker.

[8] Hesthaven, Jan S., and Warburton, Tim. 2008. *Nodal discontinuous Galerkin methods*. Texts in Applied Mathematics, vol. 54. New York: Springer. Algorithms, analysis, and applications.

[9] Gottlieb, David, and Orszag, Steven A. 1977. *Numerical analysis of spectral methods: theory and applications*. Philadelphia, PA.: Society for Industrial and Applied Mathematics. CBMS-NSF Regional Conference Series in Applied Mathematics, No. 26.

[10] Canuto, Claudio, Hussaini, M. Yousuff, Quarteroni, Alfio, and Zang, Thomas A. 1988. *Spectral methods in fluid dynamics*. New York: Springer-Verlag.

[11] Fornberg, Bengt. 1996. *A practical guide to pseudospectral methods*. Cambridge Monographs on Applied and Computational Mathematics, vol. 1. Cambridge: Cambridge University Press.

[12] Boyd, John P. 2001. *Chebyshev and Fourier spectral methods*. Second edn. Mineola, NY: Dover Publications Inc.

[13] Deville, M. O., Fischer, P. F., and Mund, E. H. 2002. *High-order methods for incompressible fluid flow*. Cambridge Monographs on Applied and Computational Mathematics, vol. 9. Cambridge: Cambridge University Press.

[14] Hesthaven, Jan S., Gottlieb, Sigal, and Gottlieb, David. 2007. *Spectral methods for time-dependent problems*. Cambridge Monographs on Applied and Computational Mathematics, vol. 21. Cambridge: Cambridge University Press.

[15] Colton, David, and Kress, Rainer. 1998. *Inverse acoustic and electromagnetic scattering theory*. Second edn. Berlin: Springer-Verlag.

[16] Kress, Rainer. 1999. *Linear integral equations*. Second edn. New York: Springer-Verlag.

[17] Bruno, Oscar P., and Reitich, Fernando. 1992. Solution of a boundary value problem for the Helmholtz equation via variation of the boundary into the complex domain. *Proc. Roy. Soc. Edinburgh Sect. A*, **122**(3-4), 317–340.

[18] Bruno, Oscar P., and Reitich, Fernando. 1993a. Numerical solution of diffraction problems: A method of variation of boundaries. *J. Opt. Soc. Am. A*, **10**(6), 1168–1175.

[19] Bruno, Oscar P., and Reitich, Fernando. 1993b. Numerical solution of diffraction problems: A method of variation of boundaries. II. Finitely conducting gratings, Padé approximants, and singularities. *J. Opt. Soc. Am. A*, **10**(11), 2307–2316.

[20] Bruno, Oscar P., and Reitich, Fernando. 1993c. Numerical solution of diffraction problems: A method of variation of boundaries. III. Doubly periodic gratings. *J. Opt. Soc. Am. A*, **10**(12), 2551–2562.

[21] Bruno, Oscar P., and Reitich, Fernando. 1994. Approximation of analytic functions: A method of enhanced convergence. *Math. Comp.*, **63**(207), 195–213.

[22] Bruno, Oscar P., and Reitich, Fernando. 1996. Calculation of electromagnetic scattering via boundary variations and analytic continuation. *Appl. Comput. Electromagn. Soc. J.*, **11**(1), 17–31.

[23] Bruno, Oscar P., and Reitich, Fernando. 1998. Boundary–variation solutions for bounded–obstacle scattering problems in three dimensions. *J. Acoust. Soc. Am.*, **104**(5), 2579–2583.

[24] Bruno, Oscar P., and Reitich, Fernando. 2001. High-order boundary perturbation methods. Pages 71–109 of: *Mathematical Modeling in Optical Science*, vol. 22. Philadelphia, PA: SIAM. Frontiers in Applied Mathematics Series.

[25] Milder, D. Michael. 1991a. An improved formalism for rough-surface scattering of acoustic and electromagnetic waves. Pages 213–221 of: *Proceedings of SPIE - The International Society for Optical Engineering (San Diego, 1991)*, vol. 1558. Bellingham, WA: Int. Soc. for Optical Engineering.

[26] Milder, D. Michael. 1991b. An improved formalism for wave scattering from rough surfaces. *J. Acoust. Soc. Am.*, **89**(2), 529–541.

[27] Milder, D. Michael, and Sharp, H. Thomas. 1991. Efficient computation of rough surface scattering. Pages 314–322 of: *Mathematical and numerical aspects of wave propagation phenomena (Strasbourg, 1991)*. Philadelphia, PA: SIAM.

[28] Milder, D. Michael, and Sharp, H. Thomas. 1992. An improved formalism for rough surface scattering. II: Numerical trials in three dimensions. *J. Acoust. Soc. Am.*, **91**(5), 2620–2626.

[29] Craig, Walter, and Sulem, Catherine. 1993. Numerical simulation of gravity waves. *J. Comput. Phys.*, **108**, 73–83.

[30] Milder, D. Michael. 1996b. Role of the admittance operator in rough-surface scattering. *J. Acoust. Soc. Am.*, **100**(2), 759–768.

[31] Milder, D. Michael. 1996a. An improved formalism for electromagnetic scattering from a perfectly conducting rough surface. *Radio Science*, **31**(6), 1369–1376.

[32] Nicholls, David P., and Reitich, Fernando. 2001a. A new approach to analyticity of Dirichlet-Neumann operators. *Proc. Roy. Soc. Edinburgh Sect. A*, **131**(6), 1411–1433.

[33] Nicholls, David P., and Reitich, Fernando. 2001b. Stability of High-Order Perturbative Methods for the Computation of Dirichlet-Neumann Operators. *J. Comput. Phys.*, **170**(1), 276–298.

[34] Nicholls, David P., and Reitich, Fernando. 2003. Analytic continuation of Dirichlet-Neumann operators. *Numer. Math.*, **94**(1), 107–146.

[35] Lamb, Horace. 1993. *Hydrodynamics*. sixth edn. Cambridge: Cambridge University Press.

[36] Zakharov, Vladimir. 1968. Stability of periodic waves of finite amplitude on the surface of a deep fluid. *J. App. Mech. Tech. Phys.*, **9**, 190–194.

[37] Rayleigh, Lord. 1907. On the dynamical theory of gratings. *Proc. Roy. Soc. London*, **A79**, 399–416.

[38] Rice, S. O. 1951. Reflection of electromagnetic waves from slightly rough surfaces. *Comm. Pure Appl. Math.*, **4**, 351–378.

[39] Nicholls, David P., and Reitich, Fernando. 2004a. Shape deformations in rough surface scattering: Cancellations, conditioning, and convergence. *J. Opt. Soc. Am. A*, **21**(4), 590–605.

[40] Nicholls, David P., and Reitich, Fernando. 2004b. Shape deformations in rough surface scattering: Improved algorithms. *J. Opt. Soc. Am. A*, **21**(4), 606–621.

[41] Malcolm, Alison, and Nicholls, David P. 2011. A field expansions method for scattering by periodic multilayered media. *J. Acoust. Soc. Am.*, **129**(4), 1783–1793.

[42] Fang, Zheng, and Nicholls, David P. 2014. An operator expansions method for computing Dirichlet–Neumann operators in linear elastodynamics. *J. Comput. Phys.*, **272**, 266–278.

[43] Nicholls, David P. 2015. A method of field expansions for vector electromagnetic scattering by layered periodic crossed gratings. *J. Opt. Soc. Am., A (to appear)*.

[44] Baker, Jr., George A., and Graves-Morris, Peter. 1996. *Padé approximants.* Second edn. Cambridge: Cambridge University Press.

[45] Bender, Carl M., and Orszag, Steven A. 1978. *Advanced mathematical methods for scientists and engineers.* New York: McGraw-Hill Book Co. International Series in Pure and Applied Mathematics.

2

High-Order Perturbation of Surfaces Short Course: Traveling Water Waves

Benjamin F. Akers

Abstract

In this contribution we discuss High-Order Perturbation of Surfaces (HOPS) methods with particular application to traveling water waves. The Transformed Field Expansion method (TFE) is discussed as a method for handling the unknown fluid domain. The procedures for computing Stokes waves and Wilton Ripples are compared. The Lyapunov-Schmidt procedure for the Wilton Ripple is presented explicitly in a simple, weakly nonlinear model equation.

2.1 Introduction

Traveling water waves have been studied for over a century, most famously by Stokes, for whom weakly-nonlinear periodic waves are now named [1–3]. In his 1847 paper, Stokes expanded the wave profile as a power series in a small parameter, the wave slope, a technique that has since become commonplace. This classic perturbation expansion, which we will refer to as the Stokes' expansion, has been applied to the water wave problem numerous times [4–9]. When the effect of surface tension is included, the expansion may be singular. This singularity, due to a resonance between a long and a short wave, was noted first by Wilton [10] and has been studied more recently in [11–15].

In these lecture notes, we explain how traveling water waves may be computed using a High-Order Perturbation of Surfaces (HOPS) approach, which numerically computes the coefficients in an amplitude-based series expansion of the free surface. For the water wave problem, a crucial aspect of any numerical approach is the method used to handle the unknown fluid domain. Popular examples include Boundary Integral Methods [16, 17], conformal mappings [18, 19], and series computations of the Dirichelet-to-Neumann operator [20, 21]. Here we discuss an alternative approach, in which the solution is expanded using the Transformed Field Expansion (TFE) method, developed in [22, 23].

The TFE method has been used to compute traveling waves on both two-dimensional (one horizontal and one vertical dimension) and three-dimensional fluids, both for planar and short-crested waves [23]. Short-crested wave solutions to the potential flow equations have been computed without surface tension [22, 24, 25] and with surface tension [26]. They have also been studied experimentally [27, 28].

The TFE method computes a Stokes expansion of the water wave to all orders. There is a long history of numerical implementation of the Stokes expansion to simulate water waves [29], with the approaches of [30, 31] and [22] of greatest relevance in the present context. However, only the TFE methods can be *rigorously* shown to converge (proof in [32]; see Figure 1 in [22] for an explicit demonstration of the ill-conditioning present in the algorithm of [30]), and thus be completely reliable for numerical simulation. In a recent paper, Wilkening and Vasan discuss this type of numerical ill-conditioning, which is due to floating-point cancellation, and show how the use of multi-precision arithmetic allows one to use the Craig-Sulem expansion to compute traveling waves (this expansion would otherwise be victim to the same ill-conditioning of Roberts et. al.) [33].

The implicit assumption of the Stokes' expansion, that solutions are analytic in the wave slope parameter ε, can also be exploited to derive weakly nonlinear model equations. In the case of shallow water waves with surface tension, well-known weakly nonlinear models include the 5th order KdV equation [34] and the KP equation [35]. Without surface tension, Boussinesq type models have been used to study weakly nonlinear short-crested waves, for example in [36]. Recently, analogues of both the 5th order KdV [37] and KP equations [38] have been derived for deep water gravity-capillary waves. In this work a weakly nonlinear model will be used to present the details required to compute a high order Stokes expansion, as in [15]. This model allows one to avoid some of the technical details of the full Euler equations, but still captures the essence of the methods required to compute Stokes waves and Wilton ripples.

In these notes, we study HOPS methods for computing water waves. We begin by discussing methods for dealing with the unknown fluid domain in the water wave problem – Taylor Series, Operator Expansions, and the TFE. Next, we discuss the Lyapanov-Schmidt reduction used to compute the series to all orders – first for Stokes waves in the TFE expansion, second for Wilton ripples in a weakly nonlinear model.

2.2 Water Waves

The widely accepted model for the motion of waves on the surface of a large body of water with constant surface tension, and in the absence viscosity, are the Euler equations

$$\phi_{xx} + \phi_{zz} = 0, \qquad\qquad\qquad z < \varepsilon\eta, \qquad (2.2.1a)$$

$$\phi_z = 0, \qquad\qquad\qquad\qquad z = -H, \qquad (2.2.1b)$$

$$\eta_t + \varepsilon\eta_x\phi_x = \phi_z, \qquad\qquad\qquad z = \varepsilon\eta, \qquad (2.2.1c)$$

$$\phi_t + \frac{\varepsilon}{2}\left(\phi_x^2 + \phi_z^2\right) + \eta - \sigma\left(\frac{\eta_{xx}}{(1 + \varepsilon^2\eta_x^2)^{3/2}}\right) = 0, \qquad z = \varepsilon\eta, \qquad (2.2.1d)$$

where η is the free-surface displacement and ϕ is the velocity potential. These equations describe the motion of an inviscid incompressible fluid undergoing an irrotational motion. System (2.2.1) has been nondimensionalized as in [23, 39]. For HOPS methods, the wave slope $\varepsilon = A/L$ is assumed to be small (A is a typical amplitude and L, the characteristic horizontal length, is chosen in the nondimensionalization so that the waves have spatial period 2π). Also, the vertical dimension has been nondimensionalized using the wavelength, so the quantity H is nondimensional ($H = h/L$). For simplicity's sake, in these notes we will consider the deep water limit $H \to \infty$. To compute traveling waves, the time dependence of (2.2.1) is prescribed, using the ansatz

$$\eta(x,t) = \eta(x+ct) \qquad \text{and} \qquad \phi(x,z,t) = \phi(x+ct,z).$$

To proceed further, one must choose a method to handle the fundamental difficulty of the water wave problem: that the problem domain is unknown.

2.3 The Fluid Domain

In this section, approaches for dealing with the vertical dependence of the water wave problem are discussed. Of course there are many other methods (for example, conformal mapping or boundary integral methods). Here the aim is to discuss boundary perturbation methods and so we will restrict our attention to three methods based on series expansions: Taylor series, Operator Expansions, and the TFE method.

2.3.1 Taylor Series

The most classic approach for the fluid domain is the one used originally by Stokes, and later by Wilton, and much later numerically by, for example, Roberts et. al. [1, 10, 24, 30]. The idea is simple, presume that the potential is analytic in the vertical dimension near $z = 0$, and Taylor expand the potential, and thus the boundary conditions, about this flat boundary. The resulting problem will have an infinite degree nonlinearity, but be posed on a half plane (or strip in finite depth). The new boundary conditions are

$$c\eta_x - \phi_z + \varepsilon\partial_x\left(\sum_{n=0}^{\infty}\frac{\varepsilon^n}{(n+1)!}\eta^{n+1}\partial_z^n\partial_x\phi\right) = 0, \qquad \text{at} \quad z=0, \qquad (2.3.1a)$$

$$c\phi_x + \eta - \sigma\partial_x\left(\frac{\partial_x\eta}{(1+\varepsilon|\partial_x\eta|^2)^{1/2}}\right) + \sum_{n=1}^{\infty}\frac{\varepsilon^n}{n!}\eta^n\partial_z^n\phi_t$$

$$+\cdots+\frac{\varepsilon}{2}\left(\sum_{n=0}^{\infty}\frac{\varepsilon^n}{n!}\eta^n\partial_z^n\partial_x\phi\right)^2 + \frac{\varepsilon}{2}\left(\sum_{n=0}^{\infty}\frac{\varepsilon^n}{n!}\eta^n\partial_z^n\phi_z\right)^2 = 0, \qquad \text{at} \quad z=0.$$

$$(2.3.1b)$$

Laplace's equation is exactly solvable for the potential in the lower half plane, given its trace at $z=0$, denoted here as $\Phi(x) = \phi(x,0)$:

$$\phi(x,z) = \mathcal{F}^{-1}\left\{\mathcal{F}\{\Phi(x)\}e^{|k|z}\right\}.$$

Thus, we can write equations (2.3.1) in terms of Φ, and replacing $\phi_z(x,0) = \mathcal{T}\Phi(x)$, effectively eliminating the vertical dependence. The operator $\mathcal{T}$ is then defined by setting $z=0$ above, or in terms of the Fourier transformed variables,

$$\hat{\phi}(k,0) = |k|\hat{\Phi}(k) = \widehat{\mathcal{T}\Phi}(k).$$

Unfortunately, this type of expansion, although effective for weakly nonlinear models (where the series are truncated at small order), are ill-conditioned when large numbers of terms are kept in the nonlinearity. Many terms in the series almost cancel, meaning that precision is lost in the result of combining these sums. To fix such numerical instability, in fixed precision storage types, one desires to avoid large-degree nonlinearities, which can be done using the TFE approach.

2.3.2 *Operator Expansions*

A popular alternative to the Taylor series approach of the previous section is to map the problem to the free surface, via the Dirichlet-to-Neumann operator (DNO). For example, one can write the water wave problem using Zakharov's canonical variables [40] $\eta(x,t)$ and

$$\xi(x,t) := \phi(x,\eta(x,t),t)$$

(the displacement and the *surface* velocity potential). This formulation was made explicit by Craig & Sulem [41], with the introduction of the DNO,

$$G(\eta)\xi := (\partial_z\phi - (\partial_x\eta)\partial_x\phi)_{z=\eta},$$

which maps Dirichlet data, ξ, to Neumann data at the interface η. In terms of this operator the evolution equations (2.2.1) can be *equivalently* stated as

$$\partial_t \eta = G(\eta)\xi, \tag{2.3.2a}$$

$$\partial_t \xi = -\eta + \sigma \eta_{xx} - A(\eta)B(\eta,\xi), \tag{2.3.2b}$$

where

$$A(\eta) = \frac{1}{2(1 + (\partial_x \eta)^2)} \tag{2.3.2c}$$

$$B(\eta,\xi) = (\partial_x \xi)^2 - (G(\eta)\xi)^2 - 2(\partial_x \eta)(\partial_x \xi)G(\eta)\xi$$

$$+ \sigma \partial_x \left(\frac{\partial_x \eta}{(1 + \varepsilon^2 (\partial_x \eta)^2)^{1/2}} - \partial_x \eta \right). \tag{2.3.2d}$$

An amplitude expansion in this setting asks one to expand the operator $G(\eta)$, and thus it is natural to refer to such a HOPS method as an operator expansion; it fundamentally relies on expansion of the DNO [42]. We refer the interested reader to [42] for the full details, including the fundamental difficulty of computing the first variation of the DNO (see also [43, 44]). Such methods can be numerically implemented, provided that one is very careful with floating-point cancellation, see [33] for discussion of a multi-precision implementation using the operator expansion of Craig & Sulem.

2.3.3 Transformed Field Expansion

The TFE method begins with a very similar idea to the Taylor series approach, solving for the potential below a fixed depth. The change is that instead of the depth being $z = 0$, we choose some depth below the minimum of the fluid interface, $z = -a$. As before, Laplace's equation can be solved below this interface, yielding a new boundary condition at $z = -a$ instead of $z = -\infty$:

$$\phi_z = \mathcal{T}\phi,$$

with $\mathcal{T}$ defined as in the previous section, simply evaluating the solution of Laplace's equation at $z = -a$ instead of $z = 0$. Although the domain for Laplace's equation is still unknown, it is now bounded and small ($-a < z < \varepsilon\eta$). The next step is to change variables, mapping the domain to a strip. Rather than conformal mapping, which would leave Laplace's equation unchanged, the TFE approach uses the transformation

$$\tilde{z} = a\left(\frac{z - \varepsilon\eta}{a + \varepsilon\eta}\right),$$

which naturally generalizes to three-dimensional fluids. In this setting, we can write Euler's equations for the transformed field $u(\tilde{x}, \tilde{z})$, using

$$u(\tilde{x}, \tilde{z}) = \phi\left(\tilde{x}, \left(\frac{a + \varepsilon\eta}{a}\right)\tilde{z} + \varepsilon\eta\right).$$

Because this transformed field u mixes x and z dependence, derivatives transform nonlinearly

$$u_{\tilde{z}} = \phi_z\left(\frac{a + \varepsilon\eta}{a}\right) \qquad u_{\tilde{x}} = \phi_x + \left(\left(\frac{a + \varepsilon\eta_x}{a}\right)\tilde{z} + \varepsilon\eta_x\right)\phi_z.$$

This transformation increases the degree of the nonlinear terms in the boundary conditions, but only to cubics. The cost of this lower degree nonlinearity is that Laplace's equation becomes inhomogeneous, and the new system is of the form

$$u_{xx} + u_{zz} = F, \qquad -a < z < 0, \tag{2.3.3a}$$

$$u_z - \mathcal{T}u = J, \qquad z = -a, \tag{2.3.3b}$$

$$c\eta_x - \phi_z = Q, \qquad z = 0, \tag{2.3.3c}$$

$$c\phi_x + \eta - \sigma\eta_{xx} = R, \qquad z = 0, \tag{2.3.3d}$$

where all the nonlinear terms are included in F, J, Q, and R, whose degree is at most cubic, and are reported in [22, 23]. Solving such an inhomogeneous equation in the vertical direction is more difficult than a homogeneous one and is typically done numerically. However the result can be shown to be numerically stable for boundary perturbation methods, due to the finite-degree nonlinearity (a stark contrast from the previous section) [32]. It is also not so expensive, since a can be chosen to be small, thus one needs only a few points when solving for the the vertical dependence.

2.4 Boundary Perturbation

In the following sections we describe how one might compute traveling solutions to (2.3.3), using a perturbation expansion, about the flat state. The overall method will be described for the water wave problem, in which we will discuss the computation of Stokes waves. The extra Lyapunov-Schmidt reduction required to compute Wilton Ripples will be presented in the case of a simpler, weakly nonlinear family of model equations, as in [15].

2.4.1 Stokes Waves

The boundary perturbation method discussed here is based on an amplitude expansion about the linear solution of the water wave problem:

$$u_{xx} + u_{zz} = 0, \qquad -a < z < 0, \qquad (2.4.1\text{a})$$

$$u_z - \mathcal{T}u = 0 \qquad z = -a, \qquad (2.4.1\text{b})$$

$$c\eta_x - \phi_z = 0, \qquad z = 0, \qquad (2.4.1\text{c})$$

$$c\phi_x + \eta - \sigma \eta_{xx} = 0, \qquad z = 0. \qquad (2.4.1\text{d})$$

The linear problem (2.4.1), can be solved exactly, and has dispersion relation

$$\omega(k)^2 = c^2 k^2 = |k|(1 + \sigma k^2).$$

Thus for each wavenumber $k \in \mathbb{R}$ there are two speeds $c = \pm c_p(k)$, where $c_p(k) = \omega(k)/k$ is the phase speed. Similarly at each speed, there are two possible wave numbers, see Figure 2.1. If we restrict to a periodic domain, $k \in \mathbb{Z}$, most of the time there is only one wavenumber, which travels at a given speed, or equivalently the two wave numbers, which travel at a given speed do not share a period. When only one wavenumber both fits in a prescribed period and travels at a given speed, the leading order solution is

$$\eta_0(x) = e^{ix} + *, \qquad (2.4.2\text{a})$$

$$u_0(x, 0) = ic_0 e^{ix} + *, \qquad (2.4.2\text{b})$$

$$c_0 = \sqrt{(1 + \sigma)}, \qquad (2.4.2\text{c})$$

in which $*$ corresponds to the complex conjugate of the preceding terms. We call the nonlinear solution for which (2.4.2) is the leading order term a Stokes' wave. The goal of boundary perturbation is now to construct a series

$$u = \sum_{n=0}^{\infty} \varepsilon^n u_n, \qquad \eta = \sum_{n=0}^{\infty} \varepsilon^n \eta_n, \qquad \text{and} \qquad c = \sum_{n=0}^{\infty} \varepsilon^n c_n,$$

with u_0, η_0, c_0 given by (2.4.2). To compute the terms in this series η_j, u_j, and c_j, which we call corrections, we first substitute the above ansatz into (2.3.3), and then collect consecutive powers of ε. At each perturbation order, one must then solve

$$\Delta u_n = F_n, \qquad -a < z < 0, \quad (2.4.3\text{a})$$

$$\partial_z u_n - \mathcal{T}u_n = J_n, \qquad z = -a, \quad (2.4.3\text{b})$$

$$c_0 \partial_x \eta_n - \partial_z u_n = Q_n - c_n \partial_x \eta_0, \qquad z = 0, \quad (2.4.3\text{c})$$

$$c_0 \partial_x u_n + (1 - \sigma \Delta_x) \eta_n = R_n - c_n \partial_x u_0, \qquad z = 0. \quad (2.4.3\text{d})$$

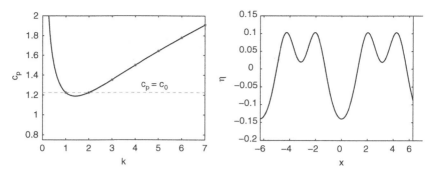

Figure 2.1. Left: The phase speed c_p for $\sigma = 1/2$, for which there is a Wilton ripple, between $k = 1$ and $k = 2$. The wave at $k = 1$ travels at the same speed as one of its harmonics (at $k = m$) only if $\sigma = 1/m$. For all other Bond numbers the solution is a Stokes' wave. Right: An example of two periods a ripple for $\sigma = 1/2$, computed in equation (2.4.6).

where the formula for F_n, J_n, Q_n, and R_n appear in [22, 23], and are known functions of η_j, u_j, c_j for $j < n$. To compute u_n, η_n from (2.4.3), one must first ensure that the right hand side is in the range of the linear operator (on the left). The linear problem has nontrivial solution, thus one must impose a solvability condition for each independent solution of the adjoint problem, which we call ψ. Stokes waves correspond to the case where the adjoint problem has a one-dimensional solution space.

The Lyapunov-Schmidt reduction for Stokes waves is quite simple. Defining $v_n = (0, 0, u_n, \eta_n)^T$ and $G_n = (F_n, J_n, Q_n, R_n)^T$, the Fredholm alternative theorem says that there are solutions of (2.4.3) if and only if

$$\langle \psi, G_n - c_n \partial_x v_0 \rangle = 0. \tag{2.4.4}$$

This equation is linear in c_n. After calculating c_n from (2.4.4), one can then invert the linear operator upon its range, calculating u_n and η_n, and proceed to the next order. This procedure is done numerically in [22, 23, 32].

2.4.2 Wilton Ripples

When the adjoint problem has more than one independent solution, the Lyapunov-Schmidt reduction required to compute solutions is more complicated. For deep water waves on a two-dimensional fluid this occurs when the Bond number $\sigma = \frac{1}{m}$, with $m \in \mathbb{N}$. For this countable set of Bond numbers, there are two nontrivial solutions to the linear problem, which one might express as

$$\eta_0 = e^{ix} + \beta_0 e^{imx} + *, \qquad \text{and} \qquad u_0(x, 0) = ic_0 e^{ix} + \beta_0 ic_0 e^{imx} + *, \tag{2.4.5}$$

both of which move at speed $c_0 = \sqrt{1 + \sigma}$. From a linear perspective, arbitrary values of β_0 are allowed. Only particular values of β_0 correspond to nonlinear solutions however. Rather than present the perturbation expansion for the full water wave problem, where the bookkeeping becomes quite arduous, we will present the expansion in a simpler model,

$$c\eta_x - \mathcal{L}\eta_x + \varepsilon(\eta^2)_x = 0, \qquad (2.4.6)$$

where $\mathcal{L}$ is a pseudo-differential operator, whose Fourier symbol is

$$\hat{\mathcal{L}}(k) = c_p(k) = \sqrt{\frac{1}{|k|} + \sigma|k|},$$

to mimic the water wave problem (we are presenting here the results of [15], for a particular choice of $\mathcal{L}$). The leading order solution will be identical to the η portion of (2.4.5). A nonlinear solution is then sought in the form

$$\eta = \sum_{n=0}^{\infty} \varepsilon^n \eta_n, \qquad \text{and} \qquad c = \sum_{n=0}^{\infty} \varepsilon^n c_n.$$

This ansatz results in a series of linear equations for η_n,

$$(c_0 - \mathcal{L})\partial_x \eta_n = -\partial_x \left(\sum_{j=0}^{n-1} \eta_j \eta_{n-1-j} \right) - \sum_{j=0}^{n-1} c_{n-j}\partial_x \eta_j. \qquad (2.4.7)$$

Just as with Stokes waves, we seek to compute the corrections (η_j, c_j) at each order, by first enforcing that the right hand side of (2.4.7) is in the range of the linear operator (on the left). In the ripple case there are two solutions to the adjoint linear problem, and we will need to enforce two solvability conditions. The speed corrections c_n are still available for one unknown at each order, but since there are now two conditions, we need a second unknown. The second unknown is a coefficient of the homogeneous solutions to (2.4.7). We choose to add these in the following way:

$$\eta_n = \eta_n^{\perp} + \beta_n e^{imx},$$

where $\eta_n^{\perp}$ is not supported at $k = 1$ or $k = m$, and is a particular solution of (2.4.7). The total correction η_n is a sum of this particular solution and a homogeneous contribution at frequency $k = m$.

The nth correction includes no contribution from the other homogeneous solution (at $k = 1$). This choice defines the total of the Fourier transform of the solution at $k = 1$ to be 1, and thus $\sum_{j=0}^{\infty} \varepsilon^j \beta_j$ is defined to be the Fourier transform of the solution at $k = m$. With these definitions, the solution is unique.

The series can be computed for all m, but for simplicity of presentation, we will only present the case considered by Wilton himself, $m = 2$. Larger values of m require similar manipulations; see a complete treatment of $m = 3$ in [15].

For $m = 2$, the character of the entire perturbation series can be determined by considering only the first three orders. The $O(\varepsilon^0)$ solution is identical to the leading order free surface of the Euler equations, here in (2.4.5), as is the leading order speed c_0. At the next order, one must impose that equation (2.4.7) is solvable, that the right hand side of (2.4.7) is orthogonal to both $\psi_j = e^{ik_j x}$:

$$\langle e^{ix}, (\eta_0^2)_x + c_1 \partial_x \eta_0 \rangle = 0,$$

$$\langle e^{2ix}, (\eta_0^2)_x + c_1 \partial_x \eta_0 \rangle = 0.$$

This can be written simply in terms of β_0 and c_1:

$$c_1 + 2\beta_0 = 0,$$

$$c_1 \beta_0 + 1 = 0.$$

Thus, $c_1 = \pm\sqrt{2}$ and $\beta_0 = \mp\sqrt{2}/2$. These numbers depend only on the nonlinearity, not the linear operator; they will be different for the full water wave problem. At later orders, the solvability conditions are

$$\left\langle e^{ix}, 2(\eta_0 \eta_{n-1})_x + c_n \partial_x \eta_0 + c_1 \partial_x \eta_{n-1} + \partial_x \left(\sum_{j=1}^{n-2} \eta_j \eta_{n-1-j} \right) + \sum_{j=1}^{n-2} c_{n-j} \partial_x \eta_j \right\rangle = 0,$$

$$\left\langle e^{2ix}, 2(\eta_0 \eta_{n-1})_x + c_n \partial_x \eta_0 + c_1 \partial_x \eta_{n-1} + \partial_x \left(\sum_{j=1}^{n-2} \eta_j \eta_{n-1-j} \right) + \sum_{j=1}^{n-2} c_{n-j} \partial_x \eta_j \right\rangle = 0.$$

In the above solvability conditions, the unknowns c_n and β_{n-1} appear linearly, and only in the first three terms in the second argument of the inner product. The resulting solvability conditions can be written as

$$\begin{pmatrix} 1 & 2 \\ \beta_0 & c_1 \end{pmatrix} \begin{pmatrix} c_n \\ \beta_{n-1} \end{pmatrix} = \vec{D}_n, \tag{2.4.8}$$

where $\vec{D}_n$ are known functions of the previous corrections.

$$\vec{D}_n = \begin{pmatrix} \left\langle e^{ix}, \partial_x \left(\sum_{j=1}^{n-2} \eta_j \eta_{n-1-j} \right) + \sum_{j=1}^{n-2} c_{n-j} \partial_x \eta_j \right\rangle \\ \left\langle e^{2ix}, \partial_x \left(\sum_{j=1}^{n-2} \eta_j \eta_{n-1-j} \right) + \sum_{j=1}^{n-2} c_{n-j} \partial_x \eta_j \right\rangle \end{pmatrix}.$$

Since the matrix on the left-hand side of (2.4.8) is invertible, this equation is solvable at all orders, and we can formally construct the solutions. To use such an approach, of course one must also consider convergence of the series, see [15] for discussion of the convergence of the above series. An example of a Wilton ripple solution for $m = 1$ is in the right panel of Figure 2.1.

Wilton ripples can be thought of a special case of a resonant interaction, where both the spatial and temporal frequencies of a set of waves sum to zero:

$$k_1 + k_2 + \cdots + k_m = 0, \qquad \omega_1 + \omega_2 + \cdots + \omega_m = 0.$$

Such nonlinear interactions are commonly referred to as triads when they contain three waves, quartets with four waves, and so on [45, 46]. Wilton ripples have wave numbers and dispersion relations satisfying such a resonant interaction, where

$$k_1 = k_2 = \cdots = k_{m-1} \qquad \text{and} \qquad k_m = mk_1,$$

along with

$$\omega_1 = \omega_2 = \cdots = \omega_{m-1} \qquad \text{and} \qquad \omega_m = m\omega_1.$$

Although these notes present only the triad ripple, $m = 1$, where the solvability conditions are quadratic, the other resonances are similar, with the exception that the solvability condition includes a cubic nonlinearity at $O(\varepsilon^2)$.

Acknowledgment

We would like to thank the Isaac Newton Institute for providing the facilities where this meeting was held, as well as Paul Milewski, David Nicholls, Tom Bridges, and Mark Groves for their excellent organizing.

References

[1] G.G. Stokes. On the theory of oscillatory waves. *Trans. Cambridge Philos. Soc.*, 8:441–455, 1847.

[2] A.D.D. Craik. The origins of water wave theory. *Ann. Rev. Fluid Mech.*, 36:1–28, 2004.

[3] A.D.D. Craik. George Gabriel Stokes on water wave theory. *Ann. Rev. Fluid Mech.*, 37:23–42, 2005.

[4] W.J. Harrison. The influence of viscosity and capillarity on waves of finite amplitude. *Proc. Lond. Math. Soc.*, 7:107–121, 1909.

[5] Rav J. C. Kamesvara. On ripples of finite amplitude. *Proc. Indian Ass. Cultiv. Sci.*, 6:175–193, 1920.

[6] H. Lamb. *Hydrodynamics*. Cambridge University Press, 1879.

[7] L.F. McGoldrick. An experiment on second-order capillary gravity resonant wave interactions. *J. Fluid Mech.*, 40:251–271, 1970.

[8] J. Reeder and M. Shinbrot. Three dimensional, nonlinear wave interaction in water of constant depth. *Non. Anal.*, 5:303–323, 1981.

[9] W.J. Pierson and Paul Fife. Some nonlinear properties of long-crested periodic waves with lengths near 2.44 centimeters. *J. Geophys. Res.*, 66:163–179, 1961.

[10] J.R. Wilton. On ripples. *Phil. Mag.*, 29:173, 1915.

[11] J.-M. Vanden-Broeck. Wilton ripples generated by a moving pressure disturbance. *J. Fluid Mech.*, 451:193–201, 2002.

[12] J.-M. Vanden-Broeck. Nonlinear gravity-capillary standing waves in water of arbitrary uniform depth. *J. Fluid Mech.*, 139:97–104, 1984.

[13] P. Christodoulides and F. Dias. Resonant capillary-gravity interfacial waves. *J. Fluid Mech.*, 265:303–343, 1994.

[14] J. Reeder and M. Shinbrot. On Wilton ripples II: Rigorous results. *Arch. Rat. Mech. Anal.*, 77:321–347, 1981.

[15] B.F. Akers and W. Gao. Wilton ripples in weakly nonlinear model equations. *Commun. Math. Sci*, 10(3):1015–1024, 2012.

[16] S.T. Grilli, P. Guyenne, and F. Dias. A fully non-linear model for three-dimensional overturning waves over an arbitrary bottom. *Int. J. Num. Meth. Fluids*, 35(7):829–867, 2001.

[17] E.I. Parau, J.-M. Vanden-Broeck, and M.J. Cooker. Nonlinear three-dimensional gravity-capillary solitary waves. *J. Fluid Mech.*, 536:99–105, 2005.

[18] Z. Wang, J.-M. Vanden-Broeck, and P.A. Milewski. Two-dimensional flexural–gravity waves of finite amplitude in deep water. *IMA J. App. Math.*, page hxt020, 2013.

[19] A.S. Fokas and A. Nachbin. Water waves over a variable bottom: a non-local formulation and conformal mappings. *J. Fluid Mech.*, 695:288–309, 2012.

[20] W. Craig and C. Sulem. Numerical simulation of gravity waves. *J. Comp. Phys.*, 108(1):73–83, 1993.

[21] B. Akers. The generation of capillary-gravity solitary waves by a surface pressure forcing. *Math. Comp. Sim.*, 82(6):958–967, 2012.

[22] D.P. Nicholls and F. Reitich. Stable, high-order computation of traveling water waves in three dimensions. *Eur. J. Mech. B/Fluids*, 25:406–424, 2006.

[23] B. Akers and D. P. Nicholls. Traveling waves in deep water with gravity and surface tension. *SIAM J. App. Math.*, 70(7):2373–2389, 2010.

[24] A.J. Roberts and L.W. Schwartz. The calculation of nonlinear short-crested gravity waves. *Phys. Fluid*, 26:2388–2392, 1983.

[25] D.P. Nicholls. Traveling water waves: Spectral continuation methods with parallel implementation. *J. Comp. Phys.*, 143:224–240, 1998.

[26] L.W. Schwartz and J.-M. Vanden-Broeck. Numerical solution of the exact equations for capillary-gravity waves. *J. Fluid Mech.*, 95:119, 1979.

[27] O. Kimmoun, H. Branger, and C. Kharif. On short-crested waves: experimental and analytical investigations. *Eur. J. Mech. B/Fluids*, 18:889–930, 1999.

[28] J. L. Hammack and D. M. Henderson. Experiments on deep-water waves with two dimensional surface patterns. *J. Offshore Mech. Arct. Eng.*, 125:48, 2003.

[29] D.P. Nicholls. Boundary perturbation methods for water waves. *GAMM-Mitt.*, 30:44–74, 2007.

[30] A. J. Roberts. Highly nonlinear short-crested water waves. *J. Fluid Mech.*, 135:301–321, 1983.

[31] T.R. Marchant and A.J. Roberts. Properties of short-crested waves in water of finite depth. *J. Austral. Math. Soc. Ser. B*, 29(1):103–125, 1987.

[32] D.P. Nicholls and J. Shen. A rigorous numerical analysis of the transformed field expansion method. *SIAM J. Num. Anal.*, 47(4):2708–2734, 2009.

[33] J. Wilkening and V. Vasan. Comparison of five methods of computing the Dirichlet-Neumann operator for the water wave problem. *arXiv preprint arXiv:1406.5226*, 2014.

[34] D.J. Korteweg and G. de Vries. On the change in form of long waves advancing in a rectangular canal and new type of long stationary waves. *Phil. Mag.*, 39:5:422–443, 1895.

[35] B.B. Kadomtsev and V.I. Petviashvili. On the stability of solitary waves in weakly dispersing media. *Sov. Phys. D.*, 15:539–541, 1970.

[36] D. R. Fuhrman and P. A. Madsen. Short-crested waves in deep water: a numerical investigation of recent laboratory experiments. *J. Fluid. Mech*, 559:391–411, 2006.

[37] B. Akers and P.A. Milewski. Model equations for gravity-capillary waves in deep water. *Stud. Appl. Math.*, 121:49–69, 2008.

[38] B. Akers and P.A. Milewski. A model equation for wavepacket solitary waves arising from capillary-gravity flows. *Stud. Appl. Math.*, 122:249–274, 2009.

[39] B. Akers and D.P. Nicholls. Spectral stability of deep two-dimensional gravity water waves: repeated eigenvalues. *SIAM J. App. Math.*, 72(2):689–711, 2012.

[40] V. Zakharov. Stability of periodic waves of finite amplitude on the surface of a deep fluid. *J. App. Mech. Tech. Phys.*, 9:190–194, 1968.

[41] W. Craig and C. Sulem. Numerical simulation of gravity waves. *J. Comp. Phys.*, 108:73–83, 1993.

[42] D.P. Nicholls. Spectral data for traveling water waves : singularities and stability. *J. Fluid Mech.*, 624:339–360, 2009.

[43] C. Fazioli and D. P. Nicholls. Parametric analyticity of functional variations of Dirichlet–Neumann operators. *Diff. Integ. Eqns.*, 21(5–6):541–574, 2008.

[44] C. Fazioli and D. P. Nicholls. Stable computation of variations of Dirichlet–Neumann operators. *J. Comp. Phys.*, 229(3):906–920, 2010.

[45] J.L. Hammack and D.M. Henderson. Resonant interactions among surface water waves. *Ann. rev. fluid mech.*, 25(1):55–97, 1993.

[46] A. D.D. Craik. *Wave Interactions and Fluid Flows*. Cambridge University Press, 1985.

3

High-Order Perturbation of Surfaces Short Course: Analyticity Theory

David P. Nicholls

Abstract

In this contribution we take up the question of convergence of the classical High-Order Perturbation of Surfaces (HOPS) schemes we introduced in the first lecture. This is intimately tied to analyticity properties of the relevant fields and Dirichlet–Neumann Operators. We show how a straightforward approach cannot succeed. However, with a simple change of variables this very method delivers not only a clear and optimal analyticity theory, but also a stable and high-order numerical scheme. We justify this latter claim with representative numerical simulations involving all three of the HOPS schemes presented thus far.

3.1 Introduction

Over the past two lectures we have derived several High-Order Perturbation of Surfaces (HOPS) schemes for the numerical simulation of (i.) solutions to boundary value problems, (ii.) surface integral operators (e.g., the Dirichlet–Neumann Operator [DNO]), and (iii.) free and moving boundary problems (e.g., traveling water waves). These HOPS methods are rapid (amounting to surface formulations accelerated by FFTs), robust (for perturbation size sufficiently small), and simple to implement (Operator Expansions is a one-line formula, while Field Expansions (FE) is two lines!). The derivation of all of these schemes is based upon the *analyticity* of the unknowns, but we have not yet specified under what conditions this is true. In this contribution we describe a straightforward framework for addressing this question that can be extended to give the most generous hypotheses known.

The first result on analyticity properties of DNOs with respect to boundary perturbations is due to Coifman & Meyer [1]. In this work the DNO was shown to be analytic as a function of Lipschitz perturbations of a line in the plane. Using a different formulation Craig, Schanz, and Sulem [2] and Craig & Nicholls [3, 4] proved analyticity of the DNO for C^1 perturbations of a hyperplane in three and general d dimensions, respectively. All of these results

32

depend upon delicate estimates of integral operators appearing in surface formulations of the problem defining the DNO.

Using a completely different approach (which we describe here), the author and F. Reitich produced a greatly simplified approach to establishing analyticity of DNOs and their related fields (at the cost of slightly less generous hypotheses, C^2 smoothness of the boundary perturbation) [5–7], which has the added benefit of generating a stabilized numerical approach. This has been generalized and extended in a number of directions, first to the Helmholtz equation by the author and Reitich [8, 9], and the author and Nigam [10, 11]. This was then extended to traveling water waves by the author and Reitich [12, 13] (see also the work of the author and Akers [14]) and the stability of these by the author [15] (see also [16–18]). On the theoretical side, the author and Hu [19, 20] showed how this very framework could be used to realize these analyticity results with the optimal smoothness requirements (Lipschitz perturbation) provided one is prepared to work with quite complicated function spaces for the field. The method was also extended to doubly perturbed domains by the author and Taber [21, 22], while the author and Fazioli generalized this to the case of *variations* of the DNO with respect to the boundary shape in [23, 24]. A rigorous numerical analysis of this algorithm for a wide array of problems was conducted by the author and Shen [25]. In this paper we revisit the proof presented in the original contribution [5], which, after many years of experience, has been further simplified and distilled.

The rest of the lecture is organized as follows. In § 3.2 we discuss a straightforward approach to an analyticity proof, which fails. In § 3.3 we outline an alternative formulation of the problem that requires a transparent boundary condition (§ 3.3.1) and a change of variables (§ 3.3.2), and results in the method of Transformed Field Expansions – TFE – (§ 4.2.1). In § 3.4 we produce an analyticity proof, which, as a side benefit, results in a stabilized numerical procedure described in § 3.5, which includes numerical tests (§ 3.5.1).

3.2 A Convergence Proof Fails

Inspired by the classical model for waves on the surface of an ideal, deep, two-dimensional fluid [5] we consider the laterally 2π-periodic Laplace problem with Dirichlet data at an irregular interface

$$\Delta v = 0 \qquad y < g(x), \tag{3.2.1a}$$

$$\partial_y v \to 0 \qquad y \to -\infty, \tag{3.2.1b}$$

$$v = \xi \qquad y = g(x). \tag{3.2.1c}$$

Supposing that $g(x) = \varepsilon f(x)$, the FE method is built upon the assumption that v depends *analytically* upon ε so that

$$v = v(x, y; \varepsilon) = \sum_{n=0}^{\infty} v_n(x, y) \varepsilon^n.$$

Following the FE philosophy, we insert this into the problem, (3.2.1), above yielding

$$\Delta v_n = 0 \qquad y < 0, \tag{3.2.2a}$$

$$\partial_y v_n \to 0 \qquad y \to -\infty, \tag{3.2.2b}$$

$$v_n = Q_n \qquad y = 0. \tag{3.2.2c}$$

In this latter system we have shown in the first lecture that

$$Q_n(x) = \delta_{n,0}\xi(x) - \sum_{m=0}^{n-1} F_{n-m}(x)\, \partial_y^{n-m} v_m(x, 0), \quad F_m(x) := \frac{f^m(x)}{m!}, \tag{3.2.3}$$

where $\delta_{n,m}$ is the Kronecker delta. We can now wonder whether one can establish analyticity of v *directly* from these recursions, (3.2.3)? The "natural" approach would be to appeal to classical elliptic theory and use the triangle inequality. Recall that, under suitable conditions, the solution of the elliptic problem, (3.2.2), at order $n > 0$ should satisfy

$$\|v_n\|_X \le C_e \|Q_n\|_Y,$$

for two function spaces (e.g., $X = H^2$ and $Y = H^{3/2}$ [26]). Now, we apply the estimate to our recursions, (3.2.3),

$$\|v_n\|_X \le C_e \left\| \delta_{n,0}\xi - \sum_{m=0}^{n-1} F_{n-m}(x)\, \partial_y^{n-m} v_m(x, 0) \right\|_Y$$

$$\le C_e \left\{ \delta_{n,0}\|\xi\|_Y + \sum_{m=0}^{n-1} \left\| F_{n-m}(x)\, \partial_y^{n-m} v_m(x, 0) \right\|_Y \right\}. \tag{3.2.4}$$

We claim that this last estimate is *useless* as the sum is *unbounded*!

Proof. The boundary condition at order n is

$$v_n(x, 0) = \delta_{n,0}\xi(x) - \sum_{m=0}^{n-1} F_{n-m}\partial_y^{n-m} v_m(x, 0).$$

Recalling, from separation of variables, that

$$v_n(x, y) = \sum_{p=-\infty}^{\infty} a_{n,p} e^{|p|y} e^{ipx},$$

we find

$$\sum_{p=-\infty}^{\infty} a_{n,p} e^{ipx} = \delta_{n,0} \sum_{p=-\infty}^{\infty} \hat{\xi}_p e^{ipx} - \sum_{m=0}^{n-1} F_{n-m} \sum_{q=-\infty}^{\infty} |q|^{n-m} a_{m,q} e^{iqx}.$$

At wavenumber p we obtain

$$a_{n,p} = \delta_{n,0} \hat{\xi}_p - \sum_{m=0}^{n-1} \sum_{q=-\infty}^{\infty} \hat{F}_{n-m,p-q} |q|^{n-m} a_{m,q}.$$

To simplify our demonstration we make the choices:

$$f(x) = \xi(x) = 2\cos(x) = e^{ix} + e^{-ix}.$$

With this we discover a number of things. First, regarding the solution at order zero, we have

$$a_{0,p} = \begin{cases} 1 & p = \pm 1, \\ 0 & p \neq \pm 1 \end{cases}.$$

Second, the powers of f satisfy

$$\begin{aligned}
f^0 &= 1, \\
f^1 &= e^{ix} + e^{-ix}, \\
f^2 &= e^{2ix} + 2 + e^{-2ix}, \\
f^3 &= e^{3ix} + 3e^{ix} + 3e^{-ix} + e^{-3ix}, \\
&\vdots \\
f^n &= e^{nix} + ne^{(n-2)ix} + \cdots + ne^{(2-n)ix} + e^{-nix},
\end{aligned}$$

so that $F_n = f^n/n! = e^{nix}/n! + \cdots + e^{-nix}/n!$. Upon defining $P_n := a_{n,n+1}$ we can show that

$$P_n = \delta_{n,0} - \sum_{m=0}^{n-1} \frac{(m+1)^{n-m}}{(n-m)!} P_m,$$

since $\hat{F}_{n,m} = 0$, $|m| > n$, and we can show (by induction) that

$$a_{n,m} = 0, \quad |m| > n+2.$$

We now appeal to the following theorem of Friedman and Reitich [27] which shows that, for this *subset* of Fourier coefficients, $P_n = a_{n,n+1}$, while the sum *converges* for ε sufficiently small, the relevant majorizing sequence (corresponding to the bound (3.2.4)) *diverges* for *any* non-zero choice of ε. $\square$

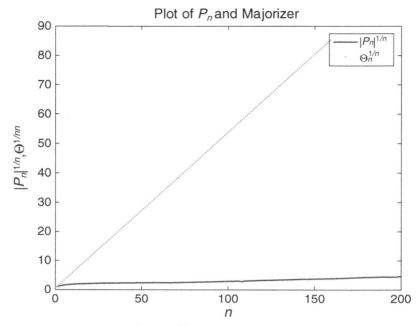

Figure 3.1. Plot of $|P_n|^{1/n}$ and $\Theta_n^{1/n}$ versus n.

Theorem 3.2.1 (Friedman & Reitich [27]). • *The sum $\sum_{n=0}^{\infty} P_n \varepsilon^n$ converges for $\varepsilon < 1/e$.*
• *Consider the majorizing sequence Θ_n (i.e., $|P_n| \le \Theta_n$) defined by*

$$\Theta_n = \delta_{n,0} + \sum_{m=0}^{n-1} \frac{(m+1)^{n-m}}{(n-m)!} \Theta_m.$$

The sum $\sum_{n=0}^{\infty} \Theta_n \varepsilon^n$ diverges for all $\varepsilon > 0$.

A graphical depiction of this result is given in Figure 3.1, which demonstrates a non-zero radius of convergence for P_n and a nonexistent domain of convergence for Θ_n.

3.3 A Change of Coordinates

One answer to the question of why this proof fails is that the "classical" HOPS schemes require the differentiation of the field (e.g., in the case of FE) and/or field trace (e.g., in the case of Operator Expansions [OE]) *across* the boundary of the problem domain, $y = g(x)$. One classical technique for analyzing free- and moving-boundary problems such as these which addresses this concern is a simple change of variables mapping the domain from the deformed geometry

$\{y < g(x)\}$ to a flat one $\{y' < 0\}$. We pursue a particular (non-conformal) choice, which, in the theory of gratings, is called the C–method [28–30] and, in atmospheric sciences, is known as σ-coordinates [31]. Now, the differentiations take place *within* the problem domain and/or at its boundary. With this strategy we realize a straightforward analyticity proof, and derive a stable and robust numerical algorithm.

3.3.1 A Transparent Boundary Condition

Before we state the change of variables we describe a domain decomposition that is very useful for HOPS methods. Once again, consider the boundary value problem (BVP) (3.2.1) generating the DNO

$$G(g)\xi = \left[\partial_y v - (\partial_x g)\partial_x v\right]_{y=g(x)}. \tag{3.3.1}$$

We choose $b > |g|_\infty$ and define the *artificial boundary* $\{y = -b\}$. It is easy to see that the BVP above, (3.2.1), is equivalent to:

$$\begin{aligned}
\Delta v &= 0 & -b < y < g(x), \\
v &= \xi & y = g(x), \\
\Delta w &= 0 & y < -b, \\
\partial_y w &\to 0 & y \to -\infty, \\
v &= w & y = -b, \\
\partial_y v &= \partial_y w & y = -b.
\end{aligned}$$

If we denote $\psi(x) := v(x, -b)$, we can solve the problem

$$\begin{aligned}
\Delta w &= 0 & y < -b, & \tag{3.3.2a} \\
w &= \psi & y = -b, & \tag{3.3.2b} \\
\partial_y w &\to 0 & y \to -\infty, & \tag{3.3.2c}
\end{aligned}$$

and, defining the (second) DNO,

$$S[\psi] := \partial_y w(x, -b),$$

see that the original BVP, (3.2.1), is equivalent to

$$\begin{aligned}
\Delta v &= 0 & -b < y < g(x), & \tag{3.3.3a} \\
v &= \xi & y = g(x), & \tag{3.3.3b} \\
\partial_y v - S[v] &= 0 & y = -b. & \tag{3.3.3c}
\end{aligned}$$

Such a strategy involves a "Transparent Boundary Condition" posed at the artificial boundary $\{y = -b\}$ and this strategy has been widely employed (see, e.g., [32–39]).

All that remains is to specify S from (3.3.2). We note that the unique, bounded, periodic solution is given by

$$w(x,y) = \sum_{p=-\infty}^{\infty} \hat{\psi}_p e^{|p|(y+b)} e^{ipx}.$$

With this we can compute

$$\partial_y w(x,y) = \sum_{p=-\infty}^{\infty} |p|\, \hat{\psi}_p e^{|p|(y+b)} e^{ipx},$$

so that

$$S[\psi] = \partial_y w(x,-b) = \sum_{p=-\infty}^{\infty} |p|\, \hat{\psi}_p e^{ipx} = |D|\,\psi,$$

which defines the order-one Fourier multiplier $|D|$ (denoted "order-one" as this operator effectively takes one derivative).

3.3.2 *The Change of Variables*

To describe the idea behind our TFE approach we fix on the problem (3.3.3), and consider the "domain flattening" change of variables

$$x' = x, \qquad y' = b\left(\frac{y-g(x)}{b+g(x)}\right),$$

which maps $\{-b < y < g(x)\}$ to $\{-b < y' < 0\}$. Note that these can be inverted by

$$x = x', \qquad y = \frac{y'(b+g(x'))}{b} + g(x').$$

Defining the transformed field

$$u(x',y') := v(x', y'(b+g(x'))/b + g(x')),$$

since

$$\frac{\partial y'}{\partial x} = b\left(\frac{(-\partial_x g)(b+g) - (y-g)(\partial_x g)}{(b+g)^2}\right)$$
$$= b\left(\frac{(-\partial_x g)(b+g) - (y'(b+g)/b)(\partial_x g)}{(b+g)^2}\right) = -(\partial_{x'} g)\left(\frac{b+y'}{b+g}\right),$$

it is not difficult to show that

$$\partial_x v = (\partial_{x'} u)(\partial_x x') + (\partial_{y'} u)(\partial_x y') = \partial_{x'} u - \frac{\partial_{x'} g}{b+g}(b+y')\partial_{y'} u$$

so that

$$(b+g)\partial_x = (b+g)\partial_{x'} - (\partial_{x'} g)(b+y')\partial_{y'}.$$

Also,

$$\partial_y v = (\partial_{x'} u)(\partial_y x') + (\partial_{y'} u)(\partial_y y') = \frac{b}{b+g}\partial_{y'} u,$$

so that

$$(b+g)\partial_y = b\partial_{y'}.$$

We summarize these as

$$M(x)\partial_x = M(x')\partial_{x'} + N(x',y')\partial_{y'}, \quad M(x)\partial_y = b\partial_{y'},$$

where

$$M(x') := b+g(x'), \quad N(x',y') := -(\partial_{x'}g)(y'+b).$$

Laplace's equation, (3.3.3a), implies

$$0 = M^2\Delta v = M^2\partial_x^2 v + M^2\partial_y^2 v$$
$$= M\partial_x [M\partial_x v] - (\partial_x M)M\partial_x v + M\partial_y [M\partial_y v].$$

Applying the change of variables yields

$$0 = \left[M\partial_{x'} + N\partial_{y'}\right]\left[M\partial_{x'}u + N\partial_{y'}u\right]$$
$$- (\partial_{x'}M)\left[M\partial_{x'}u + N\partial_{y'}u\right] + b\partial_{y'}\left[b\partial_{y'}u\right]$$
$$= M\partial_{x'}[M\partial_{x'}u] + N\partial_{y'}[M\partial_{x'}u] + M\partial_{x'}\left[N\partial_{y'}u\right] + N\partial_{y'}\left[N\partial_{y'}u\right]$$
$$- (\partial_{x'}M)M\partial_{x'}u - (\partial_{x'}M)N\partial_{y'}u + b\partial_{y'}\left[b\partial_{y'}u\right].$$

Attempting to put as many terms in "divergence form" as possible:

$$0 = \partial_{x'}\left[M^2\partial_{x'}u\right] - (\partial_{x'}M)M\partial_{x'}u + \partial_{y'}[NM\partial_{x'}u] - (\partial_{y'}N)M\partial_{x'}u$$
$$+ \partial_{x'}\left[MN\partial_{y'}u\right] - (\partial_{x'}M)N\partial_{y'}u + \partial_{y'}\left[N^2\partial_{y'}u\right] - (\partial_{y'}N)N\partial_{y'}u$$
$$- (\partial_{x'}M)M\partial_{x'}u - (\partial_{x'}M)N\partial_{y'}u + \partial_{y'}\left[b^2\partial_{y'}u\right].$$

This leads to

$$\mathrm{div}'\left[\begin{pmatrix} M^2 & MN \\ MN & b^2+N^2 \end{pmatrix}\nabla'u\right] - (\partial_{x'}g)\begin{pmatrix} M \\ N \end{pmatrix}\cdot\nabla'u = 0,$$

where we have used

$$-2\partial_{x'}M - \partial_{y'}N = -\partial_{x'}g.$$

We write this as

$$\mathrm{div}'\left[A\nabla'u\right] - (\partial_{x'}g)B\cdot\nabla'u = 0,$$

where

$$A := \begin{pmatrix} M^2 & MN \\ MN & b^2+N^2 \end{pmatrix}, \quad B := \begin{pmatrix} M \\ N \end{pmatrix}.$$

We note that

$$A = b^2 I + A_1(g) + A_2(g), \quad B = B_0 + B_1(g),$$

where

$$A_1(g) = \begin{pmatrix} 2bg & -b(y'+b)(\partial_{x'}g) \\ -b(y'+b)(\partial_{x'}g) & 0 \end{pmatrix},$$

$$A_2(g) = \begin{pmatrix} g^2 & -(y'+b)g(\partial_{x'}g) \\ -(y'+b)g(\partial_{x'}g) & (y'+b)^2(\partial_{x'}g)^2 \end{pmatrix},$$

and

$$B_0 = \begin{pmatrix} b \\ 0 \end{pmatrix}, \qquad B_1(g) = \begin{pmatrix} g \\ -(y'+b)(\partial_{x'}g) \end{pmatrix}.$$

Therefore, we can write

$$b^2 \Delta' u = F(x'; g, u), \tag{3.3.4}$$

where

$$F := -\operatorname{div}'\left[A_1 \nabla' u\right] - \operatorname{div}'\left[A_2 \nabla' u\right] + (\partial_{x'}g)B_0 \cdot \nabla' u + (\partial_{x'}g)B_1 \cdot \nabla' u, \tag{3.3.5}$$

and $F = \mathcal{O}(g)$.

The interfacial boundary condition, (3.3.3b), becomes

$$u(x',0) = v(x,g(x)) = \xi(x) = \xi(x').$$

We write the transparent boundary condition, (3.3.3c), as

$$M\partial_y v - MS[v] = 0,$$

and, noting that $u(x',-b) = v(x,-b)$, we find

$$b\partial_{y'} u - MS[u] = 0,$$

so that

$$b\partial_{y'} u - bS[u] = J(x; g, u), \tag{3.3.6}$$

where

$$J = gS[u]. \tag{3.3.7}$$

To close, we write the definition of the DNO, (3.3.1), as

$$MG(g)[\xi] = \left[M\partial_y v - (\partial_x g)M\partial_x v\right]_{y=g}.$$

In our new coordinates we have

$$MG = \left[b\partial_{y'} u - (\partial_{x'}g)M\partial_{x'} u - (\partial_{x'}g)N\partial_{y'} u\right]_{y'=0},$$

which, since $N(x',0) = -(\partial_{x'}g)b$, we rewrite as

$$bG = b\partial_{y'} u - b(\partial_{x'}g)\partial_{x'} u - g(\partial_{x'}g)\partial_{x'} u + b(\partial_{x'}g)^2\partial_{y'} u - gG,$$

or

$$bG = b\partial_{y'} u + H(x; g, u), \tag{3.3.8}$$

where

$$H(x;g,u) := -b(\partial_{x'}g)\partial_{x'}u - g(\partial_{x'}g)\partial_{x'}u + b(\partial_{x'}g)^2\partial_{y'}u - gG. \qquad (3.3.9)$$

3.3.3 Transformed Field Expansions

Gathering all of these equations and dropping the primes we find that we must seek a $(L = 2\pi)$ periodic solution of

$$b^2\Delta u = F \qquad\qquad -b < y < 0,$$
$$u = \zeta \qquad\qquad y = 0,$$
$$b\partial_y u - bS[u] = J \qquad y = -b,$$

to compute the DNO

$$bG = b\partial_y u + H.$$

Supposing that $g = \varepsilon f$, $\varepsilon \ll 1$ for now, our HOPS approach, denoted the "Transformed Field Expansions" (TFE) method, seeks solutions of the form

$$u = u(x,y;\varepsilon) = \sum_{n=0}^{\infty} u_n(x,y)\varepsilon^n, \quad G = G(\varepsilon f)[\zeta] = \sum_{n=0}^{\infty} G_n(f)[\zeta]\varepsilon^n.$$

These can be shown to satisfy

$$b^2\Delta u_n = F_n \qquad\qquad -b < y < 0 \qquad (3.3.10a)$$
$$u_n = \delta_{n,0}\zeta \qquad\qquad y = 0 \qquad (3.3.10b)$$
$$b\partial_y u_n - bS[u_n] = J_n \qquad y = -b, \qquad (3.3.10c)$$

and

$$bG_n = b\partial_y u_n + H_n. \qquad (3.3.11)$$

In these

$$F_n = -\text{div}\left[A_1(f)\nabla u_{n-1}\right] - \text{div}\left[A_2(f)\nabla u_{n-2}\right]$$
$$+ (\partial_x f)B_0 \cdot \nabla u_{n-1} + (\partial_x f)B_1(f) \cdot \nabla u_{n-2},$$

and

$$J_n = fS[u_{n-1}],$$
$$H_n = -b(\partial_x f)\partial_x u_{n-1} - fG_{n-1} - f(\partial_x f)\partial_x u_{n-2} + b(\partial_x f)^2\partial_y u_{n-2}.$$

3.4 A Convergence Proof Succeeds

We now have a recursive sequence of elliptic BVPs, (3.3.10), to solve that *can* be successfully estimated (under generous hypotheses). One can ask why these recursions are "better." An answer is that all derivatives (tangential and normal) are taken at locations *within* the problem domain. Additionally, no more than *first* derivatives of f appear, and no more than *second* powers of f (and its derivative) appear. Our recursive estimation strategy requires two (classical) elements (i.) an "Algebra Lemma" to handle products of functions, and (ii.) an "Elliptic Estimate" to bound solutions of the BVP, (3.3.10), in terms of the inhomogenous terms. The proofs of each can be found in classical texts; see, e.g., Ladyzhenskaya & Ural'tseva [26] or Evans [40].

Lemma 3.4.1. *Given an integer $s \geq 0$ and any $\sigma > 0$, there exists a constant $\mathcal{M} = \mathcal{M}(s)$ such that if $f \in C^s([0,2\pi])$, $w \in H^s[0,2\pi] \times [-b,0]$ then*

$$\|fw\|_{H^s} \leq \mathcal{M}(s) |f|_{C^s} \|w\|_{H^s},$$

and if $\tilde{f} \in C^{s+1/2+\sigma}([0,2\pi])$, $\tilde{w} \in H^{s+1/2}([0,2\pi])$ then

$$\left\|\tilde{f}\tilde{w}\right\|_{H^{s+1/2}} \leq \mathcal{M}(s) \left|\tilde{f}\right|_{C^{s+1/2+\sigma}} \|\tilde{w}\|_{H^{s+1/2}}.$$

Theorem 3.4.2. *Given an integer $s \geq 0$, if $F \in H^s([0,2\pi] \times [-b,0])$, $\xi \in H^{s+3/2}([0,2\pi])$, $J \in H^{s+1/2}([0,2\pi])$, then the unique solution of*

$$b^2 \Delta w = F \qquad\qquad -b < y < 0,$$
$$w = \xi \qquad\qquad y = 0,$$
$$b\partial_y w - bS[w] = J \qquad\qquad y = -b,$$

satisfies

$$\|w\|_{H^{s+2}} \leq C_e \left\{ \|F\|_{H^s} + \|\xi\|_{H^{s+3/2}} + \|J\|_{H^{s+1/2}} \right\},$$

for some constant $C_e = C_e(s)$.

3.4.1 The Analyticity Result

In order to establish our desired result, we begin by demonstrating that the field, u, depends analytically upon ε.

Theorem 3.4.3. *Given any integer $s \geq 0$, if $f \in C^{s+2}([0,2\pi])$ and $\xi \in H^{s+3/2}([0,2\pi])$ then $u_n \in H^{s+2}([0,2\pi] \times [-b,0])$ and*

$$\|u_n\|_{H^{s+2}} \leq KB^n,$$

for constants $K, B > 0$.

With this in hand we are able to show the following.

Theorem 3.4.4. *Given any integer* $s \geq 0$, *if* $f \in C^{s+2}([0, 2\pi])$ *and* $\xi \in H^{s+3/2}([0, 2\pi])$ *then* $G_n \in H^{s+1/2}([0, 2\pi])$ *and*

$$\|G_n\|_{H^{s+1/2}} \leq \tilde{K}B^n,$$

for constants $\tilde{K}, B > 0$.

For these we require the following inductive lemma.

Lemma 3.4.5. *Given an integer* $s \geq 0$, *if* $f \in C^{s+2}([0, 2\pi])$ *and*

$$\|u_n\|_{H^{s+2}} \leq KB^n, \quad \forall \, n < \bar{n},$$

for constants $K, B > 0$, *then there exists a constant* $\bar{C} > 0$ *such that*

$$\max\left\{\|F_{\bar{n}}\|_{H^s}, \|J_{\bar{n}}\|_{H^{s+1/2}}\right\} \leq K\bar{C}\left\{|f|_{C^{s+2}} B^{\bar{n}-1} + |f|^2_{C^{s+2}} B^{\bar{n}-2}\right\}.$$

Proof. For simplicity we focus on the term

$$F_{\bar{n}} = -\mathrm{div}\left[A_1 \nabla u_{\bar{n}-1}\right] + \cdots = -\partial_x\left[A_1^{xx}\partial_x\left[u_{\bar{n}-1}\right]\right] + \cdots,$$

where $A_1^{xx}(x) = 2bf(x)$. We can estimate

$$\|F_{\bar{n}}\|_{H^s} \leq \left\|A_1^{xx}\partial_x\left[u_{\bar{n}-1}\right]\right\|_{H^{s+1}} + \cdots \leq \mathcal{M}\left|A_1^{xx}\right|_{C^{s+1}}\|u_{\bar{n}-1}\|_{H^{s+2}} + \cdots$$

$$\leq \mathcal{M}\left(2b\,|f|_{C^{s+1}}\right)KB^{\bar{n}-1} + \cdots$$

So, if we choose

$$\bar{C} > 2b\mathcal{M},$$

then we are done. $\qquad\square$

Now we are in a position to prove Theorem 3.4.3.

Theorem 3.4.3. We work by induction in n: At order $n = 0$ we use the elliptic estimate to deduce

$$\|u_0\|_{H^{s+2}} \leq C_e\|\xi\|_{H^{s+3/2}},$$

so we set $K := C_e\|\xi\|_{H^{s+3/2}}$. Now, we suppose that the inductive estimate is valid for all $n < \bar{n}$ and use the elliptic estimate at order $\bar{n}$:

$$\|u_{\bar{n}}\|_{H^{s+2}} \leq C_e\left\{\|F_{\bar{n}}\|_{H^s} + \|J_{\bar{n}}\|_{H^{s+1/2}}\right\}.$$

Lemma 3.4.5 tells us that

$$\|u_{\bar{n}}\|_{H^{s+2}} \leq C_e 2\bar{C}K\left\{|f|_{C^{s+2}} B^{\bar{n}-1} + |f|^2_{C^{s+2}} B^{\bar{n}-2}\right\}.$$

So, we are done if we choose

$$B > \max\left\{4C_e\bar{C}, 2\sqrt{C_e\bar{C}}\right\}|f|_{C^{s+2}}.$$

$\qquad\square$

We can now prove the analyticity of the DNO.

Theorem 3.4.4. Once again, we use induction in n: At order $n = 0$ we have

$$G_0(x) = \partial_y u_0(x, 0),$$

so, from the previous theorem,

$$\|G_0\|_{H^{s+1/2}} = \|\partial_y u_0\|_{H^{s+1/2}} \leq \|u_0\|_{H^{s+2}} \leq K,$$

and we choose $\tilde{K} \geq K$. Now, we assume the inductive estimate is true for all $n < \bar{n}$ and estimate

$$\|G_{\bar{n}}\|_{H^{s+1/2}} = \|\partial_y u_{\bar{n}}\|_{H^{s+1/2}} + (1/b) \|H_{\bar{n}}\|_{H^{s+1/2}},$$

A lemma similar to Lemma 3.4.5 delivers

$$\|H_{\bar{n}}\|_{H^{s+1/2}} \leq \tilde{K}\bar{C}\left\{|f|_{C^{s+2}} B^{\bar{n}-1} + |f|_{C^{s+2}}^2 B^{\bar{n}-2}\right\}.$$

Again, we are done if we choose

$$B > \max\left\{4C_e\bar{C}, 2\sqrt{C_e\bar{C}}\right\} |f|_{C^{s+2}}.$$

$\square$

Remark We note that the theory can be extended in a number of ways. In particular to finite depth ($h < \infty$), three dimensions (by simply viewing $p \in \mathbf{Z}^2$) [5], and joint analyticity with respect to multiple boundaries [21]. Furthermore, analytic continuation can be justified by considering perturbations about a generic real-valued function $f_0(x)$ [7]. Finally variations of the DNO with respect to boundary deformations can also be shown to be analytic [23].

3.5 Stable Numerics

In light of the powerful and straightforward proof we were able to deliver for the analyticity of both the field and DNO, one can wonder about a numerical simulation based upon these TFE recursions. We recall the recursions for the u_n, (3.3.10), and G_n, (3.3.11), and the first thing we notice is the inhomogeneous nature of these, i.e. $F_n, J_n, H_n \not\equiv 0$ in general. Consequently, a method based upon separation of variables (such as FE and OE) is no longer available, thus a *volumetric* approach is mandated.

As we specify this method, we note that we have already expanded in a Taylor series in the perturbation parameter ε and this is our first "discretization." We work recursively at each perturbation order, so that once we know $\{u_0, \dots, u_{n-1}\}$ and $\{G_0, \dots, G_{n-1}\}$ we can form $\{F_n, J_n, H_n\}$. The periodic lateral boundary conditions suggest a Fourier expansion in the x variable.

Finally, at each perturbation order and each wavenumber one must solve a two-point boundary value problem. For this we choose a Chebyshev collocation method [41, 42], and thus we approximate u by

$$u^{N,N_x,N_y}(x',y';\varepsilon) := \sum_{n=0}^{N} \sum_{p=-N_x/2}^{N_x/2-1} \sum_{\ell=0}^{N_y} \hat{u}_n(p,\ell)e^{ipx'} T_\ell(2y'/b+1)\varepsilon^n,$$

where the $\hat{u}_n(p,\ell)$ are determined by a collocation approach [7, 9].

3.5.1 Numerical Tests

We now seek to validate and evaluate this third HOPS scheme, TFE, for approximating DNOs, and compare its performance against that of the previous approaches, FE and OE. Recall the exact solution we used in a previous lecture to test the FE and OE algorithms. If we choose a wavenumber, say r, and a profile $f(x)$, for a given $\varepsilon > 0$, it is easy to see that the Dirichlet data

$$\xi_r(x;\varepsilon) := v_r(x,\varepsilon f(x)) = e^{|r|\varepsilon f(x)}e^{irx},$$

generates Neumann data

$$v_r(x;\varepsilon) := \left[\partial_y v_r - \varepsilon(\partial_x f)\partial_x v_r\right](x,\varepsilon f(x))$$
$$= [|r| - \varepsilon(\partial_x f)(ir)]e^{|r|\varepsilon f(x)}e^{irx}.$$

We consider a problem with geometric and numerical parameters

$$L = 2\pi, \quad \varepsilon = 0.5, \quad b = 1, \quad f(x) = \exp(\cos(x)),$$
$$N_x = 256, \quad N_y = 64, \quad N = 16. \tag{3.5.1}$$

In Figure 3.2 we display results of our numerical experiments with the FE, OE, and TFE algorithms as N is refined from 0 to 16 using Taylor summation. We repeat this with Padé approximation in Figure 3.3 for this moderate perturbation. We notice that, while this deformation is not small, the TFE computation demonstrates it is *within* the disk of analyticity of the DNO. The instabilities in the FE and OE methods (first reported in [5]) render these methods useless with Taylor summation. We do point out that the analytic continuation technique of Padé approximation not only enhances the TFE results, but also renders the FE and OE simulations useful.

We now reconsider these calculations in the context of a much larger deformation size $\varepsilon = 1.0$ (twice as large). In Figures 3.4 and 3.5 we show results generated by the FE, OE, and TFE algorithms as N is refined from 0 to 16 using Taylor and Padé summation, respectively. As is evident from the TFE simulation, here the deformation is large enough so that we are no longer within the disk of analyticity of the DNO. While none of the HOPS algorithms

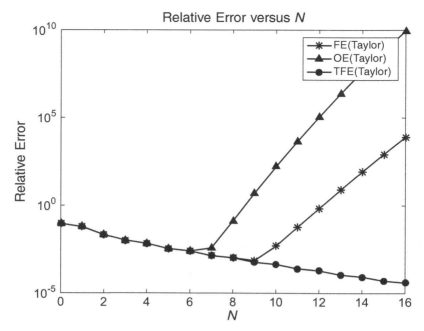

Figure 3.2. Relative error in FE, OE, and TFE algorithms with Taylor summation versus perturbation order N for configuration, (3.5.1), with moderate deformation $\varepsilon = 0.5$.

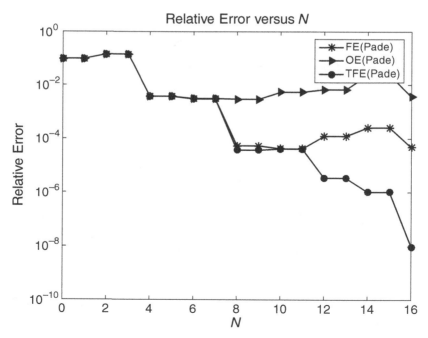

Figure 3.3. Relative error in FE, OE, and TFE algorithms with Padé summation versus perturbation order N for configuration, (3.5.1), with moderate deformation $\varepsilon = 0.5$.

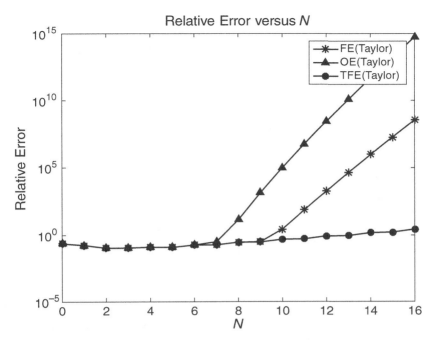

Figure 3.4. Relative error in FE, OE, and TFE algorithms with Taylor summation versus perturbation order N for configuration, (3.5.1), with large deformation $\varepsilon = 1.0$.

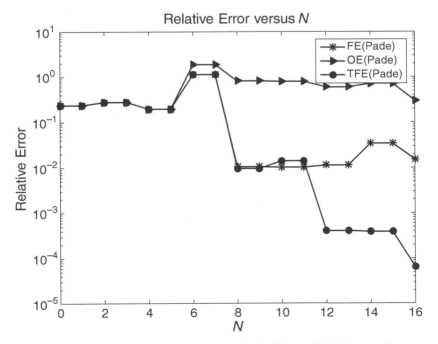

Figure 3.5. Relative error in FE, OE, and TFE algorithms with Padé summation versus perturbation order N for configuration, (3.5.1), with large deformation $\varepsilon = 1.0$.

deliver any accuracy with Taylor summation (as they are prohibited by the theory), Padé approximation allows one to access the domain of analyticity *outside* the disk of convergence.

Remark As with FE and OE, the TFE algorithm can be extended to finite depth ($h < \infty$) and three dimensions (using $p \in \mathbf{Z}^2$) [6].

Dedication. The author would like to dedicate this contribution to his wonderful daughter, Emma. Without her love and enthusiasm, life would not be near as much fun. He learned many things at the "Theory of Water Waves" programme at the Isaac Newton Institute, but none was more surprising than Emma's demonstration of how exciting it is to ride around Cambridge on the upper level (at the front, of course) of a double-decker bus.

Acknowledgments

The author gratefully acknowledges support from the National Science Foundation through grant No. DMS–1115333.

The author also would like to thank the Issac Newton Institute at the University of Cambridge for providing the facilities where this meeting was held, as well as T. Bridges, P. Milewski, and M. Groves for organizing this exciting meeting.

References

[1] Coifman, R., and Meyer, Y. 1985. Nonlinear Harmonic Analysis and Analytic Dependence. Pages 71–78 of: *Pseudodifferential operators and applications (Notre Dame, Ind., 1984)*. *Amer. Math. Soc.*, pp. 71–78.

[2] Craig, Walter, Schanz, Ulrich, and Sulem, Catherine. 1997. The Modulation Regime of Three-Dimensional Water Waves and the Davey-Stewartson System. *Ann. Inst. Henri Poincaré*, **14**, 615–667.

[3] Nicholls, David P. 1998. *Traveling Gravity Water Waves in Two and Three Dimensions*. Ph.D. thesis, Brown University.

[4] Craig, Walter, and Nicholls, David P. 2000. Traveling Two and Three Dimensional Capillary Gravity Water Waves. *SIAM J. Math. Anal.*, **32**(2), 323–359.

[5] Nicholls, David P., and Reitich, Fernando. 2001a. A new approach to analyticity of Dirichlet-Neumann operators. *Proc. Roy. Soc. Edinburgh Sect. A*, **131**(6), 1411–1433.

[6] Nicholls, David P., and Reitich, Fernando. 2001b. Stability of High-Order Perturbative Methods for the Computation of Dirichlet-Neumann Operators. *J. Comput. Phys.*, **170**(1), 276–298.

[7] Nicholls, David P., and Reitich, Fernando. 2003. Analytic Continuation of Dirichlet-Neumann Operators. *Numer. Math.*, **94**(1), 107–146.

[8] Nicholls, David P., and Reitich, Fernando. 2004a. Shape Deformations in Rough Surface Scattering: Cancellations, Conditioning, and Convergence. *J. Opt. Soc. Am. A*, **21**(4), 590–605.

[9] Nicholls, David P., and Reitich, Fernando. 2004b. Shape Deformations in Rough Surface Scattering: Improved Algorithms. *J. Opt. Soc. Am. A*, **21**(4), 606–621.

[10] Nicholls, David P., and Nigam, Nilima. 2004. Exact Non-Reflecting Boundary Conditions on General Domains. *J. Comput. Phys.*, **194**(1), 278–303.

[11] Nicholls, David P., and Nigam, Nilima. 2006. Error Analysis of a Coupled Finite Element/DtN Map Algorithm on General Domains. *Numer. Math.*, **105**(2), 267–298.

[12] Nicholls, David P., and Reitich, Fernando. 2005. On Analyticity of Traveling Water Waves. *Proc. Roy. Soc. Lond., A*, **461**(2057), 1283–1309.

[13] Nicholls, David P., and Reitich, Fernando. 2006. Rapid, Stable, High-Order Computation of Traveling Water Waves in Three Dimensions. *Eur. J. Mech. B Fluids*, **25**(4), 406–424.

[14] Akers, Benjamin F., and Nicholls, David P. 2010. Traveling Waves in Deep Water with Gravity and Surface Tension. *SIAM J. App. Math.*, **70**(7), 2373–2389.

[15] Nicholls, David P. 2007. Spectral Stability of Traveling Water Waves: Analytic Dependence of the Spectrum. *J. Nonlin. Sci.*, **17**(4), 369–397.

[16] Akers, Benjamin F., and Nicholls, David P. 2012b. Spectral Stability of Deep Two–Dimensional Gravity Water Waves: Repeated Eigenvalues. *SIAM J. App. Math.*, **72**(2), 689–711.

[17] Akers, Benjamin F., and Nicholls, David P. 2012a. Spectral Stability of Deep Two–Dimensional Gravity–Capillary Water Waves. *Stud. App. Math.*, **130**, 81–107.

[18] Akers, Benjamin, and Nicholls, David P. 2014. Spectral Stability of Finite Depth Water Waves. *Euro. J. Mech. B/Fluids*, **46**, 181–189.

[19] Hu, Bei, and Nicholls, David P. 2005. Analyticity of Dirichlet–Neumann Operators on Hölder and Lipschitz Domains. *SIAM J. Math. Anal.*, **37**(1), 302–320.

[20] Hu, Bei, and Nicholls, David P. 2010. The Domain of Analyticity of Dirichlet–Neumann Operators. *Proc. Royal Soc. Edinburgh A*, **140**(2), 367–389.

[21] Nicholls, David P., and Taber, Mark. 2008. Joint Analyticity and Analytic Continuation for Dirichlet–Neumann Operators on Doubly Perturbed Domains. *J. Math. Fluid Mech.*, **10**(2), 238–271.

[22] Nicholls, David P., and Taber, Mark. 2009. Detection of Ocean Bathymetry from Surface Wave Measurements. *Euro. J. Mech. B/Fluids*, **28**(2), 224–233.

[23] Fazioli, C., and Nicholls, David P. 2008. Parametric Analyticity of Functional Variations of Dirichlet–Neumann Operators. *Diff. Integ. Eqns.*, **21**(5–6), 541–574.

[24] Fazioli, Carlo, and Nicholls, David P. 2010. Stable Computation of Variations of Dirichlet–Neumann Operators. *J. Comp. Phys.*, **229**(3), 906–920.

[25] Nicholls, David P., and Shen, Jie. 2009. A Rigorous Numerical Analysis of the Transformed Field Expansion Method. *SIAM J. Num. Anal.*, **47**(4), 2708–2734.

[26] Ladyzhenskaya, Olga A., and Ural'tseva, Nina N. 1968. *Linear and quasilinear elliptic equations.* New York: Academic Press.

[27] Friedman, Avner, and Reitich, Fernando. 2001. Symmetry-Breaking Bifurcation of Analytic Solutions to Free Boundary Problems: An Application to a Model of Tumor Growth. *Trans. Amer. Math. Soc.*, **353**, 1587–1634.

[28] Chandezon, J., Maystre, D., and Raoult, G. 1980. A New Theoretical Method for Diffraction Gratings and its Numerical Application. *J. Opt.*, **11**(7), 235–241.

[29] Chandezon, J., Dupuis, M.T., Cornet, G., and Maystre, D. 1982. Multicoated Gratings: A Differential Formalism Applicable in the Entire Optical Region. *J. Opt. Soc. Amer.*, **72**(7), 839.

[30] Li, L., Chandezon, J., Granet, G., and Plumey, J. P. 1999. Rigorous and Efficient Grating-Analysis Method Made Easy for Optical Engineers. *Appl. Opt.*, **38**(2), 304–313.

[31] Phillips, N. A. 1957. A Coordinate System Having Some Special Advantages for Numerical Forecasting. *J. Atmos. Sci.*, **14**(2), 184–185.

[32] Johnson, Claes, and Nédélec, J.-Claude. 1980. On the Coupling of Boundary Integral and Finite Element Methods. *Math. Comp.*, **35**(152), 1063–1079.

[33] Han, Hou De, and Wu, Xiao Nan. 1985. Approximation of Infinite Boundary Condition and its Application to Finite Element Methods. *J. Comput. Math.*, **3**(2), 179–192.

[34] Keller, Joseph B., and Givoli, Dan. 1989. Exact Nonreflecting Boundary Conditions. *J. Comput. Phys.*, **82**(1), 172–192.

[35] Givoli, Dan. 1991. Nonreflecting Boundary Conditions. *J. Comput. Phys.*, **94**(1), 1–29.

[36] Givoli, Dan, and Keller, Joseph B. 1994. Special Finite Elements for Use with High-Order Boundary Conditions. *Comput. Methods Appl. Mech. Engrg.*, **119**(3–4), 199–213.

[37] Givoli, Dan. 1992. *Numerical methods for problems in infinite domains.* Studies in Applied Mechanics, vol. 33. Amsterdam: Elsevier Scientific Publishing Co.

[38] Grote, Marcus J., and Keller, Joseph B. 1995. On Nonreflecting Boundary Conditions. *J. Comput. Phys.*, **122**(2), 231–243.

[39] Givoli, D. 1999. Recent Advances in the DtN FE Method. *Arch. Comput. Methods Engrg.*, **6**(2), 71–116.

[40] Evans, Lawrence C. 1998. *Partial differential equations.* Providence, RI: American Mathematical Society.

[41] Gottlieb, David, and Orszag, Steven A. 1977. *Numerical analysis of spectral methods: theory and applications.* Philadelphia, PA.: Society for Industrial and Applied Mathematics. CBMS-NSF Regional Conference Series in Applied Mathematics, No. 26.

[42] Canuto, Claudio, Hussaini, M. Yousuff, Quarteroni, Alfio, and Zang, Thomas A. 1988. *Spectral methods in fluid dynamics.* New York: Springer-Verlag.

4

High-Order Perturbation of Surfaces Short Course: Stability of Travelling Water Waves

Benjamin F. Akers

Abstract

In this contribution we present High-Order Perturbation of Surfaces (HOPS) methods as applied to the spectral stability problem for traveling water waves. The Transformed Field Expansion method (TFE) is used for both the traveling wave and its spectral data. The Lyapunov-Schmidt reductions for simple and repeated eigenvalues are compared. The asymptotics of modulational instabilities are discussed.

4.1 Introduction

The water wave stability problem has a rich history, with great strides made in the late sixties in the work of Benjamin and Feir [1] and in the ensuing development of Resonant Interaction Theory (RIT) [2–5]. The predictions of RIT have since been leveraged heavily by numerical methods; the influential works of MacKay and Saffman [6] and McLean [7] led to a taxonomy of water wave instabilities based on RIT (Class I and Class II instabilities). The most recent review article is that of Dias & Kharif [8]; since the publication of this review a number of modern numerical stability studies have been conducted [9–13].

In these lecture notes, we explain how the spectral data of traveling water waves may be computed using a High-Order Perturbation of Surfaces (HOPS) approach, which numerically computes the coefficients in amplitude-based series expansions [14]. For the water wave problem, a crucial aspect of any numerical approach is the method used to handle the unknown fluid domain. Just as in the traveling waves lecture of this short course, numerical results will be presented from the Transformed Field Expansion (TFE) method, whose development for the spectral stability problem appears in [13, 15–17].

The TFE method computes the spectral data as a series in wave slope/ amplitude, and thus relies on analyticity of the spectral data in amplitude. A large number of studies of the spectrum have been made that do not make such

51

an assumption [9, 10, 18, 19]. On the other hand, it is known that the spectrum is analytic for all Bloch parameters at which eigenvalues are simple in the zero amplitude limit [20]. Numerically it has been observed that the spectrum is analytic in amplitude at Bloch parameters for which there are eigenvalue collisions, but that the disc of analyticity is discontinuous in Bloch parameter. This discontinuity in radius is due to modulational instabilities, as explained in [21].

These notes begin by introducing the spectral stability problem for traveling water waves. Next, the TFE formulation is described, followed by leading-order asymptotics of the spectral data. In section 4.4, the perturbation series approach is discussed, followed by the Lyapunov-Schmidt reduction for triad collisions in section 4.5. Finally, we present instabilities due to the modulation of triad collisions in section 4.6.

4.2 Spectral Stability of Water Waves

The widely accepted model for the motion of waves on the surface of a large body of water with constant surface tension, and in the absence viscosity, are the Euler equations

$$\phi_{xx} + \phi_{zz} = 0, \qquad z < \varepsilon\eta, \qquad (4.2.1a)$$

$$\phi_z = 0, \qquad z = -H \qquad (4.2.1b)$$

$$\eta_t + \varepsilon\eta_x\phi_x = \phi_z, \qquad z = \varepsilon\eta, \qquad (4.2.1c)$$

$$\phi_t + \frac{\varepsilon}{2}\left(\phi_x^2 + \phi_z^2\right) + \eta - \sigma\left(\frac{\eta_{xx}}{(1 + \varepsilon^2\eta_x^2)^{3/2}}\right) = 0, \qquad z = \varepsilon\eta, \qquad (4.2.1d)$$

where η is the free-surface displacement and ϕ is the velocity potential. These equations describe the motion of an inviscid, incompressible fluid undergoing an irrotational motion. System (4.2.1) has been nondimensionalized as in [15, 22]. We assume that the wave slope, $\varepsilon = A/L$ is small (A is a typical amplitude and L, the characteristic horizontal length, is chosen in the nondimensionalization so that the waves have spatial period 2π). Also, the vertical dimension has been nondimensionalized using the wavelength, so the quantity H is nondimensional ($H = h/L$). For simplicity's sake, in these notes we will consider the deep water limit $H \to \infty$. To compute traveling waves, the time dependence of (4.2.1) is prescribed, using the ansatz

$$\eta(x,t) = \eta(x+ct) \qquad \text{and} \qquad \phi(x,z,t) = \phi(x+ct,z).$$

In the spectral stability problem, solutions to the traveling waves problem, $\bar{\eta}, \bar{\phi}, c$, are assumed to be known. These waves are then perturbed a small amount, δ, and linearized by substituting

$$\eta(x,t) = \bar{\eta}(x-ct) + \delta\zeta(x-ct)e^{\lambda t}, \qquad \phi(x,z,t) = \bar{\phi}(x-ct,z) + \delta v(x-ct,z)e^{\lambda t},$$

and neglecting quadratic powers of δ. The resulting problem will be a non-constant coefficient, generalized spectral problem, still on an unknown domain. For amplitude expansion based methods, an effective way to handle the domain is with the TFE [23]. The TFE method is both spectrally accurate and numerically stable; see the previous lecture in this short course or [24] for a comparison of other methods in the traveling wave problem.

4.2.1 Transformed Field Expansions

The TFE method solves for the field (perturbations of the velocity potential and free surface) via an amplitude expansion of the Euler equations (4.2.1) after two transformations. First, Laplace's equation is solved exactly below a prescribed depth, $z = -a$ (via the same operator presented in part II of this short course). Second, the domain above this depth is transformed to a strip by the simple change of variables,

$$z \to a\left(\frac{z - \varepsilon\eta}{a + \varepsilon\eta}\right),$$

after which the stability problem becomes

$$v_{xx} + v_{zz} = \tilde{F}, \qquad -a < z < 0, \qquad (4.2.2a)$$

$$v_z - Tv = \tilde{J}, \qquad z = -a, \qquad (4.2.2b)$$

$$\lambda\zeta + c\zeta_x - v_z = \tilde{Q} \qquad z = 0, \qquad (4.2.2c)$$

$$\lambda v + cv_x + (1 - \sigma\partial_x^2)\zeta = \tilde{R}, \qquad z = 0. \qquad (4.2.2d)$$

The symbols $\tilde{F}, \tilde{J}, \tilde{Q}$, and $\tilde{R}$ contain all the non-constant coefficient terms, depending on $\bar{\eta}, \bar{\phi}, \zeta$, and v. This formulation yields stable, fast recursions in a Boundary Perturbation method, the formulation is discussed at length in [13, 15].

To consider the broadest class of perturbations, we will append Bloch boundary conditions to (4.2.2). If the traveling wave is of period $L = 2\pi$, the perturbations satisfy,

$$\zeta(x + 2\pi) = \zeta(x)e^{2\pi ip}$$

with similar boundary condition for v, in which $p \in \mathbb{R}$ is the Bloch, or Floquet, parameter. This decomposes the continuous spectrum of the original problem (which has periodic coefficients) to a sets of discrete eigenvalues for each value of the Bloch parameter. After this decomposition, it makes sense to construct a perturbation expansion about the flat-state eigenvalues.

To compute the spectrum at fixed Bloch parameter, all variables are expanded as a series in amplitude, both the traveling wave data (as in the traveling wave lecture of this short course) and the spectral data,

$$v = \sum_{n=0}^{\infty} v_n \varepsilon^n, \qquad \zeta = \sum_{n=0}^{\infty} \zeta_n \varepsilon^n, \qquad \text{and} \qquad \lambda = \sum_{n=0}^{\infty} \lambda_n \varepsilon^n. \qquad (4.2.3)$$

In the TFE formulation, this results in a sequence of linear problems

$$v_{n,xx} + v_{n,zz} = \tilde{F}_n(x,z), \qquad -a < z < 0, \qquad (4.2.4a)$$

$$v_{n,z} - T v_n = \tilde{J}_n(x), \qquad z = -a, \qquad (4.2.4b)$$

$$\lambda_0 \zeta_n + c_0 \zeta_{n,x} - v_{n,z} = \tilde{Q}_n(x) - \lambda_n \zeta_0, \qquad z = 0, \qquad (4.2.4c)$$

$$\lambda_0 v_n + c_0 v_{n,x} + (1 - \sigma \partial_x^2) \zeta_n = \tilde{R}_n(x) - \lambda_n v_0, \qquad z = 0. \qquad (4.2.4d)$$

The exact formula for the $\tilde{F}_n, \tilde{J}_n, \tilde{Q}_n$, and $\tilde{R}_n$ can be found in [13, 20], and we direct the motivated reader to the (tedious) details provided therein. These equations can be rapidly solved via Fourier collocation in the horizontal dimension, and an elliptic solver in the vertical dimension, for example, the Chebychev-Tau method [15]. It is in this formulation that the spectrum has been calculated to all orders, about both simple and repeated eigenvalues, in deep water and finite depth, with and without surface tension [15–17].

4.3 Leading Order Behavior

In the flat water configuration, when $n = 0$, the right hand side of (4.2.4) vanishes, $\tilde{F}_0 = \tilde{J}_0 = \tilde{Q}_0 = \tilde{R}_0 = 0$. The resulting problem is exactly solvable, with eigenvalues

$$\lambda_0 = \pm i \omega(k_j) + i c_0 \cdot k_j, \qquad (4.3.1)$$

where k_j are the wave numbers of the perturbation (including the Bloch parameter) and $\omega(k) = \sqrt{|k|(1 + \sigma |k|^2)}$ is the dispersion relation of the potential flow equations (4.2.1). The leading order solution for the eigenfunctions depends on the multiplicity of the eigenvalue in question. We will discuss these solutions separately.

4.3.1 Simple Eigenvalues

If the spectrum is simple, ignoring perturbations of the mean, so that $k_1 \neq 0$, the leading order eigenfunctions are

$$\begin{pmatrix} \zeta_0(x) \\ v_0(x,0) \end{pmatrix} = \beta_{0,1} \begin{pmatrix} 1 \\ \frac{\lambda_0 + i c_0 \cdot k_1}{|k_1|} \end{pmatrix} e^{i k_1 \cdot x} = \beta_{0,1} v_1. \qquad (4.3.2)$$

Before one can compute (ζ_n, v_n), recall that the size of an eigenfunction is not a meaningful quantity. We choose to define the size of the eigenfunctions by the size of their Fourier coefficient at k_1, here $\hat{\zeta}(k_1) = 1$. Such a choice is necessary, without which one cannot hope to uniquely compute an eigenfunction. This particular normalization of the eigenfunctions allows one to avoid including homogeneous solutions at later perturbation orders, simplifying the entire procedure.

Without collisions, the equations for λ_n are linear, and all λ_n are pure imaginary; simple eigenvalues do not lead to instability (within the radius of convergence of their series expansions). This observation motivated in-depth study of the radius of convergence of these expansions. Since the spectrum is simple almost everywhere (for almost all values of the Bond number and Bloch parameters), one might expect that the radius of convergence of these series could be used to detect instabilities, see [13]. Eigenvalue collisions however, even those of opposite Krein signature, do not always create instabilities. Next we consider the spectrum about resonant Bloch parameters, where eigenvalues collide.

4.3.2 Eigenvalue Collisions

At Bloch parameters where the flat-state ($\varepsilon = 0$) contains an eigenvalue collision, $\lambda_0(k_1) = \lambda_0(k_2)$, implying

$$\pm \omega(k_1) \pm \omega(k_2) = (k_1 - k_2)c_0.$$

Including the broadest class of perturbations, $k_j = n_j + p$ with $p \in [0,1)$ and $n_j \in \mathbb{Z}$, this condition can be rewritten as

$$k_1 - k_2 = mk_0 \quad \text{and} \quad \omega(k_1) \pm \omega(k_2) = m\omega(k_0), \tag{4.3.3}$$

where $k_0 = 1$ is the frequency of the Stokes' wave, and $m = k_1 - k_2 \in \mathbb{Z}$. Equation (4.3.3), states that the perturbations are waves whose temporal and spatial frequencies resonate with m copies of the Stokes wave (for $m + 2$ total waves). It is natural to label these resonances with the naming convention of Resonant Interaction Theory (RIT). In RIT, $m = 1$ is labeled a triad interaction, $m = 2$ is labeled a quartet, $m = 3$ is a quintet, etc. [3, 6, 7]. The leading order perturbations at repeated eigenvalues are superpositions of the eigenfunctions from the simple case,

$$\begin{pmatrix} \zeta_0(x) \\ v_0(x,0) \end{pmatrix} = \begin{pmatrix} 1 \\ \frac{\lambda_0 + ic_0 \cdot k_1}{|k_1|} \end{pmatrix} e^{ik_1 \cdot x} + \beta_{0,2} \begin{pmatrix} 1 \\ \frac{\lambda_0 + ic_0 \cdot k_2}{|k_2|} \end{pmatrix} e^{ik_2 \cdot x} = v_1 + \beta_{0,2} v_2.$$

$$\tag{4.3.4}$$

Just as for simple eigenvalues, the size of the eigenfunction is chosen so that $\hat{\zeta}(k_1) = 1$, later orders will not be supported at wavenumber k_1. From the perspective of the leading order problem, $\beta_{0,2}$ is free to take any value. At later orders, solvability conditions will allow only a discrete set of $\beta_{0,2}$, analogous to the Wilton ripple discussion earlier in this short course. The form of the solvability conditions, and the solution at later orders depends on the the the type of eigenvalue collision, which occurs at leading order, encoded by the value of m. All values of m are discussed in [15]; for brevity, we present only the triad case, $m = 1$. First we will introduce the perturbation series in a simplified notation.

4.4 Eigenvalue Perturbation

After Bloch decomposition, (4.2.2) can be thought of as a generalized eigenvalue problem,

$$(A - \lambda B)v = 0.$$

In this notation, (4.2.4) is recast

$$(A_0 - \lambda_0 B_0)v_n = \sum_{p=0}^{n-1} \left(A_{n-p} - \sum_{q=0}^{n-p} \lambda_q B_{n-p-q} \right) v_p. \tag{4.4.1}$$

To construct the solution (v, λ), requires knowledge of the solutions of the adjoint of the leading order problem,

$$(A_0 - \lambda_0 B_0)^* \psi = 0. \tag{4.4.2}$$

Simple eigenvalues have a single non-trivial solution to (4.4.2), which we label ψ_1. At a generic eigenvalue collision (of only two eigenvalues), equation (4.4.2) has two solutions ψ_1 and ψ_2. For discussion purposes, we will choose to label the solutions based on the wavenumber (so that function ψ_j has spatial dependence proportional to $e^{ik_j x}$). The eigenfunctions of the leading order problem

$$(A_0 - \lambda_0 B_0)v_0 = 0 \tag{4.4.3}$$

will be similarly labeled, so that $v_0 = v_1$, or in the case of an eigenvalue collision $v_0 = v_1 + \beta_{0,2} v_2$. As in the previous section, the eigenfunctions v_j are labeled so that they are supported at wavenumber k_j.

At every order, n, solving equation (4.4.1) requires two steps. First, one must impose that the right hand side is orthogonal to the solutions of (4.4.2). Afterward, the equation is solvable, and one may invert the linear operator against its range. For simple eigenvalues, both steps are trivial. Solvability requires enforcing

$$\lambda_n = \frac{1}{\langle \psi_1, B_0 v_0 \rangle} \left\langle \psi_1, \sum_{p=0}^{n-1} A_{n-p} v_p - \sum_{q=0}^{n-1} \sum_{p=0}^{n-q-1} \lambda_q B_{n-p-q} v_p \right\rangle.$$

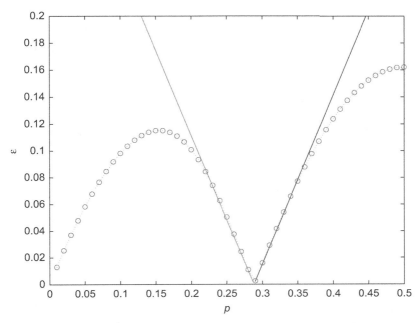

Figure 4.1. The radius of convergence of the spectral data as a function of Bloch parameter is numerically computed (circles). A triad eigenvalue collision occurs at $(p, \varepsilon) \approx (0.29, 0)$. An asymptotic prediction, equation (4.6.1), for the location of modulated instabilities, and thus loss of analyticity, is marked with the solid line.

Given the terms in the series expansion for the spectral data, one may ask "to what extent is this series summable?" The answer informs on the values of ε for which a boundary perturbation method can compute the spectrum, i.e., the radius of the disc of analyticity of the spectrum. This radius may be computed using a standard convergence test on the terms in the Taylor series (4.2.3), or by computing the Padé interpolant and finding its smallest uncancelled pole. An example of the numerically computed radius of convergence of this series as a function of Bloch parameter p is marked with circles in Figure 4.1.

4.5 Triad Instabilities

Triad instabilities arise about eigenvalue collisions $\lambda_0(k_1) = \lambda_0(k_2)$, where

$$k_1 - k_2 = \pm k_0 \qquad \text{and} \qquad \omega(k_1) \pm \omega(k_2) = \pm \omega(k_0)$$

in which k_0 is the Stokes wave, here $k_0 = 1$. Triad resonances then, are those where both the spatial and temporal frequencies of the perturbations differ by exactly those of the Stokes wave.

The boundary perturbation approach expands the traveling wave as a series, where the $O(\varepsilon^n)$ term is supported at wave numbers $|k| \leq n$. The traveling wave corrections (η_{n-j}, ϕ_{n-j}) occur as coefficients of the previous eigenfunctions (ζ_j, v_j) in the Nth right hand sides $(\tilde{F}_n, \tilde{J}_n, \tilde{Q}_n, \tilde{R}_n)$ of equation (4.2.4). In order to get non-trivial nonlinear solvability conditions, the coefficients of the wavenumber k_2 must appear in the solvability condition for k_1, thus there must be a wave interaction with $k_1 - k_2 = m$ copies of the Stokes wave. Wavenumber $k = m$ is first present in the traveling wave at $O(\varepsilon^m)$, thus for triads this first occurs at $O(\varepsilon)$. These first solvability conditions are

$$
\begin{pmatrix} \langle \psi_1, (A_1 - \lambda_0 B_1 - \lambda_1 B_0) v_1 \rangle & \langle \psi_1, (A_1 - \lambda_0 B_1 - \lambda_1 B_0) v_2 \rangle \\ \langle \psi_2, (A_1 - \lambda_0 B_1 - \lambda_1 B_0) v_1 \rangle & \langle \psi_2, (A_1 - \lambda_0 B_1 - \lambda_1 B_0) v_2 \rangle \end{pmatrix} \begin{pmatrix} \beta_{0,1} \\ \beta_{0,2} \end{pmatrix} = 0.
$$
(4.5.1)

For the water wave problem, a number of these inner products vanish. In our labeling, ψ_j and v_j are supported at k_j, and since A_1 and B_1 are supported at wavenumber $k = \pm 1$,

$$
\langle \psi_1, (A_1 - \lambda_0 B_1) v_1 \rangle = \langle \psi_1, \lambda_1 B_0 v_2 \rangle = \langle \psi_2, (A_1 - \lambda_0 B_1) v_2 \rangle = \langle \psi_2, \lambda_1 B_0 v_1 \rangle = 0.
$$

An inner product can be non-zero only if the wave numbers of the functions in the inner product sum to zero, hence the connection to RIT. The leading order eigenvalue correction is

$$
\lambda_1 = \pm \sqrt{\frac{\langle \psi_1, (A_1 - \lambda_0 B_1) v_2 \rangle \langle \psi_2, (A_1 - \lambda_0 B_1) v_1 \rangle}{\langle \psi_2, B_0 v_2 \rangle \langle \psi_1, B_0 v_1 \rangle}} = \pm \sqrt{\tau_{1,2} \tau_{2,1}}.
$$

The $\tau_{i,j}$, defined implicitly above are

$$
\tau_{i,j} = \frac{\langle \psi_i, (A_1 - \lambda_0 B_1) v_j \rangle}{\langle \psi_i, B_0 v_i \rangle}.
$$

At this order, $\beta_{0,j}$ are determined as the null vectors in (4.5.1). For deep water two-dimensional Stokes waves, labeling $k_1 = k_2 + k_0$, these inner products evaluate to

$$
\tau_{1,2} = \frac{1}{2} \left(\frac{i\omega(k_1)}{1 + \sigma k_1^2} \left((1 + \sigma)(|k_2| - k_2) \frac{\omega(k_2)}{c_0 |k_2|} + \omega^2(k_2) + c_0^2 \right) \right.
$$
$$
\left. - ik_1 \left(\frac{\omega(k_2)}{|k_2|} k_2 + c_0 \right) \right),
$$
$$
\tau_{2,1} = \frac{1}{2} \left(\frac{i\omega(k_2)}{1 + \sigma |k_2|^2} \left((1 + \sigma)(-|k_1| - k_1) \frac{\omega(k_1)}{c_0 |k_1|} + \omega^2(k_1) + c_0^2 \right) \right.
$$
$$
\left. - ik_2 \left(\frac{\omega(k_1)}{|k_1|} k_1 + c_0 \right) \right).
$$

From the above formulae, we see that the $\tau_{i,j}$ are pure imaginary, and stability is determined by the sign of the product $\tau_{1,2} \tau_{2,1}$.

At all later orders, the corrections to the perturbations are decomposed into two parts

$$v_n = v_{n,p} + v_{n,h}$$

where the $v_{n,p}$ are particular solutions, chosen to be orthogonal to the null vectors v_1 and v_2. The $v_{n,h} = \beta_{n,2}v_2$ are the the homogeneous solutions of (4.4.1) at $n = 0$, whose coefficients $\beta_{n,2}$ are set by solvability (as are the λ_n). This decomposition was unnecessary for simple eigenvalues, as our choice of normalization of the eigenfunctions set $v_{n,h} = 0$ for $n > 0$. The general order, $n \geq 2$, corrections solve

$$(A_0 - \lambda_0 B_0)v_n = -\sum_{j=1}^{n}\left(A_j - \sum_{k=0}^{j}\lambda_k B_{j-k}\right)v_{n-j}. \tag{4.5.2}$$

To solve for v_n in equation (4.5.2), one must first impose that the right hand side is in the range of the linear operator, $(A_0 + \lambda_0 B_0)$. At triads for $n \geq 2$ the resulting solvability conditions are linear,

$$\begin{pmatrix} \langle \psi_1, (A_1 - \lambda_1 B_0 - \lambda_0 B_1)v_2 \rangle & \langle \psi_1, B_0 v_0 \rangle \\ \langle \psi_2, (A_1 - \lambda_1 B_0 - \lambda_0 B_1)v_2 \rangle & \langle \psi_2, B_0 v_0 \rangle \end{pmatrix}\begin{pmatrix} \beta_{n-1,2} \\ \lambda_n \end{pmatrix}$$

$$= -\begin{pmatrix} \langle \psi_1, (A_1 - \lambda_1 B_0 - \lambda_0 B_1)v_{n-1,p} \rangle \\ \langle \psi_2, (A_1 - \lambda_1 B_0 - \lambda_0 B_1)v_{n-1,p} \rangle \end{pmatrix}$$

$$+ \begin{pmatrix} \langle \psi_1, \left(A_n - \sum_{j=0}^{n-1}\lambda_j B_{n-j}\right)v_0 \rangle \\ \langle \psi_2, \left(A_n - \sum_{j=0}^{n-1}\lambda_j B_{n-j}\right)v_0 \rangle \end{pmatrix}$$

$$+ \begin{pmatrix} \langle \psi_1, \sum_{j=2}^{n-1}\left(A_j - \sum_{k=0}^{j}\lambda_k B_{j-k}\right)v_{n-j} \rangle \\ \langle \psi_2, \sum_{j=2}^{n-1}\left(A_j - \sum_{k=0}^{j}\lambda_k B_{j-k}\right)v_{n-j} \rangle \end{pmatrix}. \tag{4.5.3}$$

Thus to compute the general term, one must impose (4.5.3), then solve (4.5.2) for $v_{n,p}$. The complete correction at order $O(\varepsilon^n)$, is not known until the solvability conditions are imposed at $O(\varepsilon^{n+1})$, at which point we have determined $v_{n,h}$.

The above series expansions, and similar expansions about quartets, quintets, etc., compute the spectrum at any *fixed* values of the Bloch parameter. This approach works well both far from, and exactly at, resonant Bloch parameters. It does not perform well in the neighborhood of resonant Bloch parameters, the radius of convergence of the series shrinks as it approaches resonant configurations, see Figure 4.1. This is due to the change in form of the series at resonant Bloch parameters, where flat state eigenvalues collide, and the linear operator $(A_0 - \lambda_0 B_0)$ has two-dimensional kernel. To study the spectrum near, but not exactly at, resonant Bloch parameters, one must include the effect of modulation, by allowing the Bloch parameter to depend on amplitude.

4.6 Modulations of Triads

In this section we refer to modulational instabilities as those whose Bloch parameter depends on amplitude. Often the term modulational instability is used to refer only to the long-wave, Benjamin-Feir instability. Our use of the term modulation is consistent with its meaning in the Benjamin-Feir setting, as the expansion in Bloch parameter can be thought of as including the effects of waves modulated by their sidebands. For brevity reasons, we will discuss only modulational instabilities of triads; for quartets and Benjamin-Feir see [21].

To compute the modulated spectrum, first notice that the operators A and B, and thus the eigenfunctions and eigenvalues (v, λ), are all functions of the Bloch parameter p. To compute modulational instabilities, we couple to equation (4.2.3), an amplitude expansion for the Bloch parameter

$$p = p_0 + \varepsilon p_1 + \varepsilon^2 p_2 + \cdots.$$

This expansion introduces an extra unknown at each order in ε. Since the series was already solvable, the Bloch parameter corrections p_j cannot be determined by equation (4.4.1). What results instead is the functional dependence of the spectrum on the Bloch parameter corrections.

Triad instabilities arise at $O(\varepsilon)$; non-trivial modulation of these instabilities can also be recovered at this order. The modulational contribution (the variations of the operators A_j, B_j with respect to frequency) appear in the triad solvability conditions. These modulations occur in the form of perturbations of the phase speed and group velocity. For a general triad, the $\tau_{i,j} \neq 0$, and the first non-zero correction to the flat-state spectrum is

$$\lambda_1 = -i\left(c_0 - \frac{c_g(k_1) + c_g(k_2)}{2}\right)p_1 \pm \frac{1}{2}\sqrt{\tau_{1,2}\tau_{2,1} - ((c_g(k_2) - c_g(k_1))p_1)^2},$$

where $c_g(k_j) = \omega_k(k_j)$ is the group velocity vector at wave number k_j.

Both τ_j are pure imaginary, so if $\tau_{2,1}\tau_{1,2} > 0$, then there is a band of p_1 where instabilities occur, which includes the non-modulated case $p_1 = 0$. Instabilities exist within the symmetric interval

$$|p_1| < \frac{\sqrt{\tau_{1,2}\tau_{2,1}}}{|c_g(k_1) - c_g(k_2)|}. \tag{4.6.1}$$

The boundaries of this interval are marked by the solid straight lines in Figure 4.1. In this figure, we see that the boundaries of the region where modulated instabilities occur predicts well the radius of convergence of a non-modulational expansion of the spectrum. On the other hand the largest triad instabilities are the non-modulational ones; the triads in the band of instabilities where λ_1 has the largest real part are at non-modulational, at $p_1 = 0$.

The effect of modulation can be considered in the absence of triads. If there is no triad interaction, then $\tau_{i,j} = 0$, and there can be no instability at $O(\varepsilon)$. There may be instability at later orders, with its scaling and character depending on the degree of the resonance. The cases of quartets and the Benjamin-Feir instability, a four-eigenvalue collision, are discussed in detail in [21]. The lesson to be learned here is that non-modulational expansions of the spectra lose their analyticity at asymptotically small locations, which can be predicted using a modulational expansion of the spectral data.

In these notes, we focus on the leading order asymptotics of the spectrum. When implementing such an expansion to all orders a number of details become important. For example, one must consider convergence of the modulational expansion of the operators (in frequency space in addition to amplitude space). The effects of floating point cancellations become crucially important. The TFE method was derived to elegantly deal with such cancellations, alternatively one may use extended precision for intermediate computations – see [24]. The cost of such an expansion is also a factor, which can be significantly reduced by solving for the corrections recursively, see for example [22].

Acknowledgments

I would like to thank the Isaac Newton Institute for providing the facilities where this meeting was held, as well as Paul Milewski, David Nicholls, Tom Bridges and Mark Groves for their excellent organizing.

References

[1] Benjamin, T. B., and Feir, J.E. 1967. The disintegration of wave trains on deep water. *Journal of Fluid Mechanics*, **27**, 417–430.

[2] Phillips, O.M. 1960. On the dynamics of unsteady gravity waves of finite amplitude. Part 1. The elementary interactions. *Journal of Fluid Mechanics*, **9**, 193–217.

[3] Hammack, J. L., and Henderson, D. M. 2003. Experiments on deep-water waves with two dimensional surface patterns. *Journal of Offshore Mechanics and Arctic Engineering*, **125**, 48.

[4] Craik, A. D.D. 1985. *Wave Interactions and Fluid Flows*. Cambridge University Press.

[5] Benney, DJ. 1962. Non-linear gravity wave interactions. *Journal of Fluid Mechanics*, **14**(4), 577–584.

[6] MacKay, R. S., and Saffman, P.G. 1986. Stability of water waves. *Proceedings of the Royal Society of London. A*, **406**, 115–125.

[7] McLean, J. 1982. Instabilities of finite-amplitude gravity waves on water of finite depth. *Journal of Fluid Mechanics*, **114**, 331–341.

[8] Dias, F., and Kharif, C. 1999. Nonlinear gravity and gravity-capillary waves. *Annual Review of Fluid Mechanics*, **31**, 301–346.

[9] Tiron, R., and Choi, W. 2012. Linear stability of finite-amplitude capillary waves on water of infinite depth. *Journal of Fluid Mechanics*, **696**, 402–422.

[10] Francius, M., and Kharif, C. 2006. Three-dimensional instabilities of periodic gravity waves in shallow water. *Journal of Fluid Mechanics*, **561**, 417–437.

[11] Deconinck, B., and Trichtchenko, O. 2014. Stability of periodic gravity waves in the presence of surface tension. *European Journal of Mechanics-B/Fluids*, **46**, 97–108.

[12] Deconinck, B., and Oliveras, K. 2011. The instability of periodic surface gravity waves. *J. Fluid Mech.*, **675**, 141–167.

[13] Nicholls, D.P. 2009. Spectral data for traveling water waves : singularities and stability. *Journal of Fluid Mechanics*, **624**, 339–360.

[14] Nicholls, D.P., and Reitich, F. 2006. Stable, high-order computation of traveling water waves in three dimensions. *European Journal of Mechanics B/Fluids*, **25**, 406–424.

[15] Akers, B., and Nicholls, D.P. 2012. Spectral stability of deep two-dimensional gravity water waves: repeated eigenvalues. *SIAM Journal on Applied Mathematics*, **72**(2), 689–711.

[16] Akers, B., and Nicholls, D. P. 2013. Spectral stability of deep two-dimensional gravity capillary water waves. *Studies in Applied Mathematics*, **130**, 81–107.

[17] Akers, B., and Nicholls, D. P. 2014. The spectrum of finite depth water waves. *European Journal of Mechanics - B/Fluids*, **46**, 181–189.

[18] Longuet-Higgins, M.S. 1978. The instabilities of gravity waves of finite amplitude in deep water. I. Superharmonics. *Proceedings of the Royal Society of London. A. Mathematical and Physical Sciences*, **360**(1703), 471–488.

[19] Ioualalen, M., and Kharif, C. 1994. On the subharmonic instabilities of steady three-dimensional deep water waves. *Journal of Fluid Mechanics*, **262**, 265–291.

[20] Nicholls, D. P. 2007. Spectral stability of traveling water waves: analytic dependence of the spectrum. *Journal of Nonlinear Science*, **17**(4), 369–397.

[21] Akers, B.F. 2014. Modulational instabilities of periodic traveling waves in deep water. *preprint*.

[22] Akers, B., and Nicholls, D. P. 2010. Traveling waves in deep water with gravity and surface tension. *SIAM Journal on Applied Mathematics*, **70**(7), 2373–2389.

[23] Nicholls, D.P., and Reitich, F. 2005. On analyticity of traveling water waves. *Proceedings of the Royal Society of London. A*, **461**(2057), 1283–1309.

[24] Wilkening, J., and Vasan, V. 2014. Comparison of five methods of computing the Dirichlet-Neumann operator for the water wave problem. *arXiv preprint arXiv:1406.5226*.

5

A Novel Non-Local Formulation of Water Waves

Athanassios S. Fokas & Konstantinos Kalimeris

Abstract

An introduction to the new formulation of the water wave problem on the basis of the unified transform is presented. The main presentation is on the three-dimensional irrotational water wave problem with surface tension forces included. Examples considered are the doubly-periodic case, the linear case, the case of a variable bottom, and the case of non-zero vorticity.

5.1 Introduction

There have been numerous important developments in the study of surface water waves that date back to the classical works of Stokes and his contemporaries in the nineteenth century. A new reformulation of this problem was presented in [1]. This reformulation is based on the so-called *unified transform* or the Fokas method, which provides a novel approach for the analysis of linear and integrable nonlinear boundary value problems [2, 3].

This chapter is organised as follows: section 5.2 presents the novel formulation of the 3D water waves in the case of a flat bottom. This formulation is used in section 5.3 for the derivation of the associated linearized equations as well as a 2D Boussinesq equation and the KP equation. The case of the 2D periodic water waves is discussed in section 5.4. 3D water waves in a variable bottom are discussed in section 5.5. Finally, the case of 2D water waves with constant vorticity are considered in section 5.6.

A non-local formulation governing two fluids bounded above either by a rigid lid or a free surface is presented in [4]. The case of three fluids bounded above by a rigid lid is considered in [5]. A hybrid of the novel formulation and an approach based on conformal mappings is presented in [6].

5.2 3D Water Waves with Flat Bottom

Let the domain Ω_f (where the subscript f denotes flat bottom) be defined by

$$\Omega_f = \left\{ -\infty < x_j < \infty, \ j = 1,2; \ -h < y < \eta(x_1,x_2,t); \ t > 0 \right\}, \qquad (5.2.1)$$

where η denotes the free surface of the water. One of the major difficulties of the problem of water waves is the fact that η is unknown.

Let ϕ denote the velocity potential. The two unknown functions $\eta(x_1,x_2,t)$ and $\phi(x_1,x_2,y,t)$ satisfy the following equations:

$$\Delta\phi = 0 \quad \text{in } \Omega_f, \qquad (5.2.2a)$$

$$\phi_y = 0 \quad \text{on } y = -h, \qquad (5.2.2b)$$

$$\eta_t + \phi_{x_1}\eta_{x_1} + \phi_{x_2}\eta_{x_2} = \phi_y \quad \text{on } y = \eta, \qquad (5.2.2c)$$

$$\phi_t + \frac{1}{2}\left(\phi_{x_1}^2 + \phi_{x_2}^2 + \phi_y^2\right) + g\eta = \frac{\sigma}{\rho}\nabla\cdot\left(\frac{\nabla\eta}{\sqrt{1+|\nabla\eta|^2}}\right) \quad \text{on } y = \eta, \quad (5.2.3)$$

where $\nabla = \left(\partial_{x_1},\partial_{x_2}\right)$, g is the gravitational acceleration, σ and ρ denote the constant surface tension and density respectively, and h is the constant unperturbed fluid depth. Equation (5.2.2b) states that the velocity normal to the flat bottom surface defined by $y = -h$, vanishes. Equation (5.2.2c), the so-called kinematic condition, implies that fluid particles on the free surface remain on the free surface, whereas equation (5.2.3), the so-called dynamic boundary condition (or Bernoulli's equation), implies continuity of pressure across the free surface. We assume that η, as well as the derivatives of ϕ, vanish as $x_1^2 + x_2^2 \to \infty$.

Equations (5.2.2) define a well-posed boundary value problem for the Laplace equation in the domain Ω_f. Equation (5.2.3) provides the additional information required for the determination of ϕ and η, since the function η appearing in equations (5.2.2) is unknown.

Our strategy for obtaining a novel reformulation of equations (5.2.2) and (5.2.3) is the following: Laplace's equation is linear, thus we can construct an associated *global relation*, which plays a crucial role in the unified transform. Using this relation and appropriate transformations, it is possible to obtain a novel non-local equation coupling η and q, where q denotes the value of ϕ on the free surface, i.e.,

$$q(x_1,x_2,t) = \phi(x_1,x_2,\eta(x_1,x_2,t),t), \quad -\infty < x_j < \infty, \quad j = 1,2, \ t > 0. \qquad (5.2.4)$$

The novel non-local equation to be derived below, together with (5.2.3), where ϕ is written in terms of q, constitute two equations describing surface water waves. These two equations for the two unknown functions $\eta(x_1,x_2,t)$ and

$q(x_1, x_2, t)$ are the following:

$$\int_{-\infty}^{\infty} \int_{-\infty}^{\infty} e^{ik_1 x_1 + ik_2 x_2} \left\{ i\eta_t \cosh\left[k(\eta + h)\right] + (\mathbf{k} \cdot \nabla q) \frac{\sinh[k(\eta + h)]}{k} \right\} dx_1 dx_2 = 0,$$

$$k_1, k_2 \text{ real, } t > 0 \qquad (5.2.5)$$

and

$$q_t + \frac{1}{2}|\nabla q|^2 + g\eta - \frac{(\eta_t + \nabla q \cdot \nabla \eta)^2}{2(1 + |\nabla \eta|^2)} = \frac{\sigma}{\rho} \nabla \cdot \left(\frac{\nabla \eta}{\sqrt{1 + |\nabla \eta|^2}} \right),$$

$$-\infty < x_j < \infty, \quad t > 0, \qquad (5.2.6)$$

where,

$$\mathbf{k} = (k_1, k_2), \quad \nabla = (\partial_{x_1}, \partial_{x_2}), \quad k = \sqrt{k_1^2 + k_2^2}. \qquad (5.2.7)$$

In what follows we will derive (5.2.5) and (5.2.6). We start with the latter equation and note that the definition (5.2.4) implies

$$\phi_{x_j} + \phi_y \eta_{x_j} = q_{x_j}, \quad j = 1, 2. \qquad (5.2.8)$$

These two equations and equation (5.2.2c) provide a system of 3 linear equations for the three functions $\phi_{x_1}, \phi_{x_2}, \phi_y$. The solution of this system yields the following expressions:

$$\phi_{x_1} = \frac{(1 + \eta_{x_2}^2) q_{x_1} - \eta_{x_1} \eta_{x_2} q_{x_2} - \eta_{x_1} \eta_t}{1 + |\nabla \eta|^2}, \qquad (5.2.9a)$$

$$\phi_{x_2} = \frac{(1 + \eta_{x_1}^2) q_{x_2} - \eta_{x_1} \eta_{x_2} q_{x_1} - \eta_{x_2} \eta_t}{1 + |\nabla \eta|^2}, \qquad (5.2.9b)$$

$$\phi_y = \frac{\eta_t + \eta_{x_1} q_{x_1} + \eta_{x_2} q_{x_2}}{1 + |\nabla \eta|^2}. \qquad (5.2.9c)$$

These equations, and the occurrence of extensive simplifications, yield

$$\phi_{x_1}^2 + \phi_{x_2}^2 + \phi_y^2 = \frac{\left(1 + \eta_{x_2}^2\right) q_{x_1}^2 + \left(1 + \eta_{x_1}^2\right) q_{x_2}^2 + \eta_t^2 - 2\eta_{x_1} \eta_{x_2} q_{x_1} q_{x_2}}{1 + |\nabla \eta|^2}.$$

Substituting the expression $\left(\phi_{x_1}^2 + \phi_{x_2}^2 + \phi_y^2\right)$ in equation (5.2.3) and also replacing ϕ_t by $q_t - \eta_t \phi_y$, where ϕ_y is given by equation (5.2.9c), we find an equation involving q and η. Simplifying this equation we find (5.2.6).

For the derivation of (5.2.5) we will use the following fact: Suppose that the functions $\phi(x_1, x_2, y)$ and $\psi(x_1, x_2, y)$ satisfy Laplace's equation (5.2.2a) in a domain Ω. Let $(\hat{N}_1, \hat{N}_2, \hat{N}_3)$ be the unit vector normal to the surface $\partial \Omega$ in the outward direction. Then,

$$\int_{\partial\Omega} \left[\hat{N}_1 \left(\phi_y \psi_{x_1} + \psi_y \phi_{x_1} \right) + \hat{N}_2 \left(\phi_y \psi_{x_2} + \psi_y \phi_{x_2} \right) \right.$$

$$\left. + \hat{N}_3 \left(\phi_y \psi_y - \phi_{x_1} \psi_{x_1} - \phi_{x_2} \psi_{x_2} \right) \right] dS = 0, \qquad (5.2.10)$$

where dS is the surface element on $\partial\Omega$.

Indeed, since both ϕ and ψ satisfy Laplace's equation, we have

$$\phi_y \left(\psi_{x_1 x_1} + \psi_{x_2 x_2} + \psi_{yy} \right) + \psi_y \left(\phi_{x_1 x_1} + \phi_{x_2 x_2} + \phi_{yy} \right) = 0.$$

This identity can be rewritten in the following form:

$$\left(\phi_y \psi_{x_1} + \psi_y \phi_{x_1} \right)_{x_1} + \left(\phi_y \psi_{x_2} + \psi_y \phi_{x_2} \right)_{x_2} + \left(\phi_y \psi_y - \phi_{x_1} \psi_{x_1} - \phi_{x_2} \psi_{x_2} \right)_y = 0, \qquad (5.2.11)$$

hence the divergence theorem implies (5.2.10).

Using for ψ the particular solution

$$E(x_1, x_2, k_1, k_2) = e^{i(k_1 x_1 + k_2 x_2) + ky}, \quad k = \pm\sqrt{k_1^2 + k_2^2}, \qquad (5.2.12)$$

as well as choosing Ω to be Ω_f, equation (5.2.10) becomes

$$\int_{\partial\Omega_f} E \left[\hat{N}_1 \left(ik_1 \phi_y + k\phi_{x_1} \right) + \hat{N}_2 \left(ik_2 \phi_y + k\phi_{x_2} \right) \right.$$

$$\left. + \hat{N}_3 \left(k\phi_y - ik_1 \phi_{x_1} - ik_2 \phi_{x_2} \right) \right] dS = 0. \qquad (5.2.13)$$

On the bottom we have

$$dS = dx_1 \, dx_2, \quad \hat{N}_1 = \hat{N}_2 = 0, \quad \hat{N}_3 = -1, \quad y = -h, \quad \phi_y = 0,$$

thus, on the bottom the integrand of (5.2.13) is given by

$$e^{i(k_1 x_1 + k_2 x_2) - kh} \left(ik_1 \phi_{x_1} + ik_2 \phi_{x_2} \right).$$

On the free surface we have

$$\hat{N}_j \, dS = N_j \, dx_1 \, dx_2, \quad j = 1,2,3 \quad N_1 = -\eta_{x_1}, \quad N_2 = -\eta_{x_2}, \quad N_3 = 1, \quad y = \eta,$$

thus, on the free surface the integrand of (5.2.13) is given by

$$e^{i(k_1 x_1 + k_2 x_2) + k\eta} \left[k \left(\phi_y - \phi_{x_1} \eta_{x_1} - \phi_{x_2} \eta_{x_2} \right) - ik_1 \left(\phi_{x_1} + \phi_y \eta_{x_1} \right) \right.$$

$$\left. - ik_2 \left(\phi_{x_2} + \phi_y \eta_{x_2} \right) \right].$$

The three terms in the above expression appearing in parentheses equal η_t, q_{x_1}, q_{x_2}, respectively (see (5.2.2c) and (5.2.8)). Thus, assuming that the quantity

$\sqrt{\phi_{x_1}^2 + \phi_{x_2}^2 + \phi_y^2}$ decays on the sides of Ω_f, equation (5.2.13) yields the following *global relation*:

$$\int_{-\infty}^{\infty}\int_{-\infty}^{\infty} e^{i(k_1 x_1 + k_2 x_2)} \left\{ e^{kh} \left(k\eta_t - ik_1 q_{x_1} - ik_2 q_{x_2} \right) + \right.$$

$$\left. + e^{-kh} \left[ik_1 \phi_{x_1} (x_1, x_2, -h, t) + ik_2 \phi_{x_2} (x_1, x_2, -h, t) \right] \right\} dx_1 dx_2 = 0. \quad (5.2.14)$$

This equation is valid for both $\sqrt{k_1^2 + k_2^2}$ and $-\sqrt{k_1^2 + k_2^2}$. Thus, we supplement (5.2.14) with the equation obtained from (5.2.14) by replacing k with $-k$. Multiplying these two equations by $\exp[kh]$ and $\exp[-kh]$ respectively, and then subtracting the resulting equations we find (5.2.5).

Remark 5.1   1. The identity (5.2.11), in contrast to the classical Green's identity for Laplace's equation, has the advantage that it involves only *derivatives* of ϕ and not ϕ itself. This is important for our purposes, since although the derivatives of ϕ decay on the sides of the domain Ω_f, ϕ itself may not decay.

2. The function q was first introduced in the classical paper of Zakharov [7], where it was shown that there exists a Hamiltonian formulation of water waves, and (η, q) are canonically conjugate coordinates.

3. Equation (5.2.5) provides the *summation* of the infinite series formulated and analysed by Craig and Sulem in [8].

4. In the case of one-dimensional water waves, equation (5.2.5) takes the following simple form:

$$\int_{-\infty}^{\infty} e^{ikx} \{ i\eta_t \cosh k[\eta + h] + q_x \sinh k[\eta + h] \} dx = 0, \quad k \text{ real}, \quad t > 0. \quad (5.2.15)$$

Furthermore, in this case under the additional assumption of zero surface tension, equation (5.2.6) becomes

$$q_t + \frac{1}{2} \left(q_{x_1} \right)^2 + g\eta - \frac{\left[\eta_t + q_{x_1} \eta_{x_1} \right]^2}{2 \left[1 + \left(\eta_{x_1} \right)^2 \right]} = 0, \quad -\infty < x_1 < \infty, \quad t > 0. \quad (5.2.16)$$

5. In the case of traveling waves, i.e., in the case where

$$\eta(x,t) = \eta(x - ct), \quad q(x,t) = q(x - ct),$$

equation (5.2.16) becomes

$$cq' + \frac{1}{2}(q')^2 + g\eta - \frac{(c\eta' + q'\eta')^2}{2[1 + (\eta')^2]} = 0,$$

where prime denotes differentiation. It was noted by Deconinck and Oliveras in [9] that the above equation is a quadratic equation for q':

$$(q')^2 + 2cq' + 2g\eta[1 + (\eta')^2] - c^2(\eta')^2 = 0.$$

Hence,

$$q' = -c + \sqrt{[1 + (\eta')^2](c^2 - 2g\eta)},$$

which immediately implies the well known estimate

$$\eta \leq \frac{c^2}{2g}. \tag{5.2.17}$$

5.3 The Linear Limit, A Two-Dimensional Boussinesq and the KP

In order to rewrite the basic equations (5.2.5) and (5.2.6) in a nondimensional form, we first replace in these equations all dependent and independent variables by prime variables, and then make the following substitutions:

$$x_1' = lx_1, \quad x_2' = \frac{l}{\gamma}x_2, \quad k_1' = \frac{k_1}{l}, \quad k_2' = \frac{\gamma}{l}k_2, \quad t' = \frac{l}{c_0}t,$$

$$q' = \frac{gla}{c_0}q, \quad \eta' = a\eta, \tag{5.3.1}$$

where $c_0 = \sqrt{gh}$ and l, l/γ are typical length scales (such as wavelengths) in the x_1-, x_2-directions respectively. Introducing the dimensionless parameters

$$\varepsilon = \frac{a}{h}, \quad \mu = \frac{h}{l}, \tag{5.3.2}$$

equations (5.2.5) and (5.2.6) yield the following nondimensional equations:

$$\int_{-\infty}^{\infty} \int_{-\infty}^{\infty} \left\{ i\eta_t \cosh[k\mu(1 + \varepsilon\eta)] + \frac{1}{\mu}\left(\frac{k_1}{k}q_{x_1} + \gamma^2\frac{k_2}{k}q_{x_2}\right) \right.$$

$$\times \left. \sinh[k\mu(1 + \varepsilon\eta)] \right\} dxdx_2, \quad k_1, k_2 \text{ real}, \quad t > 0, \tag{5.3.3a}$$

$$q_t + \eta + \frac{\varepsilon}{2}\left[(q_{x_1})^2 + \gamma^2(q_{x_2})^2\right] - \frac{1}{2}\varepsilon\mu^2\frac{\left[\eta_t + \varepsilon\left(q_{x_1}\eta_{x_1} + \gamma^2 q_{x_2}\eta_{x_2}\right)\right]^2}{1 + (\varepsilon\mu)^2\left[(\eta_{x_1})^2 + \gamma^2(\eta_{x_2})^2\right]}$$

$$= \tilde{\sigma} \mu^2 \left\{ \frac{\partial}{\partial x_1} \frac{\eta_{x_1}}{\sqrt{1 + (\varepsilon \mu)^2 \left[\left(\eta_{x_1} \right)^2 + \gamma^2 \left(\eta_{x_2} \right)^2 \right]}} \right.$$

$$\left. + \gamma^2 \frac{\partial}{\partial x_2} \frac{\eta_{x_2}}{\sqrt{1 + (\varepsilon \mu)^2 \left[\left(\eta_{x_1} \right)^2 + \gamma^2 \left(\eta_{x_2} \right)^2 \right]}} \right\},$$

$$- \infty < x_j < \infty, \quad j = 1, 2, \quad t > 0, \tag{5.3.3b}$$

where

$$k = \sqrt{k_1^2 + \gamma^2 k_2^2}, \quad \tilde{\sigma} = \frac{\sigma}{\rho g h^2}. \tag{5.3.4}$$

Using equations (5.3.3), it is straightforward to derive various approximations.

The Linear Limit

In the linear limit $|\eta|$, $|q_t|$, and $|q_{x_j}|$, $j = 1, 2$, are small. Thus, letting $\varepsilon = 0$ in equations (5.3.3) we find the following linear equations for q and η:

$$\int_{-\infty}^{\infty} \int_{-\infty}^{\infty} e^{i(k_1 x_1 + k_2 x_2)} \left\{ i\eta_t \cosh[k\mu] + \frac{1}{\mu} \left(\frac{k_1}{k} q_{x_1} + \gamma^2 \frac{k_2}{k} q_{x_2} \right) \right.$$

$$\left. \times \sinh[k\mu] \right\} dx_1 dx_2 = 0, \quad k_1, k_2 \text{ real } t > 0, \tag{5.3.5a}$$

$$q_t + \eta = \tilde{\sigma} \mu^2 \left(\eta_{x_1 x_1} + \gamma^2 \eta_{x_2 x_2} \right), \quad -\infty < x_j < \infty, \quad j = 1, 2, \quad t > 0. \tag{5.3.5b}$$

Let the symbol $\wedge$ on top of a variable denote its two-dimensional Fourier transform. Equation (5.3.5a) and the Fourier transform of (5.3.5b) yield the following equations:

$$i\hat{\eta}_t + \frac{1}{\mu} \left[\frac{k_1}{k} \left(\hat{q}_{x_1} \right) + \gamma^2 \frac{k_2}{k} \left(\hat{q}_{x_2} \right) \right] \tanh[k\mu] = 0, \tag{5.3.6a}$$

$$\hat{q}_t + \hat{\eta} = \tilde{\sigma} \mu^2 \left[\left(\hat{\eta}_{x_1 x_1} \right) + \gamma^2 \left(\hat{\eta}_{x_2 x_2} \right) \right], \quad k_1, k_2, \text{ real, } t > 0. \tag{5.3.6b}$$

The definition of the Fourier transform implies

$$\hat{q}_{x_j} = -ik_j \hat{q}, \quad \hat{\eta}_{x_j x_j} = -k_j^2 \hat{\eta}, \quad j = 1, 2,$$

thus equations (5.3.6) become

$$\hat{\eta}_t - \frac{k}{\mu} \tanh[k\mu] \hat{q} = 0,$$

$$\hat{q}_t + (1 + \tilde{\sigma} \mu^2 k^2) \hat{\eta} = 0, \quad k_1, k_2, \text{ real, } t > 0.$$

Hence,

$$\hat{\eta}_{tt} + \left\{ \frac{k}{\mu}(1 + \tilde{\sigma}\mu^2 k^2)\tanh[k\mu] \right\} \hat{\eta} = 0, \quad k_1, k_2, \text{ real, } t > 0. \tag{5.3.7}$$

A Two-Dimensional Boussinesq and the KP Equations

In the next approximation we neglect terms of order $O(\mu^4)$, $O(\varepsilon^2)$, $O(\mu^2\varepsilon)$. We employ the expansions

$$\cosh[k\mu(1+\varepsilon\eta)] \sim 1 + \mu^2 k^2, \quad \sinh[k\mu(1+\varepsilon\eta)] \sim \mu k + \frac{\mu^3 k^3}{6} + \varepsilon\mu k\eta,$$

and use integration by parts to replace k_j with $i\partial_{x_j}$, $j = 1,2$. Using this procedure, after tedious but straightforward calculations, equations (5.3.3) yield the following Boussinesq type equations:

$$\left(1 - \frac{\mu^2}{2}\tilde{\Delta}\right)\eta_t + \left(\tilde{\Delta} - \frac{\mu^2}{6}\tilde{\Delta}^2\right)q + \varepsilon\left(\eta_{x_1}q_{x_1} + \gamma^2\eta_{x_2}q_{x_2}\right) + \varepsilon\eta\tilde{\Delta}q = 0,$$

$$\tag{5.3.8a}$$

$$\eta = -q_t - \frac{\varepsilon}{2}\left[(q_{x_1})^2 + \gamma^2(q_{x_2})^2\right] + \tilde{\sigma}\mu^2\tilde{\Delta}\eta, \quad -\infty < x_j < \infty, \quad t > 0,$$

$$\tag{5.3.8b}$$

where

$$\tilde{\Delta} = \partial_{x_1}^2 + \gamma^2\partial_{x_2}^2. \tag{5.3.9}$$

Equation (5.3.8b) implies

$$\eta \sim -\left(1 + \tilde{\sigma}\mu^2\tilde{\Delta}\right)q_t - \frac{\varepsilon}{2}\left[(q_{x_1})^2 + \gamma^2(q_{x_2})^2\right]. \tag{5.3.10a}$$

Substituting this equation into (5.3.8a) we obtain

$$\left[1 + \left(\tilde{\sigma} - \frac{1}{2}\right)\mu^2\tilde{\Delta}\right]q_{tt} - \left(\tilde{\Delta} - \frac{\mu^2}{6}\tilde{\Delta}^2\right)q$$

$$+ \varepsilon\left(2q_{x_1}q_{x_1t} + 2\gamma^2 q_{x_2}q_{x_2t} + q_t\tilde{\Delta}q\right) = 0. \tag{5.3.10b}$$

To the leading order approximation, $q_{tt} \sim \tilde{\Delta}q$, thus equation (5.3.10b) is asymptotically equivalent to the following equation:

$$q_{tt} - \tilde{\Delta}q + \left(\tilde{\sigma} - \frac{1}{3}\right)\mu^2\tilde{\Delta}^2 q + \varepsilon\partial_t\left[(q_{x_1})^2 + \gamma^2(q_{x_2})^2\right] + \varepsilon q_t\tilde{\Delta}q = 0. \tag{5.3.11}$$

In the case of zero surface tension, an equation that is asymptotically equivalent to (5.3.11) was derived by Benney and Luke in [10].

If

$$\gamma = O(\mu), \quad \varepsilon = O(\mu^2),$$

then (5.3.11) becomes

$$q_{tt} - q_{xx} + \left(\tilde{\sigma} - \frac{1}{3} \right) \mu^2 q_{x_1 x_1 x_1} - \gamma^2 q_{x_2 x_2} + \varepsilon \left(2q_{x_1} q_{x_1 t} + q_t q_{x_1 x_1} \right) = 0. \quad (5.3.12)$$

Letting

$$\xi = x_1 - T, \quad y = x_2, \quad T = \varepsilon t$$

and using

$$\partial_t = -\partial_\xi + \varepsilon \partial_T, \quad \partial_{x_1} = \partial_\xi, \quad \partial_{x_1} = \partial_y,$$

(5.3.12) becomes

$$2\varepsilon q_{T\xi} + \left(\frac{1}{3} - \tilde{\sigma} \right) \mu^2 q_{\xi\xi\xi\xi} + \gamma^2 q_{yy} + 3\varepsilon q_\xi q_{\xi\xi} = 0. \quad (5.3.13)$$

Letting

$$W = q_\xi, \quad \varepsilon = \mu^2 = \gamma^2 \quad (5.3.14)$$

and taking the derivative with respect to ξ, (5.3.13) becomes the well known Kadomtsev-Petviashvili (KP) equation

$$2W_{T\xi} + \left(\frac{1}{3} - \tilde{\sigma} \right) W_{\xi\xi\xi} + W_{yy} + 3(WW_\xi)_\xi = 0. \quad (5.3.15)$$

Using (5.3.14) but neglecting variations in the y-direction, equation (5.3.13) becomes the celebrated Korteweg-deVries (KdV) equation,

$$2W_T + \left(\frac{1}{3} - \tilde{\sigma} \right) W_{\xi\xi\xi} + 3\varepsilon WW_\xi = 0. \quad (5.3.16)$$

Remark 5.2   KP with dominant surface tension (namely $\tilde{\sigma} > \frac{1}{3}$), supports lumps, i.e., localized soliton-type solutions. It is shown in [1] that for sufficiently large surface tension, it is possible to compute numerically lumps for equations (5.3.11).

5.4 The Periodic Problem in Two Dimensions

Let the domain Ω_p (where the subscript p denotes periodic) be defined by

$$\Omega_p = \{-L < x < L, ; \quad -h < y < \eta(x,t); \quad t > 0\}, \quad (5.4.1)$$

where η denotes the free boundary of the water.

In this case the two unknown functions $\eta(x,t)$ and $\phi(x,y,t)$ satisfy the following equations:

$$\Delta\phi = 0 \quad \text{in } \Omega_p, \tag{5.4.2a}$$

$$\phi_y = 0 \quad \text{on } y = -h, \tag{5.4.2b}$$

$$\eta_t + \phi_x\eta_x = \phi_y \quad \text{on } y = \eta, \tag{5.4.2c}$$

$$\phi_t + \frac{1}{2}\left(\phi_x^2 + \phi_y^2\right) + g\eta = \frac{\sigma}{\rho}\partial_x\left(\frac{\eta_x}{\sqrt{1+\eta_x^2}}\right) \quad \text{on } y = \eta, \tag{5.4.3}$$

where g is the gravitational acceleration, σ and ρ denote the constant surface tension and density respectively, and h is the constant unperturbed fluid depth. We assume that η and ϕ are 2L-periodic functions in x. As a consequence, their derivatives are also periodic.

Now equation (5.2.5) takes the form

$$\int_{-L}^{L} e^{ikx}\{i\eta_t\cosh[k(\eta+h)] + q_x\sinh[k(\eta+h)]\}\,dx = 0, \quad t > 0, \tag{5.4.4}$$

where $k = k_n = (n+\frac{1}{2})\frac{\pi}{L}$, $n \in \mathbb{Z}$. Furthermore, in this case under the additional assumption of zero surface tension we find that equation (5.2.16) is satisfied for $x \in [-L,L]$. Note that (5.4.4) is an identity for $k = 0$, due to the periodicity of η.

The steps of the derivation of (5.4.4) are the same as for the derivation of (5.2.5) but now we get two extra integrals at the left and right side of Ω_p. In what follows we show that the sum of these two integrals is zero, provided that we choose $k = k_n = (n+\frac{1}{2})\frac{\pi}{L}$, $n \in \mathbb{Z}$. Indeed, the integral equation (5.2.13) takes the form

$$\int_{\partial\Omega_p} e^{ikx+ky}\left(\hat{N}_1 - i\hat{N}_2\right)\left(\phi_x + i\phi_y\right)dS = 0, \tag{5.4.5}$$

where $(\hat{N}_1, \hat{N}_2)$ is the unit normal vector to the curve $\partial\Omega_p$, in the outward direction.

Hence, on the right side, where $\hat{N}_1 = 1$, $\hat{N}_2 = 0$ and $x = L$, we get

$$\int_{-h}^{\eta(L)} e^{ikL+ky}\left(\phi_x(L) + i\phi_y(L)\right)dy. \tag{5.4.6}$$

Furthermore on the left side, where $\hat{N}_1 = -1$, $\hat{N}_2 = 0$ and $x = -L$, we get

$$-\int_{-h}^{\eta(-L)} e^{-ikL+ky}(-1)\left(\phi_x(-L) + i\phi_y(-L)\right)dy. \tag{5.4.7}$$

Due to the periodicity condition, the sum of the above integrals vanishes provided that $e^{ikL} + e^{-ikL} = 0$, or equivalently when $k = k_n$.

Moreover, the supplementary second integral equation is given by

$$\int_{\partial\Omega_p} e^{ikx-ky} \left(\hat{N}_1 + i\hat{N}_2 \right) (\phi_x - i\phi_y) \, dS = 0,$$

which along with (5.4.5) and the same procedure followed in the beginning of the chapter gives the global relation (5.4.4).

The construction of the analogue equation in three (2+1) dimensions is straightforward.

5.5 The Case of a Variable Bottom

Let the domain Ω_v (where the subscript v denotes variable bottom) be defined by

$$\Omega_v = \left\{ -\infty < x_j < \infty, \ j = 1, 2; \ -(h + H(x_1, x_2)) < y < \eta(x_1, x_2, t); \ t > 0 \right\},$$
$$(5.5.1)$$

where the function H has sufficient smoothness and decay. The associated global relation is an equation similar with (5.2.13) where $\partial\Omega_f$ is replaced by $\partial\Omega_v$.

The bottom is now defined by the surface

$$y = -h - H(x_1, x_2),$$

hence on this surface

$$\hat{N}_j dS = N_j dx_1 dx_2, \ j = 1, 2, \ N_1 = -H_{x_1}, \ N_2 = -H_{x_2}, \ N_3 = -1.$$

Thus the relevant integrand of (5.2.13) of the global relation becomes

$$e^{i(k_1 x_1 + k_2 x_2) - kh - kH} \left[-H_{x_1} \left(ik_1\phi_y + k\phi_{x_1} \right) - H_{x_2} \left(ik_2\phi_y + k\phi_{x_2} \right) \right.$$
$$\left. - \left(k\phi_y - ik_1\phi_{x_1} - ik_2\phi_x \right) \right], \qquad (5.5.2)$$

where ϕ is evaluation at $\{x_1, x_2, -(y + H), t\}$. The bracket appearing in (5.5.2) simplifies to

$$ik_1 \left(\phi_{x_1} - \phi_y H_{x_1} \right) + ik_2 \left(\phi_{x_2} - \phi_y H_{x_2} \right) - k \left(\phi_y + \phi_{x_1} H_{x_1} + \phi_{x_2} H_{x_2} \right). \quad (5.5.3)$$

Let Φ denote the value of ϕ on the bottom, i.e.,

$$\Phi(x_1, x_2, t) = \phi(x_1, x_2, -h - H(x_1, x_2), t). \qquad (5.5.4)$$

Then

$$\Phi_{x_j} = \phi_{x_j} - \phi_y H_{x_j}, \ j = 1, 2.$$

Thus, the first two parentheses appearing in (5.5.3) equal Φ_{x_1} and Φ_{x_2}. Furthermore, since $g + H$ is independent of t, the third parentheses appearing in (5.5.3) vanishes (compare with (5.2.2c) where η is replaced with $-y - H$). Hence, in analogy with (5.2.14) we now find the equation

$$\int_{-\infty}^{\infty} \int_{-\infty}^{\infty} e^{i(k_1 x_1 + k_2 x_2)} \left\{ e^{k(\eta + h)} \left[k\eta_t - ik_1 q_{x_1} - ik_2 q_{x_2} \right] \right.$$

$$\left. + e^{-k_1 t} \left[ik_1 \Phi_{x_1} + ik_2 \Phi_{x_2} \right] \right\} dx_1 dx_2 = 0, \quad k_1, k_2 \text{ real,} \quad t > 0.$$

By replacing k with $-k$ we get a supplementary equation. By adding and subtracting these two equations we obtain the following system of two non-local equations:

$$\int_{-\infty}^{\infty} \int_{-\infty}^{\infty} e^{ik_1 x_1 + ik_2 x_2} \left\{ i\eta_t \cosh\left[k(\eta + h)\right] + (\mathbf{k} \cdot \nabla q) \frac{\sinh[k(\eta + h)]}{k} \right.$$

$$\left. + (\mathbf{k} \cdot \nabla \Phi) \frac{\sinh[kH]}{k} \right\} dx_1 dx_2 = 0, \tag{5.5.5a}$$

$$\int_{-\infty}^{\infty} \int_{-\infty}^{\infty} e^{ik_1 x_1 + ik_2 x_2} \left\{ i\eta_t \sinh\left[k(\eta + h)\right] + \frac{(\mathbf{k} \cdot \nabla q)}{k} \cosh[k(\eta + h)] \right.$$

$$\left. - \frac{(\mathbf{k} \cdot \nabla \Phi)}{k} \cosh[kH] \right\} dx_1 dx_2 = 0, \quad k_1, k_2 \text{ real,} \quad t > 0. \tag{5.5.5b}$$

Equations (5.5.5) and Bernoulli's equation (5.2.6) are three equations for the three unknown functions

$$\eta(x_1, x_2, t), \quad q(x_1, x_2, t), \quad \Phi(x_1, x_2, t).$$

Conservation Laws The functions

$$\exp[i(k_1 x_1 + k_2 x)], \quad \cosh[k(\eta + h)], \quad \sinh[k(\eta + h)], \quad \sinh[kh], \quad \cosh[kH]$$

appearing in equations (5.5.5) can be expanded in terms of power series involving k_1^m and k_2^n, m, n positive integers. Thus, by equating the coefficients of k_1^m and of k_2^n to zero, it is straightforward to obtain a series of interesting integral relations. In particular, using

$$\frac{\sinh[k(\eta + h)]}{k} = \eta + h + O\left((k_1^2 + k_2^2)^{\frac{3}{2}}\right),$$

$$\frac{\sinh[kH]}{k} = H + O\left((k_1^2 + k_2^2)^{\frac{3}{2}}\right),$$

equation (5.5.5a) yields:

$$\partial_t \int_{-\infty}^{\infty} \int_{-\infty}^{\infty} \eta \, dx_1 dx_2 = 0, \tag{5.5.6}$$

$$\partial_t \int_{-\infty}^{\infty} \int_{-\infty}^{\infty} x_j \eta \, dx_1 dx_2 = \int_{-\infty}^{\infty} \int_{-\infty}^{\infty} \left[q_{x_j}(\eta + h) + H\Phi_{x_j} \right] dx_1 dx_2 = 0, \quad j = 1, 2.$$

$$\tag{5.5.7}$$

Similarly equation (5.5.5b) yields:

$$\int_{-\infty}^{\infty} \int_{-\infty}^{\infty} \left(q_{x_j} - \Phi_{x_j}\right) dx_1 dx_2 = 0, \quad j = 1, 2. \tag{5.5.8}$$

Equation (5.5.6) expresses the conservation of mass. Also if $H = 0$, the RHS of equation (5.5.7) can be related to the momentum of the field P_j, $j = 1, 2$, defined by

$$P_j = \int_{-\infty}^{\infty} \int_{-\infty}^{\infty} \left(\int_{-h}^{\eta} \rho \phi_{x_j}(x_1, x_2, y, t) dy\right) dx_1 dx_2, \quad j = 1, 2.$$

Since the momentum is a conserved quantity, it follows that for $H = 0$ the RHS of equation (5.5.7) is constant. Hence, equation (5.5.7) implies that the LHS of (5.5.7) that defines the center of mass grows linearly in time.

Boussinesq Type Equation

It is shown in [6] that in the weakly nonlinear, weakly dispersive regime, the system of the three equations (5.2.6), (5.5.5a) and (5.5.5b), can be reduced to a system of two equations. Indeed, supplementing equations (5.3.1) with the equations

$$y' = hy, \quad H' = h(1 + H), \quad \Phi' = \frac{ac_0 \Phi}{\mu},$$

it is shown in [6] that, in analogy with equations (5.3.8a) and (5.3.8b), we now have the following equations:

$$\left(1 + \frac{\mu^2}{2} \tilde{\Delta}\right) \eta_t + \left(\tilde{\Delta} - \frac{\mu^2}{6} \tilde{\Delta}^2\right) q + \varepsilon \left(\eta_{x_1} q_{x_1} + \gamma^2 \eta_{x_2} q_{x_2}\right) + \varepsilon \eta \tilde{\Delta} q$$

$$+ \tilde{\nabla} \cdot H \tilde{\nabla} q + \mu^2 (\tilde{\nabla} \cdot H \tilde{\nabla} \Psi) = 0 \tag{5.4.9a}$$

and

$$\eta = -q_t - \frac{\varepsilon}{2} \left(q_{x_1}^2 + \gamma^2 q_{x_2}^2\right), \tag{5.5.9b}$$

where $\tilde{\Delta}$ is defined by (5.3.9), $\tilde{\nabla}$ is defined by

$$\tilde{\nabla} = \left(\partial_{x_1}, \gamma \partial_{x_2}\right)$$

and Ψ is expressed in terms of η and q by the equation

$$\Psi = -\eta_t - \frac{\mu^2}{2} \tilde{\Delta} q + \frac{1}{2} \tilde{\nabla} \cdot H^2 \tilde{\nabla} q. \tag{5.5.9c}$$

It is also shown in [6] that by combining the non-local formulation of [1] with conformal mappings, it is possible in the two-dimensional case to obtain a system of two equations *without* any asymptotic approximations.

5.6 The Case of Constant Vorticity in Two Dimensions

Let $u(x,y,t)$ and $v(x,y,t)$ denote the horizontal and vertical components of the velocity. It is shown in [11] that for the two-dimensional case with constant vorticity, i.e., for the case

$$w = v_x - u_y = \gamma, \quad \gamma \text{ constant,}$$

the basic equations are as follows:

$$\Delta\varphi = 0 \text{ in } \Omega_v$$

$$(\varphi_x - \gamma y)H_x + \varphi_y = 0 \text{ on } y = -h_0 + H,$$

$$\eta_t + (\varphi_x - \gamma y)\eta_x - \varphi_y = 0 \text{ on } y = \eta,$$

$$\varphi_t - \gamma\,\partial_x^{-1}\eta_t + \frac{1}{2}\varphi_y^2 + \frac{1}{2}(\varphi_x - \gamma y)^2 + gy = \frac{\sigma}{\rho}\nabla_x \cdot \frac{\nabla_x\eta}{\sqrt{1 + (\nabla_x\eta)^2}}.$$

A novel non-local formulation of the above equations, analogous to the two-dimensional version of equations (5.5.5), can be found in [11].

The analogue of the estimate (5.2.17) is now

$$\eta \leq \frac{c^2}{2g} + \frac{\sigma}{2\rho g}\frac{d}{dz}\left(\frac{\eta'}{\sqrt{1 + (\eta')^2}}\right).$$

Remark 5.3   By using the parametrizations

$$x_j = X_j(s,t), \quad j = 1,2; \quad y = Y(s,t),$$

where $s \in \mathbb{R}$ is the arc-length, it is straightforward but cumbersome to bypass the assumption that $\eta(x_1, x_2, t)$ is a single valued function [11].

Acknowledgments

A.S.F. acknowledges the support of EPSRC, UK.

References

[1] M.J. Ablowitz, A.S. Fokas and Z.H. Musslimani, On a new non-local formulation of water waves, J. Fluid Mech., 562, 313, 2006.

[2] A.S. Fokas, A unifed transform method for solving linear and certain nonlinear PDEs, Proc. R. Soc. Lond. A, 453, 1411-1443, 1997.

[3] A.S. Fokas, A unifed approach to boundary value problems, SIAM, 78, 2008.

[4] T.S. Haut and M.J. Ablowitz, A reformulation and applications of interfacial fluids with a free surface. J. Fluid Mech., 631, 375-396, 2009.

[5] D. Burini, S. De Lillo and D. Skouteris, On a coupled system of shallow water equations admitting travelling wave solutions, submitted.

[6] A.S. Fokas and A. Nachbin, Water waves over a variable bottom: a non-local formulation and conformal mapping, J. Fluid Mech., 695, 288–309, 2012.

[7] V.E. Zakharov, Stability of periodic waves of finite amplitude on the surface of a deep fluid, J. Appl. Mech. Tech. Phys., 2, 190, 1968.

[8] W. Craig and C. Sulem, Numerical simulation of gravity waves, J. Comput. Phys., 108(1), 73-83, 1993.

[9] B. Deconinck and K. Oliveras, The instability of periodic surface gravity waves. J. Fluid Mech., 675, 141-167, 2011.

[10] D.J. Benney and J.C. Luke, Interactions of permanent waves of finite amplitude, J. Maths and Phys., 43, 455, 1964.

[11] A. Ashton and A.S. Fokas, A non-local formulation of rotational water waves, J. Fluid. Mech., 689 (1), 2011.

6

The Dimension-Breaking Route to Three-Dimensional Solitary Gravity-Capillary Water Waves

Mark D. Groves

Abstract

It is well known that the water-wave problem has small-amplitude line solitary-wave solutions, which to leading order are described by solutions of the Korteweg-deVries equation (for strong surface tension) or the nonlinear Schrödinger equation (for weak surface tension). The three-dimensional generalisations of these model equations (the Kadomtsev-Petviashvili equation and Davey-Stewartson system) also admit *periodically modulated solitary waves*, which have a solitary-wave profile in the direction of propagation and are periodic in the transverse direction. This article describes an existence theory for three-dimensional periodically modulated solitary-wave solutions of the water-wave problem; they emanate from the line solitary waves in a *dimension-breaking bifurcation*. The key to these results is a formulation of the water-wave problem as an evolutionary system in which the transverse horizontal variable plays the role of time, a careful study of the purely imaginary spectrum of the operator obtained by linearising the evolutionary system at a line solitary wave, and an application of an infinite-dimensional version of the classical Lyapunov centre theorem.

6.1 Introduction

In this article we consider the three-dimensional irrotational flow of a perfect fluid of unit density subject to the forces of gravity and surface tension. The fluid is bounded below by a rigid horizontal bottom $\{y = 0\}$ and above by a free surface $\{y = h + \eta(x, z, t)\}$. Our primary interest is in travelling waves moving without change of shape and with constant speed $c > 0$ from left to right in the x-direction. In a moving frame of reference in which they are stationary the equations of motion for travelling waves are Laplace's equation

$$\phi_{xx} + \phi_{yy} + \phi_{zz} = 0, \qquad 0 < y < 1 + \eta, \tag{6.1.1}$$

78

for the Eulerian velocity potential ϕ describing the flow, with boundary conditions

$$\phi_y = 0, \qquad\qquad y = 0, \qquad\qquad (6.1.2)$$

$$\phi_y = -\eta_x + \eta_x\phi_x + \eta_z\phi_z, \qquad y = 1+\eta, \qquad (6.1.3)$$

and

$$-\phi_x + \frac{1}{2}(\phi_x^2 + \phi_y^2 + \phi_z^2) + \alpha\eta$$

$$-\beta\left[\frac{\eta_x}{\sqrt{1+\eta_x^2+\eta_z^2}}\right]_x - \beta\left[\frac{\eta_z}{\sqrt{1+\eta_x^2+\eta_z^2}}\right]_z = 0, \qquad y = 1+\eta. \quad (6.1.4)$$

Here we have introduced dimensionless variables, choosing h as length scale and h/c as time scale; the parameters α and β are defined in terms of the Froude and Bond numbers $F = c/\sqrt{gh}$ and $\tau = \sigma/gh^2$ by the formulae $\alpha = 1/F^2$ and $\beta = \tau/F^2$ (so that $\beta = \tau\alpha$), in which g is the acceleration due to gravity and σ is the coefficient of surface tension. A *solitary wave* is a nontrivial solution (η,ϕ) of (6.1.1)–(6.1.4) with $\eta(x,z) \to 0$ as $|x| \to \infty$.

A brief review of the classical weakly nonlinear theory for solitary waves is instructive.

Strong surface tension: Figure 6.1 shows the linear dispersion relation for a sinusoidal travelling wave train with wave number μ; for each fixed value $\tau_0 > \frac{1}{3}$ of τ the dispersion curve has a unique minimum at $(\mu, \alpha^{-1}) = (0, 1)$. Working in the strong surface-tension parameter regime

$$(\beta_0, \alpha_0) \in C_s, \quad \beta = \beta_0, \quad \alpha = \alpha_0 + \varepsilon^2, \qquad (6.1.5)$$

where $C_s = \{(\beta, \alpha) : \beta > \frac{1}{3}, \alpha = 1\}$, and substituting

$$\eta(x,z) = \varepsilon^2\zeta(\varepsilon x, \varepsilon^2 z)$$

into equations (6.1.1)–(6.1.4), one finds that to leading order ζ satisfies the Kadomtsev-Petviashvili ('KP-I') equation

$$\left((\beta_0 - \tfrac{1}{3})\zeta_{XX} - \zeta - \tfrac{3}{2}\zeta^2\right)_{XX} - \zeta_{ZZ} = 0, \qquad (6.1.6)$$

where $X = \varepsilon x$, $Z = \varepsilon^2 z$ (see Ablowitz & Clarkson [1, §1.2]). In the special case of z-independent waves this equation reduces to the Korteweg-deVries equation

$$(\beta_0 - \tfrac{1}{3})\zeta_{XX} - \zeta - \tfrac{3}{2}\zeta^2 = 0.$$

Mark D. Groves

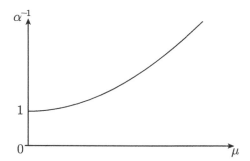

Figure 6.1. The linear dispersion relation for a sinusoidal travelling wave train
with wave number μ and $\tau = \tau_0$, where $\tau_0 > 1/3$.

The function

$$\zeta^\star_{\mathrm{KdV}}(X) = -\operatorname{sech}^2\left(\frac{X}{2(\beta_0 - \frac{1}{3})^{1/2}}\right) \tag{6.1.7}$$

is a symmetric solitary-wave solution of the Korteweg-deVries equation and
hence a *line* solitary-wave solution to the Kadomtsev-Petviashvili equation.
The Kadomtsev-Petviashvili equation also has a family of *periodically modu-
lated* solitary-wave solutions, which have qualitatively the same profile as $\zeta^\star_{\mathrm{KdV}}$
in X and are periodic in Z (see Tajiri & Murakami [2] for explicit formulae).
These waves emanate from the line solitary waves in a *dimension-breaking
bifurcation* – a phenomenon in which a spatially inhomogeneous solution of
a partial differential equation emerges from a solution that is homogeneous
in one or more spatial dimensions (Haragus & Kirchgässner [3]) – and are
sketched in Figure 6.2.
Weak surface tension: Figure 6.3 shows the linear dispersion relation for a
sinusoidal travelling wave train with wave number μ; for each fixed value $\tau_0 \in$
$(0, \frac{1}{3})$ of τ the dispersion curve has a unique minimum at $(\mu, \alpha^{-1}) = (\mu_0, \alpha_0^{-1})$.
(The relationship between α_0, $\beta_0 = \tau_0 \alpha_0$ and μ_0 can be expressed in the form

$$\beta_0 = -\frac{1}{2}\operatorname{cosech}^2 \mu_0 + \frac{1}{2\mu_0}\coth \mu_0, \qquad \alpha_0 = \frac{\mu_0^2}{2}\operatorname{cosech}^2 \mu_0 + \frac{\mu_0}{2}\coth \mu_0,$$

which defines a curve C_{w} in the (β, α)-plane parameterised by $\mu \in (0, \infty)$.)
Working in the weak surface-tension parameter regime

$$(\beta_0, \alpha_0) \in C_{\mathrm{w}}, \quad \beta = \beta_0, \quad \alpha = \alpha_0 + \varepsilon^2 \tag{6.1.8}$$

and substituting

$$\eta(x,z) = \frac{1}{2}\varepsilon\zeta(\varepsilon x, \varepsilon z)\mathrm{e}^{\mathrm{i}\mu_0 x} + \frac{1}{2}\varepsilon\overline{\zeta(\varepsilon x, \varepsilon z)}\mathrm{e}^{-\mathrm{i}\mu_0 x}$$

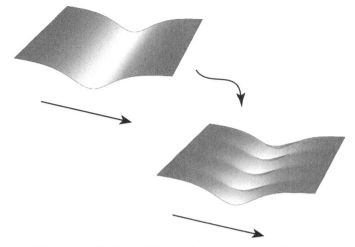

Figure 6.2. The periodically modulated solitary wave on the right emerges from the line solitary wave on the left in a dimension-breaking bifurcation. (In unscaled coordinates these waves have small amplitude and long length scales in both horizontal directions.)

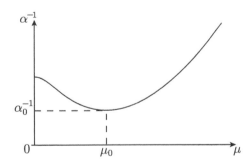

Figure 6.3. The linear dispersion relation for a sinusoidal travelling wave train with wave number μ and $\tau = \tau_0$, where $\tau_0 \in (0, 1/3)$.

into equations (6.1.1)–(6.1.4), one finds that to leading order ζ satisfies the elliptic-elliptic Davey-Stewartson system

$$\zeta - A_1 \zeta_{XX} - A_2 \zeta_{ZZ} - A_3 |\zeta|^2 \zeta + 4A_4 x \zeta \psi_X = 0, \qquad (6.1.9)$$

$$-(1 - \alpha_0^{-1})\psi_{XX} - \psi_{ZZ} - A_4(|\zeta|^2)_X = 0, \qquad (6.1.10)$$

where $X = \varepsilon x$, $Z = \varepsilon z$; the coefficients are given by the formulae

$$A_1 = \beta_0 + (1 - \mu_0 \coth(\mu_0)) \operatorname{cosech}^2(\mu_0),$$

$$A_2 = \mu_0^{-1} \coth \mu_0,$$

$$A_3 = -\frac{\mu_0^3}{8\sigma^3}\left(\frac{(1-\sigma^2)(9-\sigma^2)a_0+\beta_0\mu_0^2(3-\sigma^2)(7-\sigma^2)}{a_0\sigma^2-\beta_0\mu_0^2(3-\sigma^2)}+8\sigma^2\right.$$
$$\left.-\frac{2\mu_0}{a_0\sigma}(1-\sigma^2)^2-3\beta_0\mu_0\sigma^3\right),$$

$$A_4 = \frac{\mu_0(a_0\sinh(2\mu_0)+\mu_0)}{4a_0\sinh^2(\mu_0)},$$

where $\sigma = \tanh(\mu_0) = \mu_0(a_0+\beta_0\mu_0^2)^{-1}$ (see Djordjevic & Redekopp [4], Ablowitz & Segur [5], noting the misprint in equation (2.24d), and Sulem & Sulem [6, §11.1.1]). In the special case of z-independent waves these equations reduce to the focussing cubic nonlinear Schrödinger equation

$$\zeta - A_1\partial_X^2\zeta - A_5|\zeta|^2\zeta = 0,$$

where $A_5 = A_3 + 4(1-a_0^{-1})^{-1}A_4^2 > 0$.

The functions

$$\zeta_{\text{NLS}}^\star(X) = \pm\left(\frac{2}{A_5}\right)^{1/2}\text{sech}\left(\frac{X}{A_1^{1/2}}\right) \tag{6.1.11}$$

are symmetric solitary-wave solutions of the cubic nonlinear Schrödinger and hence line solitary-wave solutions of the Davey-Stewartson system. The Davey-Stewartson system also has a family of periodically modulated solitary-wave solutions that have qualitatively the same profile as $\zeta_{\text{NLS}}^\star$ in X and are periodic in Z (see Groves, Sun & Wahlén [7], noting that explicit formulae are not available). These waves emanate from the line solitary waves in a dimension-breaking bifurcation and are sketched in Figure 6.4.

The weakly nonlinear theory predicts the existence of line and periodically modulated solitary-wave solutions to the water-wave problem (6.1.1)–(6.1.4), and these predictions have been confirmed by rigorous mathematics.

- A family of line solitary waves $(\eta_{\text{KdV}}^\star, \phi_{\text{KdV}}^\star)_\varepsilon$ in the strong surface-tension regime (6.1.5) were discovered by Amick and Kirchgässner [8] (see also Kirchgässner [9] and Sachs [10]); they are symmetric, monotonically decaying waves of depression given asymptotically by

$$\eta_{\text{KdV}}^\star(x) = \varepsilon^2\zeta_{\text{KdV}}^\star(\varepsilon x) + O(\varepsilon^4 e^{-\rho\varepsilon|x|}),$$

for some $\rho > 0$ (the corresponding asymptotic formula for $\phi_{\text{KdV}}^\star$ is given in the Appendix).

- Two families of line solitary waves $(\eta_{\text{NLS}}^\star, \phi_{\text{NLS}}^\star)_\varepsilon$ in the weak surface-tension regime (6.1.8) were discovered by Iooss and Kirchgässner [11] (see also Dias & Iooss [12] and Iooss & Pérouème [13]); they are wave packets consisting of a decaying envelope that modulates an underlying periodic

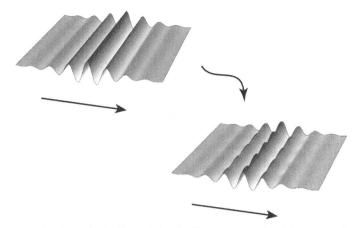

Figure 6.4. The periodically modulated solitary wave on the right emerges from the line solitary wave on the left in a dimension-breaking bifurcation. (In unscaled coordinates these waves have small amplitude and long length scales in both horizontal directions.)

wave train with wave number μ_0, and are given asymptotically by

$$\eta_{\text{NLS}}^{\star}(x) = \varepsilon\zeta_{\text{NLS}}^{\star}(\varepsilon x)\cos(\mu_0 x) + O(\varepsilon^2 \mathrm{e}^{-\rho\varepsilon|x|})$$

for some $\rho > 0$ (more precise asymptotic formulae for $\eta_{\text{NLS}}^{\star}$ and $\phi_{\text{NLS}}^{\star}$ are given in the Appendix).

The corresponding existence theories for periodically modulated solitary waves were given by Groves, Haragus & Sun [14] (strong surface tension) and Groves, Sun & Wahlén [15] (weak surface tension). In the present article we review the dimension-breaking theory developed in those references, reworking the older theory (for strong surface tension) with the more efficient methods introduced in the newer theory (for weak surface tension) to create a unified approach to the two cases.

Our existence proof is based upon *spatial dynamics*, a framework for studying stationary boundary-value problems by treating them as evolution equations in which an unbounded spatial variable plays the role of time. The method was pioneered by Kirchgässner [9] for two-dimensional travelling water waves and is in particular the basis of the existence theories for the Amick-Kirchgässner and Iooss-Kirchgässner line solitary waves described earlier (C_s and C_w correspond to the curves C_4 and C_2 in Kirchgässner's bifurcation diagram). It was extended to the three-dimensional water-wave setting by Groves & Mielke [16] and has since been used to construct a wide variety of three-dimensional gravity-capillary waves (see Groves & Haragus [17]). We use the method by formulating (6.1.1)–(6.1.4) as a reversible

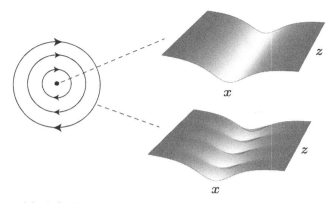

Figure 6.5. A family of periodic solutions surrounding a nontrivial equilibrium solution to (6.1.12) in its phase space (left) corresponds to a dimension-breaking bifurcation of a branch of periodically modulated solitary waves from a line solitary wave (right).

evolutionary equation

$$u_z = f(u), \tag{6.1.12}$$

where u belongs to a space $\mathcal{X}$ of functions that converge to zero as $x \to \pm\infty$ and $f : \mathcal{D}(f) \subseteq \mathcal{X} \to \mathcal{X}$ is an unbounded, densely defined vector field. Observe that the nonzero equilibria of equation (6.1.12) are precisely the line solitary-wave solutions of the water-wave problem; of particular interest in this respect are the equilibria $u^\star$ corresponding to $(\eta^\star_{\mathrm{KdV}}, \phi^\star_{\mathrm{KdV}})$ or $(\eta^\star_{\mathrm{NLS}}, \phi^\star_{\mathrm{NLS}})$. A solution to equation (6.1.12) of the form

$$u(x,z) = u^\star(x) + u'(x,z), \tag{6.1.13}$$

where u' has small amplitude and is periodic in z, corresponds to a periodically modulated solitary wave that is generated by a dimension-breaking bifurcation from a line solitary wave (see Figure 6.5). Accordingly, we construct a family of small-amplitude periodic solutions to the equation

$$u'_z = f(u^\star_\varepsilon + u') \tag{6.1.14}$$

for each sufficiently small value of ε. Explicit formulae for equations (6.1.12) and (6.1.14) (together with the definitions of suitable function spaces in which to study them) are presented in Sections 6.2.1 and 6.2.2.

Small-amplitude periodic solutions of reversible evolutionary equations are classically found using the Lyapunov centre theorem, which asserts the existence of a family of these solutions with frequency near ω_0 provided that $\pm i\omega_0$ are nonresonant imaginary eigenvalues of the corresponding linearised system. The theorem remains true for infinite-dimensional systems with finite-dimensional linear central subspaces under additional hypotheses on the decay rate of the resolvent operator along the imaginary axis (Devaney [18]). In the present setting the central subspace is however infinite-dimensional due to the presence of essential spectrum at the origin (a feature typical of spatial dynamics formulations for problems in unbounded domains). It was pointed out by Iooss [19] that this difficulty does not automatically rule out an application of the reversible Lyapunov centre theorem. Denoting the linear and nonlinear parts of the reversible vector field in question by respectively L and N, one finds upon examining the proof of the Lyapunov centre theorem that it is not required that L is invertible on the whole of phase space, rather merely on the range of N ('the Iooss condition at the origin'). We thus arrive at the following generalisation of the reversible Lyapunov centre theorem, which we refer to as the *Lyapunov-Iooss theorem*.

Theorem 6.1.1. *Consider the differential equation*

$$v_\tau = L(v) + N(v), \tag{6.1.15}$$

in which $v(\tau)$ belongs to a Banach space $\mathcal{X}$. Suppose that $\mathcal{Y}$, $\mathcal{Z}$ are further Banach spaces with the properties that

(i) *$\mathcal{Z}$ is continuously embedded in $\mathcal{Y}$ and continuously and densely embedded in $\mathcal{X}$,*

(ii) *$L : \mathcal{Z} \subseteq \mathcal{X} \to \mathcal{X}$ is a closed linear operator,*

(iii) *there is an open neighbourhood $\mathcal{U}$ of the origin in $\mathcal{Y}$ such that $L \in \mathcal{L}(\mathcal{Y}, \mathcal{X})$ and $N \in C^3_{\mathrm{b,u}}(\mathcal{U}, \mathcal{X})$ (and hence $N \in C^3_{\mathrm{b,u}}(\mathcal{U} \cap \mathcal{Z}, \mathcal{X})$) with $N(0) = 0$, $\mathrm{d}N[0] = 0$.*

Suppose further that

(iv) *equation (6.1.15) is reversible: there exists an involution $S \in \mathcal{L}(\mathcal{X}, \mathcal{X})$ with $SLv = -LSv$ and $SN(v) = -N(Sv)$ for all $v \in \mathcal{U}$,*

and that the following spectral hypotheses are satisfied:

(v) *$\pm i\omega_0$ are nonzero simple eigenvalues of L;*

(vi) *$in\omega_0 \in \rho(L)$ for $n \in \mathbb{Z} \setminus \{-1, 0, 1\}$;*

(vii) *$\|(L - in\omega_0 I)^{-1}\|_{\mathcal{X} \to \mathcal{X}} = o(1)$ and $\|(L - in\omega_0 I)^{-1}\|_{\mathcal{X} \to \mathcal{Z}} = O(1)$ as $n \to \infty$;*

(viii) *for each $v^\dagger \in \mathcal{U}$ the equation*

$$Lv = -N(v^\dagger)$$

has a unique solution $v \in \mathcal{Y}$ and the mapping $v^\dagger \mapsto v$ belongs to $C^3_{\mathrm{b,u}}(\mathcal{U}, \mathcal{Y})$.

Under these hypotheses there exist an open neighbourhood $\mathcal{N}$ of the origin in $\mathbb{R}$ and a continuously differentiable branch $\{(v(s), \omega(s))\}_{s \in S}$ of reversible,

$2\pi/\omega(s)$-*periodic solutions in* $C^1_{\mathrm{per}}(\mathbb{R}, \mathcal{Y} \oplus \mathcal{X}) \cap C_{\mathrm{per}}(\mathbb{R}, \mathcal{Y} \oplus \mathcal{Z})$ *to* (6.1.15) *with amplitude* $O(|s|)$. *Here the direct sum refers to the decomposition of a function into its mode 0 and higher-order Fourier components, the subscript 'per' indicates a* $2\pi/\omega(s)$-*periodic function and* $\omega(s) = \omega_0 + O(|s|^2)$.

Our main task is now to study the purely imaginary spectrum of the linearisation L of the vector field on the right-hand side of (6.1.14). Motivated by the weakly nonlinear theory, we use the decomposition $\eta = \eta_1 + \eta_2$, where the supports of the Fourier transforms of η_1 and η_2 lie respectively in

$$
S = \begin{cases} [-\delta, \delta], & (\beta_0, \alpha_0) \in C_s, \\ [-\mu_0 - \delta, -\mu_0 + \delta] \cup [\mu_0 - \delta, \mu_0 + \delta], & (\beta_0, \alpha_0) \in C_w, \end{cases}
$$

and $\mathbb{R} \setminus S$, and write

$$
\eta_1(x) = \begin{cases} \varepsilon^2 \zeta(\varepsilon x), & (\beta_0, \alpha_0) \in C_s, \\ \dfrac{1}{2} \varepsilon \zeta(\varepsilon x) e^{i\mu_0 x} + \dfrac{1}{2} \varepsilon \overline{\zeta(\varepsilon x)} e^{-i\mu_0 x}, & (\beta_0, \alpha_0) \in C_w. \end{cases}
$$

We find, after a straightforward but lengthy calculation, that the pair $\pm i\varepsilon^r k$, $k > 0$, where

$$
r = \begin{cases} 2, & (\beta_0, \alpha_0) \in C_s, \\ 1, & (\beta_0, \alpha_0) \in C_w, \end{cases}
$$

are eigenvalues of L if and only if a reduced operator $\mathcal{B}_{\varepsilon,k} = \mathcal{B}_{0,k} + \mathcal{R}_{\varepsilon,k}$ has a zero eigenvalue; here $\mathcal{B}_{0,k}$ is the operator obtained by linearising the left-hand side of the Kadomtsev-Petviashvili equation (6.1.6) or Davey-Stewartson system (6.1.9), (6.1.10) at its line solitary wave and taking the Fourier transform with respect to z, and $\mathcal{R}_{\varepsilon,k}(\zeta)$ is a remainder term, which is $O(\varepsilon^{1/r})$ in an appropriate sense.

A comparison with well-studied operators in mathematical physics shows that $\mathcal{B}_{0,0} = \mathcal{B}_{0,k} - k^2 I$ has precisely one simple negative eigenvalue $-k_0^2$, so that $\mathcal{B}_{0,k}$ has a simple zero eigenvalue precisely when $k = k_0$, and we deduce from a spectral perturbation argument that $\mathcal{B}_{\varepsilon,k}$ has a simple zero eigenvalue at a unique value k_ε, which satisfies the estimate $|k_0 - k_\varepsilon| = O(\varepsilon^{1/r})$. A similar argument is used to verify the Iooss condition at the origin. Applying the above reduction to the equation

$$
Lv = -N(v^\dagger),
$$

where N is the nonlinear part of the vector field on the right-hand side of (6.1.14), we obtain the reduced equation

$$
\mathcal{C}_\varepsilon(\zeta) = \zeta^\dagger,
$$

where C_0 is known explicitly and $C_\varepsilon - C_0$ is $O(\varepsilon^{1/r})$ in an appropriate sense. The operator C_0 is invertible (in the class of symmetric functions) and it follows that the reduced equation is solvable.

The precise sense in which the spectral results for $\mathcal{B}_{\varepsilon,k}$ and $\mathcal{C}_{\varepsilon,k}$ indicated above are equivalent to the spectral hypotheses in the Lyapunov-Iooss theorem is stated in Section 6.2.3 (see Theorems 6.2.3 and 6.2.4), where the application of the theorem is discussed in detail. The derivation of the reduced equations is deferred to Section 6.3, while their spectral analysis is presented in Section 6.4. Altogether we obtain the following existence result for periodically modulated solitary waves.

Theorem 6.1.2. *There exist an open neighbourhood $\mathcal{N}_\varepsilon$ of the origin in $\mathbb{R}$ and a family of periodically modulated solitary waves $\{(\eta_s(x,z),\phi_s(x,y,z))\}_{s\in\mathcal{N}_\varepsilon}$, which emerges from the line solitary wave in a dimension-breaking bifurcation. Here*

$$\eta_s(x,z) = \eta_\varepsilon^\star(x) + \eta_s^\star(x,z),$$

in which $\eta_s^\star(\cdot,\cdot)$ has amplitude $O(|s|)$ and is even in both arguments and periodic in its second with frequency $\varepsilon^2 k_\varepsilon + O(|s|^2)$; the positive number k_ε satisfies $|k_\varepsilon - k_0| = O(\varepsilon)$.

6.2 Dimension-Breaking Phenomena

6.2.1 Spatial Dynamics

In this section we formulate equations (6.1.1)–(6.1.4) as an evolutionary equation in which the unbounded horizontal direction z plays the role of time. This spatial dynamics formulation is obtained using a physical argument based upon the observation that equations (6.1.1)–(6.1.4) follow from the formal variational principle

$$\delta \int_{-\infty}^\infty \int_{-\infty}^\infty \left\{ \int_0^{1+\eta} \left(-\phi_x + \frac{1}{2}(\phi_x^2 + \phi_y^2 + \phi_z^2) \right) dy \right. $$
$$\left. + \frac{1}{2}\alpha\eta^2 + \beta\left(\sqrt{1+\eta_x^2+\eta_z^2} - 1\right) \right\} d(x,z) = 0, \quad (6.2.1)$$

where the variation is taken in (η,ϕ). A more satisfactory version of this variational principle is obtained using the change of variable

$$\phi(x,y,z) = \Phi(x,y',z), \qquad y = y'(1+\eta(x,z)), \qquad (6.2.2)$$

which transforms the variable fluid domain D_η into the fixed domain $\mathbb{R} \times (0,1) \times \mathbb{R}$; we drop the primes for notational convenience. One obtains

Hamilton's principle that

$$\delta \mathcal{L} = 0, \qquad \mathcal{L} = \int_{-\infty}^{\infty} L(\eta, \Phi, \eta_z, \Phi_z) \, dz,$$

in which

$$L(\eta, \Phi, \eta_z, \Phi_z) = \int_{\mathbb{R}} \left\{ \frac{1}{2} \alpha \eta^2 + \beta \left[\sqrt{1 + \eta_x^2 + \eta_z^2} - 1 \right] \right\} dx$$

$$+ \int_{\Sigma} \left(- \left[\Phi_x - \frac{y \eta_x \Phi_y}{1 + \eta} \right] + \frac{1}{2} \left[\Phi_x - \frac{y \eta_x \Phi_y}{1 + \eta} \right]^2 + \frac{\Phi_y^2}{2(1 + \eta)^2} \right.$$

$$+ \frac{1}{2} \left[\Phi_z - \frac{y \eta_z \Phi_y}{1 + \eta} \right]^2 \right) (1 + \eta) \, dy \, dx$$

and $\Sigma = \mathbb{R} \times (0, 1)$. The next step is to carry out a Legendre transform. Define new variables ω and ξ by the formulae

$$\omega = \frac{\delta L}{\delta \eta_z} = - \int_0^1 \left(\Phi_z - \frac{y \eta_z \Phi_y}{1 + \eta} \right) y \Phi_y \, dy + \frac{\beta \eta_z}{\sqrt{1 + \eta_x^2 + \eta_z^2}}, \qquad (6.2.3)$$

$$\xi = \frac{\delta L}{\delta \Phi_z} = \left(\Phi_z - \frac{y \eta_z \Phi_y}{1 + \eta} \right) (1 + \eta), \qquad (6.2.4)$$

in which the variational derivatives are taken formally in respectively $L^2(\mathbb{R})$ and $L^2(\Sigma)$, and set

$$H(\eta, \omega, \Phi, \xi)$$

$$= \int_{\Sigma} \xi \Phi_z \, dy \, dx + \int_{\mathbb{R}} \omega \eta_z \, dx - L(\eta, \Phi, \eta_z, \Phi_z)$$

$$= \int_{\Sigma} \left\{ (1 + \eta) \Phi_x - y \eta_x \Phi_y + \frac{\xi^2 - \Phi_y^2}{2(1 + \eta)} - \frac{(1 + \eta)}{2} \left(\Phi_x - \frac{y \eta_x \Phi_y}{1 + \eta} \right)^2 \right\} dy \, dx$$

$$+ \int_{\mathbb{R}} \left\{ -\frac{1}{2} \alpha \eta^2 + \beta - (\beta^2 - W^2)^{1/2} (1 + \eta_x^2)^{1/2} \right\} dx, \qquad (6.2.5)$$

where

$$W = \omega + \frac{1}{1 + \eta} \int_0^1 y \Phi_y \xi \, dy.$$

The above procedure suggests that the equations

$$\eta_z = \frac{\delta H}{\delta \omega}, \qquad \omega_z = -\frac{\delta H}{\delta \eta}, \qquad \Phi_z = \frac{\delta H}{\delta \xi}, \qquad \xi_z = -\frac{\delta H}{\delta \Phi}$$

formally represent Hamilton's equations for a spatial dynamics formulation of the hydrodynamic problem as a Hamiltonian system. An explicit calculation shows that these equations are given by

$$\eta_z = W \left(\frac{1 + \eta_x^2}{\beta^2 - W^2} \right)^{1/2}, \tag{6.2.6}$$

$$\omega_z = \frac{W}{(1+\eta)^2} \left(\frac{1+\eta_x^2}{\beta^2 - W^2} \right)^{1/2} \int_0^1 y \Phi_y \xi \, dy - \left[\eta_x \left(\frac{\beta^2 - W^2}{1+\eta_x^2} \right)^{1/2} \right]_x$$

$$+ \int_0^1 \left\{ \frac{\xi^2 - \Phi_y^2}{2(1+\eta)^2} + \frac{1}{2} \left(\Phi_x - \frac{y\eta_x \Phi_y}{1+\eta} \right)^2 + \left[\left(\Phi_x - \frac{y\eta_x \Phi_y}{1+\eta} \right) y \Phi_y \right]_x \right.$$

$$\left. + \left(\Phi_x - \frac{y\eta_x \Phi_y}{1+\eta} \right) \frac{y\eta_x \Phi_y}{1+\eta} \right\} dy + \alpha\eta - \Phi_x|_{y=1}, \tag{6.2.7}$$

$$\Phi_z = \frac{\xi}{(1+\eta)} + \frac{Wy\Phi_y}{1+\eta} \left(\frac{1+\eta_x^2}{\beta^2 - W^2} \right)^{1/2}, \tag{6.2.8}$$

$$\xi_z = -\frac{\Phi_{yy}}{1+\eta} - \left[(1+\eta) \left(\Phi_x - \frac{y\eta_x \Phi_y}{1+\eta} \right) \right]_x + \left[\left(\Phi_x - \frac{y\eta_x \Phi_y}{1+\eta} \right) y\eta_x \right]_y$$

$$+ \frac{W(y\xi)_y}{1+\eta} \left(\frac{1+\eta_x^2}{\beta^2 - W^2} \right)^{1/2}, \tag{6.2.9}$$

where

$$W = \omega + \frac{1}{1+\eta} \int_0^1 y\Phi_y\xi \, dy,$$

and are accompanied by the boundary conditions

$$\Phi_y = 0, \quad y = 0, \tag{6.2.10}$$

$$\Phi_y = -y\eta_x + y\eta_x \Phi_x + \frac{\eta\Phi_y}{1+\eta} - \frac{y^2\eta_x^2\Phi_y}{1+\eta} + \frac{Wy\xi}{1+\eta} \left(\frac{1+\eta_x^2}{\beta^2 - W^2} \right)^{1/2}, \quad y = 1, \tag{6.2.11}$$

which emerge as a consequence of the integration by parts necessary to compute the variational derivative with respect to Φ. We write Hamilton's equations as

$$u_z = f(u) \tag{6.2.12}$$

where $u = (\eta, \omega, \Phi, \xi)$, $f(u) = (f_1(u), f_2(u), f_3(u), f_4(u))$ and $f_1(u), \ldots, f_4(u)$ are defined by the right-hand sides of (6.2.6)–(6.2.9). Similarly, we write the boundary conditions as

$$\Phi_y = B(u) \quad \text{on } y = 0, 1, \tag{6.2.13}$$

where $B(u)$ is defined by the right-hand side of equation (6.2.11); note the helpful identity

$$f_4(\eta, \omega, \Phi, \xi) = -[(1+\eta)\Phi_x - y\eta_x \Phi_y]_x + [-\Phi_y + y\eta_x + B(\eta, \omega, \Phi, \xi)]_y. \tag{6.2.14}$$

In keeping with our study of dimension breaking from line solitary waves we henceforth choose $(\beta_0, \alpha_0) \in C_s$ or $(\beta_0, \alpha_0) \in C_w$ and write

$$\beta = \beta_0, \qquad \alpha = \alpha_0 + \varepsilon^2.$$

To identify an appropriate functional-analytic setting for these equations, define

$$X^s = H^{s+1}(\mathbb{R}) \times H^s(\mathbb{R}) \times H^{s+1}(\Sigma) \times H^s(\Sigma), \qquad s \geq 0,$$

where Σ is the strip $\mathbb{R} \times (0,1)$, let M be an open neighbourhood of the origin in X^1 contained in the set

$$\{u \in X^1 : |W(x)| < \beta, \eta(x) > -1 \text{ for all } x \in \mathbb{R}\}$$

(so that M is a manifold domain of X^0) and note that $B, f : M \to H^1(\Sigma)$ are analytic mappings (see Bagri & Groves [20, Proposition 2.1]). Equation (6.2.12) therefore constitutes an evolutionary equation in the infinite-dimensional phase space X^0 with

$$\mathcal{D}(f) = \{u \in M : \Phi_y = B(u) \text{ on } y = 0, 1\}.$$

Furthermore, f is the Hamiltonian vector field for the Hamiltonian system (X^0, Ω, H), where the position-independent symplectic form Ω on X^0 is given by

$$\Omega(u_1, u_2) = \int_{\mathbb{R}} (\omega_2 \eta_1 - \eta_2 \omega_1) \, dx + \int_\Sigma (\xi_2 \Phi_1 - \Phi_2 \xi_1) \, dy \, dx$$

and $H : M \to \mathbb{R}$ is the analytic function defined by

$$H(u) = \int_\Sigma \left\{ (1+\eta)\Phi_x - y\eta_x \Phi_y + \frac{\xi^2 - \Phi_y^2}{2(1+\eta)} - \frac{(1+\eta)}{2}\left(\Phi_x - \frac{y\eta_x \Phi_y}{1+\eta}\right)^2 \right\} dy \, dx$$

$$+ \int_{\mathbb{R}} \left\{ -\frac{1}{2}(\alpha_0 + \varepsilon^2)\eta^2 + \beta_0 - (\beta_0^2 - W^2)^{1/2}(1+\eta_x^2)^{1/2} \right\} dx$$

(see Groves, Haragus & Sun [14, §2]). This Hamiltonian system is *reversible*, that is, Hamilton's equations are invariant under the transformation

$$z \mapsto -z, \quad u \mapsto S(u),$$

where the *reverser* $S : X^s \to X^s$ is defined by $S(\eta, \omega, \Phi, \xi) = (\eta, -\omega, \Phi, -\xi)$.

Equation (6.2.12) is also invariant under the reflection $R : X^s \to X^s$ given by

$$R(\eta(x), \omega(x), \Phi(x,y), \xi(x,y)) = (\eta(-x), \omega(-x), -\Phi(-x,y), -\xi(-x,y)),$$

and we seek solutions that are invariant under this symmetry by replacing X^s by

$$X_r^s := X^s \cap \mathrm{Fix}\, R$$

$$= H_e^{s+1}(\mathbb{R}) \times H_e^s(\mathbb{R}) \times H_o^{s+1}(\Sigma) \times H_o^s(\Sigma),$$

where

$$H^s_e(\mathbb{R}) = \{w \in H^s(\mathbb{R}) : w(x) = w(-x) \text{ for all } x \in \mathbb{R}\},$$

$$H^s_o(\Sigma) = \{w \in H^s(\Sigma) : w(x,y) = -w(-x,y) \text{ for all } (x,y) \in \Sigma\},$$

(with corresponding definitions for $H^s_o(\mathbb{R})$ and $H^s_e(\Sigma)$) and M by $M_r := M \cap$ Fix R. It is also possible to extend M_r by replacing X^1_r in its definition with the extended function space $X^1_\star$, where

$$X^s_\star = H^{s+1}_e(\mathbb{R}) \times H^s_e(\mathbb{R}) \times H^{s+1}_{\star,o}(\Sigma) \times H^s_o(\Sigma)$$

and

$$H^{s+1}_{\star,0}(\Sigma) = \{w \in L^2_{loc}(\Sigma) : w_x, w_y \in H^s(\Sigma),$$
$$w(x,y) = -w(-x,y) \text{for all } (x,y) \in \Sigma\}$$

(a Banach space with norm $\|u\|_{\star,s+1} := (\|u_x\|^2_s + \|u_y\|^2_s)^{1/2}$); note the relationship $M_r = M_\star \cap X^1_r$ between M_r and its extension $M_\star$. This feature allows one to consider solutions to (6.2.12), (6.2.13) whose Φ-component is not evanescent; in particular line solitary waves fall into this category.

Line solitary waves are equilibrium solutions of equations (6.2.12), (6.2.13) of the form $u^\star = (\eta^\star, 0, \Phi^\star, 0) \in M_\star$, where $\eta^\star : \mathbb{R} \to \mathbb{R}$ and $\Phi^\star : \Sigma \to \mathbb{R}$ are smooth functions whose derivatives are all $O(\varepsilon)$ (see the Appendix), and we seek solutions of the form

$$u = u^\star + u', \qquad u' \in M'_\star,$$

where $M'_\star = \{u' \in X^1_r : u_\star + u' \in M_\star\}$. Substituting this *Ansatz* into (6.2.12), (6.2.13) and again dropping the prime for notational convenience, we find that

$$u_z = f(u^\star + u) \tag{6.2.15}$$

with boundary conditions

$$\Phi_y = B_1(\eta, \Phi) + B_{nl}(\eta, \omega, \Phi, \xi) \quad \text{on } y = 0, 1, \tag{6.2.16}$$

where

$$B_1(\eta, \Phi) = dB[\eta^\star, 0, \Phi^\star, 0](\eta, \omega, \Phi, \xi)$$

$$= y(-\eta_x + \eta^\star_x \Phi_x + \Phi^\star_x \eta_x)$$

$$+ \frac{\eta^\star \Phi_y}{1 + \eta^\star} + \frac{\Phi^\star_y \eta}{(1 + \eta^\star)^2} + \frac{y^2 (\eta^\star_x)^2 \Phi^\star_y \eta}{(1 + \eta^\star)^2} - \frac{y^2 (\eta^\star_x)^2 \Phi_y}{1 + \eta^\star} - \frac{2y^2 \eta^\star_x \Phi^\star_y \eta_x}{1 + \eta^\star}$$

and

$$B_{nl}(\eta, \omega, \Phi, \xi) = B(\eta^\star + \eta, \omega, \Phi^\star + \Phi, \xi) - B(\eta^\star, 0, \Phi^\star, 0) - B_1(\eta, \Phi).$$

A small-amplitude periodic solution of (6.2.15), (6.2.16) corresponds to a periodically modulated solitary wave in the form of a three-dimensional perturbation of a line solitary wave.

6.2.2 Boundary Conditions

Equations (6.2.15), (6.2.16) cannot be studied using standard methods for evolutionary equations because of the nonlinear boundary conditions. This difficulty is handled using a change of variable which leads to an equivalent problem in a linear space. For $(\eta, \omega, \Phi, \xi)$ in M_r or $M_\star$ define $Q(\eta, \omega, \Phi, \xi) = (\eta, \omega, \Gamma, \xi)$, where

$$\Gamma = \Phi + \Theta_y$$

and $\Theta \in H^3(\Sigma)$ is the unique solution of the boundary-value problem

$$-\Theta_{xx} - \Theta_{yy} + B_1(0, \Theta_y) = B_{nl}(\eta, \omega, \Phi, \xi), \qquad 0 < y < 1,$$

$$\Theta = 0, \qquad\qquad\qquad y = 0, 1$$

(the linear operator on the left-hand side of this boundary-value problem is uniformly strongly elliptic with smooth coefficients); note in particular that

$$- \Phi_y + B_1(\eta, \Phi) + B_{nl}(\eta, \omega, \Phi, \xi) = -\Gamma_y + B_1(\eta, \Gamma) - \Theta_{xx}. \qquad (6.2.17)$$

Lemma 6.2.1.

(i) *The mapping Q is a near-identity, analytic diffeomorphism from a neighbourhood M_r of the origin in X_r^1 onto a neighbourhood $\tilde{M}$ of the origin in X_r^1.*

(ii) *For each $u \in M_\star$ the operator $dQ[u]$ also defines an isomorphism $\widehat{dQ}[u] : X_r^0 \to X_r^0$, and the operators $\widehat{dQ}[u], \widehat{dQ}[u]^{-1} \in \mathcal{L}(X_r^0, X_r^0)$ depend analytically on $u \in M_r$.*

(iii) *Statements (i) and (ii) also hold when M_r, $\tilde{M}_r$ are replaced by $M_\star$, $\tilde{M}_\star$, where $M_r = M_\star \cap X_r^1$ and $\tilde{M}_r = \tilde{M}_\star \cap X_r^1$.*

The above change of variable transforms (6.2.15) into the equation

$$v_z = Lv + N(v) \qquad (6.2.18)$$

with boundary conditions

$$\Gamma_y = B_1(\eta, \Gamma) \quad \text{on } y = 0, 1 \qquad (6.2.19)$$

for the variable $v = Q(u)$, in which $L := d\tilde{f}[0] = df[u^\star]$ and $N := \tilde{f} - L$ are the linear and nonlinear parts of the transformed vector field

$$\tilde{f}(v) := \widehat{dQ}[Q^{-1}(v)](f(u^\star + Q^{-1}(v))).$$

Here

$$\mathcal{D}(L) = Y_\star^1 := \left\{ (\eta, \omega, \Gamma, \xi) \in X_\star^1 : \Phi_y = B_1(\eta, \Gamma) \text{ on } y = 0, 1 \right\}$$

and $\mathcal{D}(N)$ is the neighbourhood $U_\star := \tilde{M}_\star \cap Y_\star^1$ of the origin in $Y_\star^1$, and we may replace $\tilde{M}_\star$, $Y_\star^1$, $U_\star$ by $\tilde{M}_r$, Y_r^1, U_r. The linear operator L is given by the explicit formula

$$L \begin{pmatrix} \eta \\ \omega \\ \Gamma \\ \xi \end{pmatrix} = \begin{pmatrix} \dfrac{\omega}{\beta_0} + h_1(\omega,\xi) \\ (\alpha_0 + \varepsilon^2)\eta - \Gamma_x|_{y=1} - \beta_0\eta_{xx} + h_2(\eta,\Gamma) \\ \xi + H_1(\omega,\xi) \\ -\Gamma_{xx} - \Gamma_{yy} + H_2(\eta,\Gamma), \end{pmatrix},$$

where

$$h_1(\omega,\xi) = \frac{(1+(\eta_x^\star)^2)^{1/2}}{\beta_0}\left(\omega + \frac{1}{1+\eta^\star}\int_0^1 y\Phi_y^\star\xi\,dy\right) - \frac{\omega}{\beta_0},$$

$$h_2(\eta,\Gamma) = \beta_0\eta_{xx} - \beta_0\left[\frac{\eta_x}{(1+(\eta_x^\star)^2)^{3/2}}\right]_x$$

$$+ \int_0^1 \Bigg\{ \Phi_x^\star\Gamma_x - \frac{\Phi_y^\star\Gamma_y}{(1+\eta^\star)^2} + \frac{(\Phi_y^\star)^2\eta}{(1+\eta^\star)^3} - \frac{y^2(\eta_x^\star)^2\Phi_y^\star\Gamma_y}{(1+\eta^\star)^2}$$

$$- \frac{y^2\eta_x^\star(\Phi_y^\star)^2\eta_x}{(1+\eta^\star)^2} + \frac{y^2\eta_x^\star(\Phi_y^\star)^2\eta}{(1+\eta^\star)^3} + \left[y\Phi_y^\star\Gamma_x + y\Phi_x^\star\Gamma_y\right.$$

$$\left. - \frac{2y^2\eta_x^\star\Phi_y^\star\Gamma_y}{1+\eta^\star} - \frac{y^2(\Phi_y^\star)^2\eta_x}{1+\eta^\star} + \frac{y^2(\Phi_y^\star)^2\eta_x^\star\eta}{(1+\eta^\star)^2}\right]_x\Bigg\}\,dy,$$

$$H_1(\omega,\xi) = -\frac{\eta^\star\xi}{1+\eta^\star} + \frac{(1+(\eta_x^\star)^2)^{1/2}}{\beta_0(1+\eta^\star)}\left(\omega + \frac{1}{1+\eta^\star}\int_0^1 y\Phi_y^\star\xi\,dy\right)y\Phi_y^\star,$$

$$H_2(\eta,\Gamma) = (F_1(\eta,\Gamma))_x + (F_3(\eta,\Gamma))_y$$

and

$$F_1(\eta,\Gamma) = -\eta^\star\Gamma_x - \Phi_x^\star\eta + y\Phi_y^\star\eta_x + y\eta_x^\star\Gamma_y,$$

$$F_3(\eta,\Gamma) = y\eta_x + B_1(\eta,\Gamma).$$

Furthermore, the change of variable preserves the reversibility of the evolutionary equation.

6.2.3 Existence Results

We construct periodically modulated solitary waves by applying the Lyapunov-Iooss theorem (Theorem 6.1.1) to (6.2.18), (6.2.19), taking $\mathcal{X} = X_r^0$, $\mathcal{Y} = Y_r^\star$, $\mathcal{Z} = Y_r^1$ and $\mathcal{U} = U_\star$ (and of course $\tau = z$ and $S(\eta,\omega,\Gamma,\xi) = (\eta,-\omega,\Gamma,-\xi)$). The spectral hypotheses are verified by studying the resolvent equations

$$(L - i\lambda I)u = u^\dagger, \qquad \lambda \in \mathbb{R},$$

for L, that is

$$\frac{\omega}{\beta_0} + h_1(\omega, \xi) = i\lambda\eta + \eta^\dagger, \tag{6.2.20}$$

$$(\alpha_0 + \varepsilon^2)\eta - \Gamma_x|_{y=1} - \beta_0\eta_{xx} + h_2(\eta, \Gamma) = i\lambda\omega + \omega^\dagger, \tag{6.2.21}$$

$$\xi + H_1(\omega, \xi) = i\lambda\Gamma + \Gamma^\dagger, \tag{6.2.22}$$

$$-\Gamma_{xx} - \Gamma_{yy} + (F_1(\eta, \Gamma))_x + (F_3(\eta, \Gamma))_y = i\lambda\xi + \xi^\dagger \tag{6.2.23}$$

with

$$\Gamma_y = 0 \qquad \text{on } y = 0, \tag{6.2.24}$$

$$\Gamma_y = -\eta_x + F_3(\eta, \Gamma) \qquad \text{on } y = 1; \tag{6.2.25}$$

since L is real and anticommutes with the reverser S it suffices to examine non-negative values of λ, real values of η, Γ, $\omega^\dagger$, $\xi^\dagger$ and imaginary values of ω, ξ, $\eta^\dagger$, $\Gamma^\dagger$.

The weakly nonlinear theory given in Section 6.1 suggests that the support of the Fourier transform $\hat{\eta} = \mathcal{F}[\eta]$ of η is concentrated near the origin for strong surface tension and near wavenumbers $\mu = \pm\mu_0$ for weak surface tension. We therefore decompose it into the sum of the functions

$$\eta_1 = \chi(D)\eta := \mathcal{F}^{-1}[\chi\hat{\eta}], \qquad \eta_2 = (1 - \chi(D))\eta := \mathcal{F}^{-1}[(1 - \chi)\hat{\eta}],$$

where

$$\chi(\mu) = \begin{cases} \chi_0(\mu), & (\beta_0, \alpha_0) \in C_s, \\ \chi_0(\mu - \mu_0) + \chi_0(\mu + \mu_0), & (\beta_0, \alpha_0) \in C_w, \end{cases} \tag{6.2.26}$$

χ_0 is the characteristic function of the interval $[-\delta, \delta]$ and δ is a small positive number, and use the *Ansatz*

$$\eta_1(x) = \begin{cases} \varepsilon^2\zeta(\varepsilon x), & (\beta_0, \alpha_0) \in C_s, \\ \dfrac{1}{2}\varepsilon\zeta(\varepsilon x)e^{i\mu_0 x} + \dfrac{1}{2}\varepsilon\overline{\zeta(\varepsilon x)}e^{-i\mu_0 x}, & (\beta_0, \alpha_0) \in C_w. \end{cases}$$

Observe that $\eta_1 \in \chi(D)L_e^2(\mathbb{R})$ is equivalent to the requirement that ζ is real and lies in $\chi_0(\varepsilon D)L_e^2(\mathbb{R})$ for $(\beta_0, \alpha_0) \in C_s$ and ζ is complex and lies in $\chi_0(\varepsilon D)L_c^2(\mathbb{R})$ for $(\beta_0, \alpha_0) \in C_w$, where

$$H_c^s(\mathbb{R}) = \{u \in H^s(\mathbb{R}) : u(-x) = \overline{u(x)} \text{ for all } x \in \mathbb{R}\}, \quad s \geq 0.$$

In Section 6.3 we convert (6.2.20)–(6.2.25) into a linear equation of Kadomtsev-Petviashvili or Davey-Stewartson type by determining η_2, ω, Γ and ξ as functions of η_1 and computing a reduced equation that determines η_1. The results of these calculations are summarised in the following theorems; note that for notational simplicity we henceforth omit the suffix 'KP' or 'DS' on operators (and their associated function spaces) when stating results that apply to both.

Theorem 6.2.2. *Fix* $\lambda^\star > 0$. *There exists a constant* $k_{\max} > 0$ *such that equations (6.2.20)–(6.2.25) have a unique solution* $u \in Y_r^1$ *for each* $\lambda \in [\varepsilon^r k_{\max}, \lambda^\star]$ *and each* $u^\dagger \in X_r^0$.

Theorem 6.2.3.

(i) *Suppose that* $(\beta_0, \alpha_0) \in C_s$ *and* $\lambda = \varepsilon^2 k$, *where* $0 < k \leq k_{\max}$. *There exist* $\zeta_{\varepsilon,k}^\dagger \in \mathcal{L}(X_r^0, \chi_0(\varepsilon D)L_e^2(\mathbb{R}))$ *and an injection* $\check{u}_{\varepsilon,k}^{KP} \in \mathcal{L}(\chi_0(\varepsilon D)L_e^2(\mathbb{R}) \times X_r^0, Y_r^1)$ *such that* $u \in Y_r^1$ *solves the resolvent equations*

$$(L - \mathrm{i}\lambda I)u = u^\dagger$$

if and only if $u = \check{u}_{\varepsilon,k}^{KP}(\zeta, u^\dagger)$ *for some solution of the reduced equation*

$$\mathcal{B}_{\varepsilon,k}^{KP}(\zeta) = \zeta_{\varepsilon,k}^\dagger(u^\dagger).$$

The operator $\mathcal{B}_{\varepsilon,k}^{KP} : \mathcal{D}_{\mathcal{B}}^{KP} \subseteq W^{KP} \to W^{KP}$ *is given by*

$$\mathcal{B}_{\varepsilon,k}^{KP}(\zeta) = -\zeta_{xx} + (\beta_0 - \tfrac{1}{3})\zeta_{xxxx} + k^2\zeta - 3(\zeta\zeta_{KdV}^\star)_{xx} - \mathcal{R}_{\varepsilon,k}^{KP}(\zeta),$$

where $W^{KP} = L_e^2(\mathbb{R})$, $\mathcal{D}_{\mathcal{B}}^{KP} = H_e^4(\mathbb{R})$ *and* $\mathcal{R}_{\varepsilon,k}^{KP} \in \mathcal{L}(H_e^4(\mathbb{R}), L_e^2(\mathbb{R}))$ *satisfies the estimate*

$$\|\mathcal{R}_{\varepsilon,k}^{KP}(\zeta)\|_0 \leq c\varepsilon\|\zeta\|_3;$$

each solution ζ *of the reduced equation satisfies* $\zeta \in \chi_0(\varepsilon D)H_e^4(\mathbb{R})$. *In particular,* $L - \mathrm{i}\varepsilon^2 kI$ *is (semi-)Fredholm if* $\mathcal{B}_{\varepsilon,k}^{KP}$ *is (semi-)Fredholm and the kernels of* $L - \mathrm{i}\varepsilon^2 kI$ *and* $\mathcal{B}_{\varepsilon,k}^{KP}$ *have the same dimension.*

(ii) *Suppose that* $(\beta_0, \alpha_0) \in C_w$ *and* $\lambda = \varepsilon k$, *where* $0 < k \leq k_{\max}$. *There exist* $\zeta_{\varepsilon,k}^\dagger \in \mathcal{L}(X_r^0, \chi_0(\varepsilon D)L_c^2(\mathbb{R}))$ *and an injection* $\check{u}_{\varepsilon,k}^{DS} \in \mathcal{L}(\chi_0(\varepsilon D)L_c^2(\mathbb{R}) \times X_r^0, Y_r^1)$ *such that* $u \in Y_r^1$ *solves the resolvent equations*

$$(L - \mathrm{i}\lambda I)u = u^\dagger$$

if and only if $u = \check{u}_{\varepsilon,k}^{DS}(\zeta, u^\dagger)$ *for some solution of the reduced equation*

$$\mathcal{B}_{\varepsilon,k}^{DS}(\zeta, \psi) = (\zeta_{\varepsilon,k}^\dagger(u^\dagger), 0).$$

The operator $\mathcal{B}_{\varepsilon,k}^{DS} : \mathcal{D}_{\mathcal{B}}^{DS} \subseteq W^{DS} \to W^{DS}$ *is given by*

$$\mathcal{B}_{\varepsilon,k}^{DS}(\zeta, \psi)$$
$$= \begin{pmatrix} A_2^{-1}\zeta - A_2^{-1}A_1\zeta_{xx} + k^2\zeta - A_2^{-1}(A_3 + A_5)\zeta_{NLS}^{\star 2}\zeta - A_2^{-1}A_3\zeta_{NLS}^{\star 2}\bar{\zeta} + 4A_2^{-1}A_4\zeta_{NLS}^\star\psi_x - \mathcal{R}_{\varepsilon,k}^{DS}(\zeta) \\ -(1 - a_0^{-1})\psi_{xx} + k^2\psi - 2A_4\mathrm{Re}(\zeta_{NLS}^\star\zeta)_x, \end{pmatrix}$$

where $W^{\mathrm{DS}} = L^2_{\mathrm{e}}(\mathbb{R}) \times L^2_{\mathrm{o}}(\mathbb{R})$, $\mathcal{D}^{\mathrm{DS}}_{\mathcal{B}} = H^2_{\mathrm{e}}(\mathbb{R}) \times H^2_{\mathrm{o}}(\mathbb{R})$ *and* $\mathcal{R}^{\mathrm{DS}}_{\varepsilon,k} \in$
$\mathcal{L}(H^2_{\mathrm{c}}(\mathbb{R}), L^2_{\mathrm{c}}(\mathbb{R}))$ *satisfies the estimate*

$$\|\mathcal{R}^{\mathrm{DS}}_{\varepsilon,k}(\zeta)\|_0 \leq c\varepsilon\|\zeta\|_1;$$

each solution (ζ, ψ) *of the reduced equation satisfies* $\zeta \in \chi_0(\varepsilon D)H^2_{\mathrm{c}}(\mathbb{R})$.
In particular, $L - \mathrm{i}\varepsilon k I$ is (semi-)Fredholm if $\mathcal{B}^{\mathrm{DS}}_{\varepsilon,k}$ is (semi-)Fredholm and the kernels of $L - \mathrm{i}\varepsilon k I$ and $\mathcal{B}^{\mathrm{DS}}_{\varepsilon,k}$ have the same dimension.

Theorem 6.2.4. *Define* $Z^0_{\mathrm{r}} = H^1_{\mathrm{e}}(\mathbb{R}) \times L^2_{\mathrm{e}}(\mathbb{R}) \times H^1_{\mathrm{o}}(\Sigma) \times H^1_{\mathrm{e}}(\Sigma) \times \mathring{H}^1_{\mathrm{o}}(\Sigma)$,
where

$$\mathring{H}^1_{\mathrm{o}}(\Sigma) = \{u \in H^1_{\mathrm{o}}(\Sigma) : u|_{y=0} = u|_{y=1} = 0\}.$$

(i) *Suppose that* $(\beta_0, \alpha_0) \in C_{\mathrm{s}}$. *There exist* $\zeta^\dagger_{\varepsilon,0} \in \mathcal{L}(Z^0_{\mathrm{r}}, \chi_0(\varepsilon D)L^2_{\mathrm{e}}(\mathbb{R}))$ *and an injection* $\breve{u}^{\mathrm{KP}}_{\varepsilon,0} \in \mathcal{L}(\chi_0(\varepsilon D)L^2_{\mathrm{e}}(\mathbb{R}) \times Z^0_{\mathrm{r}}, Y^1_\star)$ *such that* $u \in Y^1_\star$ *solves*

$$Lu = u^\dagger,$$

where $\xi^\dagger = -(\xi^\dagger_1)_x - (\xi^\dagger_3)_y$, *if and only if* $u = \breve{u}^{\mathrm{KP}}_{\varepsilon,0}(\zeta, \eta^\dagger, \omega^\dagger, \Gamma^\dagger, \xi^\dagger_1, \xi^\dagger_3)$
for some solution of the reduced equation

$$\mathcal{C}^{\mathrm{KP}}_\varepsilon(\zeta) = \zeta^\dagger_{\varepsilon,0}(\eta^\dagger, \omega^\dagger, \Gamma^\dagger, \xi^\dagger_1, \xi^\dagger_3).$$

The operator $\mathcal{C}^{\mathrm{KP}}_\varepsilon : H^2_{\mathrm{e}}(\mathbb{R}) \subseteq L^2_{\mathrm{e}}(\mathbb{R}) \to L^2_{\mathrm{c}}(\mathbb{R})$ *is given by*

$$\mathcal{C}^{\mathrm{KP}}_\varepsilon(\zeta) = -\zeta + (\beta_0 - \tfrac{1}{3})\zeta_{xx} - 3\zeta\zeta^\star_{\mathrm{KdV}} - \mathcal{R}^{\mathrm{KP}}_{\varepsilon,0}(\zeta)$$

and $\mathcal{R}^{\mathrm{KP}}_{\varepsilon,0} \in \mathcal{L}(H^2_{\mathrm{c}}(\mathbb{R}), L^2_{\mathrm{c}}(\mathbb{R}))$ *satisfies the estimate*

$$\|\mathcal{R}^{\mathrm{KP}}_{\varepsilon,0}(\zeta)\|_0 \leq c\varepsilon\|\zeta\|_1;$$

each solution of the reduced equation lies in $\chi_0(\varepsilon D)H^2_{\mathrm{e}}(\mathbb{R})$.

(ii) *Suppose that* $(\beta_0, \alpha_0) \in C_{\mathrm{w}}$. *There exist* $\zeta^\dagger_{\varepsilon,0} \in \mathcal{L}(Z^0_{\mathrm{r}}, \chi_0(\varepsilon D)L^2_{\mathrm{c}}(\mathbb{R}))$ *and an injection* $\breve{u}^{\mathrm{DS}}_{\varepsilon,0} \in \mathcal{L}(\chi_0(\varepsilon D)L^2_{\mathrm{c}}(\mathbb{R}) \times Z^0_{\mathrm{r}}, Y^1_\star)$ *such that* $u \in Y^1_\star$ *solves*

$$Lu = u^\dagger,$$

where $\xi^\dagger = -(\xi^\dagger_1)_x - (\xi^\dagger_3)_y$, *if and only if* $u = \breve{u}^{\mathrm{DS}}_{\varepsilon,0}(\zeta, \eta^\dagger, \omega^\dagger, \Gamma^\dagger, \xi^\dagger_1, \xi^\dagger_3)$
for some solution of the reduced equation

$$\mathcal{C}^{\mathrm{DS}}_\varepsilon(\zeta) = \zeta^\dagger_{\varepsilon,0}(\eta^\dagger, \omega^\dagger, \Gamma^\dagger, \xi^\dagger_1, \xi^\dagger_3).$$

The operator $\mathcal{C}^{\mathrm{DS}}_\varepsilon : H^2_{\mathrm{c}}(\mathbb{R}) \subseteq L^2_{\mathrm{c}}(\mathbb{R}) \to L^2_{\mathrm{c}}(\mathbb{R})$ *is given by*

$$\mathcal{C}^{\mathrm{DS}}_\varepsilon(\zeta) = A^{-1}_2\zeta - A^{-1}_2 A_1\zeta_{xx} - 2A^{-1}_2 A_5\zeta^{\star 2}_{\mathrm{NLS}}\zeta - A^{-1}_2 A_5\zeta^{\star 2}_{\mathrm{NLS}}\overline{\zeta} - \mathcal{R}^{\mathrm{DS}}_{\varepsilon,0}(\zeta)$$

and $\mathcal{R}^{\mathrm{DS}}_{\varepsilon,0} \in \mathcal{L}(H^2_{\mathrm{c}}(\mathbb{R}), L^2_{\mathrm{c}}(\mathbb{R}))$ *satisfies the estimate*

$$\|\mathcal{R}^{\mathrm{DS}}_{\varepsilon,0}(\zeta)\|_0 \leq c\varepsilon\|\zeta\|_1;$$

each solution of the reduced equation lies in $\chi_0(\varepsilon D)H^2_{\mathrm{c}}(\mathbb{R})$.

We study the spectrum of $\mathcal{B}_{\varepsilon,k}$ in Section 6.4, establishing the following result, whose corollary relates it to the operator L.

Lemma 6.2.5. *For each sufficiently small $k_{\min} > 0$ there exists a unique number $k_\varepsilon \in [k_{\min}, k_{\max}]$ with the following properties.*

(i) *The operator $\mathcal{B}_{\varepsilon,k} : \mathcal{D}_\mathcal{B} \subseteq W \to W$ is an isomorphism for each $k \in [k_{\min}, k_{\max}] \setminus \{k_\varepsilon\}$.*
(ii) *The operator $\mathcal{B}_{\varepsilon,k_\varepsilon} : \mathcal{D}_\mathcal{B} \subseteq W \to W$ is Fredholm with index 0 and has a one-dimensional kernel.*

Corollary. *The imaginary number $i\varepsilon^r k$ belongs to $\rho(L)$ for each $k \in [k_{\min}, k_{\max}] \setminus \{k_\varepsilon\}$, while $i\varepsilon^r k_\varepsilon$ is a simple eigenvalue of L.*

Proof. It remains only to show that the eigenvalue $i\varepsilon^r k_\varepsilon$ is algebraically simple. Observe that $\Omega(Lu_1, u_2) = -\Omega(u_1, Lu_2)$ and in particular that $\Omega((L - i\varepsilon^r k_\varepsilon)u_1, u_2) = -\Omega(u_1, (L + i\varepsilon^r k_\varepsilon)u_2)$ for $u_1, u_2 \in \mathcal{D}(L)$. It follows that $\Omega(f, \bar{u}_\varepsilon) = 0$ is a necessary condition for $f \in X_r^0$ to lie in the range of $L - i\varepsilon^r k_\varepsilon I$, where u_ε is an eigenvector corresponding to the eigenvalue $i\varepsilon^r k_\varepsilon$. Using the formulae

$$\omega = \frac{i\varepsilon^r k_\varepsilon \beta_0 \eta_\varepsilon}{(1 + (\eta_x^\star)^2)^{1/2}} - i\varepsilon^r k_\varepsilon \int_0^1 y\Phi_y^\star \Gamma_\varepsilon \, dy + \frac{i\varepsilon^r k_\varepsilon \eta_\varepsilon}{1 + \eta^\star} \int_0^1 y\Phi_y^\star \, dy,$$
$$\xi = (1 + \eta^\star)i\varepsilon^r k_\varepsilon \Gamma_\varepsilon - i\varepsilon^r k_\varepsilon \eta_\varepsilon y\Phi_y^\star$$

(see equations (6.3.1), (6.3.2)), we find that

$$\Omega(u_\varepsilon, \bar{u}_\varepsilon) = -2i\varepsilon^r k_\varepsilon \beta_0 \int_\mathbb{R} |\eta_\varepsilon|^2 \, dx - 2i\varepsilon^r k_\varepsilon \int_\Sigma |\Gamma_\varepsilon|^2 \, dy \, dx$$
$$+ O(\varepsilon^{2r}(\|\eta_\varepsilon\|_0^2 + \|\Gamma_\varepsilon\|_0^2))$$
$$\neq 0,$$

so that u_ε does not lie in the range of $L - i\varepsilon k_\varepsilon I$. $\qquad\square$

The resolvent decay estimate is a consequence of the next lemma.

Lemma 6.2.6. *There exists a constant $\lambda^\star > 0$ such that each solution $u \in Y_r^1$ of equations (6.2.20)–(6.2.25) with $u^\dagger \in X_r^0$ satisfies the estimate*

$$\|u\|_{X^1}^2 + \lambda^2 \|u\|_{X^0}^2 \le c\|u^\dagger\|_{X^0}^2 \qquad (6.2.27)$$

whenever $\lambda \ge \lambda^\star$.

Proof. Standard *a priori* resolvent estimates for constant-coefficient operators yield

$$\|\eta\|_2^2 + \|\omega\|_1^2 + \|\Phi\|_2^2 + \|\xi\|_1^2 + \lambda^2(\|\eta\|_1^2 + \|\omega\|_0^2 + \|\Phi\|_1^2 + \|\xi\|_0^2)$$
$$\leq c(\|h_1 + \eta^\dagger\|_1^2 + \|h_2 + \omega^\dagger\|_0^2 + \|H_1 + \Gamma^\dagger\|_1^2 + \|F_{1x} + F_{3y} + \xi^\dagger\|_0^2$$
$$+ \|F_3|_{y=1}\|_{1/2}^2 + |\lambda| \|F_3|_{y=1}\|_0^2)$$

for sufficiently large values of λ (e.g. see Groves & Mielke [16, Lemma 3.4]). The assertion follows from this result and the estimates

$$\|h_1\|_1, \|h_2\|_0, \|H_1\|_1, \|F_{1x}\|_0, \|F_{3y}\|_0 \leq c\varepsilon(\|\eta\|_2^2 + \|\omega\|_1^2 + \|\Phi\|_2^2 + \|\xi\|_1^2),$$
$$\|F_3|_{y=1}\|_0 \leq c\varepsilon(\|\eta\|_1 + \|\Phi_x|_{y=1}\|_0 + \|\Phi_y|_{y=1}\|_0),$$
$$\|F_3|_{y=1}\|_{1/2} \leq c\varepsilon(\|\eta\|_{3/2} + \|\Phi_x|_{y=1}\|_{1/2} + \|\Phi_y|_{y=1}\|_{1/2}).$$

$\square$

Corollary. *The operator* $L - \mathrm{i}\lambda I$ *is an isomorphism* $\mathcal{D}(L) \to X_r^0$ *whenever* $\lambda \geq \lambda^\star$.

Proof. Theorem 6.2.3 shows that the resolvent set of L is nonempty, and this operator is therefore closed. It follows that $L - \mathrm{i}\lambda I$ is also closed, and one finds from the *a priori* estimate (6.2.27) that it is injective and has closed range for $\lambda \geq \lambda^\star$, so that it is semi-Fredholm. Since $L - \mathrm{i}\lambda I$ is a continuous function of λ with values in $\mathcal{L}(\mathcal{D}(L), X_r^0)$ its index is constant and hence equal to the index of $L - \mathrm{i}\lambda^\star I$, which is zero (Theorem 6.2.2). An injective Fredholm operator with index zero is an isomorphism. $\square$

It remains to establish the Iooss condition at the origin.

Lemma 6.2.7. *The operators* $\mathcal{C}_\varepsilon^{\mathrm{KP}} : H_e^2(\mathbb{R}) \to L_e^2(\mathbb{R})$ *and* $\mathcal{C}_\varepsilon^{\mathrm{DS}} : H_c^2(\mathbb{R}) \to L_c^2(\mathbb{R})$ *are isomorphisms.*

Proposition 6.2.8 *For each* $\eta^\dagger \in H_e^1(\mathbb{R})$, $\omega^\dagger \in L_e^2(\mathbb{R})$, $\Gamma^\dagger \in H_o^s(\Sigma)$ *and*

$$\xi^\dagger = -(\xi_1^\dagger)_x - (\xi_3^\dagger)_y,$$

where $\xi_1^\dagger \in H_e^1(\Sigma)$ *and* $\xi_3^\dagger \in \mathring{H}_o^1(\Sigma)$, *the equation*

$$Lu = u^\dagger$$

has a unique solution $u \in Y_\star^1$, *which satisfies the estimate*

$$\|u\|_{Y_\star^1} \leq c(\|\eta^\dagger\|_1 + \|\omega^\dagger\|_0 + \|\Gamma^\dagger\|_1 + \|\xi_1^\dagger\|_1 + \|\xi_3^\dagger\|_1).$$

Lemma 6.2.9. *For each* $v^\dagger \in U_\star$ *the equation*

$$Lv = -N(v^\dagger)$$

has a unique solution $v \in Y_\star^1$, *which satisfies the estimate*

$$\|v\|_{Y_\star^1} \le c\|v^\dagger\|_{Y_\star^1}.$$

Proof. Choose $v^\dagger \in U_\star$ and write $u^\dagger = Q^{-1}(v^\dagger)$.

Recall that

$$g(v) = \widehat{dQ}[u](f(u^\star + u)), \qquad u = Q^{-1}(v)$$

and that Q, $\widehat{dQ}[u]$ do not alter the fourth component of their arguments. The fourth component of $g(v^\dagger)$ is therefore given by

$$
\begin{aligned}
g_4(v^\dagger) = {} & f_4(u^\star + u^\dagger) \\
= {} & -[\Gamma_x^\dagger + (\eta^\star + \eta^\dagger)\Gamma_x^\dagger + \Phi_x^\star \eta^\dagger - y(\eta_x^\star + \eta_x^\dagger)\Gamma_y^\dagger - y\Phi_y^\star \eta_x^\dagger]_x \\
& + [-\Gamma_y^\dagger + B_1(\eta^\dagger, \Gamma^\dagger) + y\eta_x^\dagger]_y,
\end{aligned}
$$

in which the first line follows from (6.2.14), the second from the fact that $f_4(u^\star) = 0$ and the definitions of B_1, B_{nl}, the third from (6.2.17) and the fourth from the identity $\Gamma_{xx}^\dagger = \Phi_{xx}^\dagger + \Theta_{xxy}^\dagger$. According to the definition $N(v) = g(v) - Lv$ the fourth component of $N(v^\dagger)$ is

$$
\begin{aligned}
N_4(v^\dagger) = {} & f_4(u^\star + u^\dagger) - L_4 v^\dagger \\
= {} & -[F_1(\eta^\dagger, \Gamma^\dagger) + (\eta^\star + \eta^\dagger)\Gamma_x^\dagger + \Phi_x^\star \eta^\dagger - y(\eta_x^\star + \eta_x^\dagger)\Gamma_y^\dagger - y\Phi_y^\star \eta_x^\dagger]_x.
\end{aligned}
$$

Applying Proposition 6.2.8 with

$$\xi_1^\dagger = -\left(F_1(\eta^\dagger, \Gamma^\dagger) + (\eta^\star + \eta^\dagger)\Gamma_x^\dagger + \Phi_x^\star \eta^\dagger - y(\eta_x^\star + \eta_x^\dagger)\Gamma_y^\dagger - y\Phi_y^\star \eta_x^\dagger\right), \quad \xi_3^\dagger = 0,$$

one finds that the equation

$$Lv = -N(v^\dagger)$$

has a unique solution $v \in Y_\star^1$, which satisfies

$$\|v\|_{Y_\star^1} \le c(\|N_1(v^\dagger)\|_1 + \|N_2(v^\dagger)\|_0 + \|N_3(v^\dagger)\|_1 + \|\xi_1^\dagger\|_1).$$

The assertion follows by combining this estimate with

$$\|\xi_1^\dagger\|_1 \le c(\|\eta^\dagger\|_2 + \|\nabla\Gamma^\dagger\|_1 + \|\nabla\Gamma^\dagger\|_1) \le c\|v^\dagger\|_{Y_\star^1}$$

and

$$\|N(v^\dagger)\|_{X^0} \le c\|v^\dagger\|_{Y_\star^1}. \qquad \square$$

Remark We have not examined the portion of $\sigma(L)$ which lies in the interval $(0, i\varepsilon^r k_{min})$ since this information is not required for an application of the Lyapunov-Iooss theorem. It can however be demonstrated that this interval lies in the resolvent set of L (see Groves, Haragus & Sun [14] and Groves, Sun & Wahlén [15]).

6.3 Derivation of the Reduced Equation

6.3.1 The Reduction Method

We begin by eliminating ω, Φ and ξ from (6.2.20)–(6.2.25) and deriving a single, equivalent equation for η. Equations (6.2.20) and (6.2.22) clearly yield the explicit formulae

$$\omega = \frac{\beta_0}{(1+(\eta_x^\star)^2)^{1/2}}(\eta^\dagger + i\lambda\eta) - \frac{1}{1+\eta^\star}\int_0^1 y\Phi_y^\star\xi \, dy, \qquad (6.3.1)$$

$$\xi = (1+\eta^\star)(\Gamma^\dagger + i\lambda\Gamma) - y\Phi_y^\star(\eta^\dagger + i\lambda\eta). \qquad (6.3.2)$$

for ω and Φ, and substituting these formulae into (6.2.23)–(6.2.25), we obtain the boundary-value problem

$$-\hat{\Gamma}_{yy} + q^2\hat{\Gamma} = \hat{F}^\dagger(u^\dagger) - i\mu\hat{F}_1(\eta, \Gamma) - i\lambda\hat{F}_2(\eta, \Gamma)$$

$$- (\hat{F}_3(\eta, \Gamma))_y, \qquad 0 < y < 1, \qquad (6.3.3)$$

$$\hat{\Gamma}_y = 0 \qquad \text{on } y = 0, \qquad (6.3.4)$$

$$\hat{\Gamma}_y = -i\mu\hat{\eta} + \hat{F}_3(\eta, \Gamma) \qquad \text{on } y = 1 \qquad (6.3.5)$$

for the Fourier transform $\hat{\Gamma}(\mu, y)$ of $\Gamma(x, y)$, where $q = \sqrt{\mu^2 + \lambda^2}$,

$$F_2(\eta, \Gamma) = -i\lambda\eta^\star\Gamma + i\lambda y\Phi_y^\star\eta,$$

$$F^\dagger(u^\dagger) = \xi^\dagger + i\lambda(1+\eta^\star)\Gamma^\dagger - i\lambda y\Phi_y^\star\eta^\dagger$$

and $\xi^\dagger = -\xi_{1x}^\dagger - \xi_{3y}^\dagger$ for $\lambda = 0$. On the other hand, substituting (6.3.1), (6.3.2) into (6.2.21) yields

$$(\alpha_0 + \varepsilon^2)\eta - \Gamma_x|_{y=1} - \beta_0\eta_{xx} + h_2(\eta, \Gamma)$$

$$= \omega^\dagger + i\lambda \frac{\beta_0}{(1+(\eta_x^\star)^2)^{1/2}}(\eta^\dagger + i\lambda\eta)$$

$$- \frac{i\lambda}{1+\eta^\star}\int_0^1 y\Phi_y^\star[(1+\eta^\star)(\Gamma^\dagger + i\lambda\Gamma) - y\Phi_y^\star(\eta^\dagger + i\lambda\eta)] \, dy,$$

which we write as

$$g_\varepsilon(D, \lambda)\eta = \mathcal{N}(\eta, \Gamma, u^\dagger), \qquad (6.3.6)$$

where

$$g_\varepsilon(\mu, \lambda) = \alpha_0 + \varepsilon^2 + \beta_0 q^2 - \frac{\mu^2}{q^2} q \coth q$$

and

$$\mathcal{N}(\eta, \Gamma, u^{\dagger})$$

$$= \omega^{\dagger} + i\lambda\beta_0 \frac{\eta^{\dagger}}{(1 + (\eta_x^{\star})^2)^{1/2}} + \Gamma_x|_{y=1} - \mathcal{F}^{-1}\left[\frac{\mu^2}{q^2} q \coth q \,\hat{\eta}\right]$$

$$- \beta_0\left(\eta_{xx} - \left[\frac{\eta_x}{(1 + (\eta_x^{\star})^2)^{3/2}}\right]_x\right) - \lambda^2\beta_0\left(\frac{1}{(1 + (\eta_x^{\star})^2)^{1/2}} - 1\right)\eta$$

$$- \frac{i\lambda}{1 + \eta^{\star}} \int_0^1 [-i\lambda y^2(\Phi_y^{\star})^2\eta - y^2(\Phi_y^{\star})^2\eta^{\dagger} + i\lambda y\Phi_y^{\star}\Gamma$$

$$+ y\Phi_y^{\star}\Gamma^{\dagger} + i\lambda y\Phi_y^{\star}\eta^{\star}\Gamma + y\Phi_y^{\star}\eta^{\star}\Gamma^{\dagger}]\,dy$$

$$- \int_0^1 \left\{\Phi_x^{\star}\Gamma_x - \frac{\Phi_y^{\star}\Gamma_y}{(1 + \eta^{\star})^2} + \frac{(\Phi_y^{\star})^2\eta}{(1 + \eta^{\star})^3} - \frac{y^2(\eta_x^{\star})^2\Phi_y^{\star}\Gamma_y}{(1 + \eta^{\star})^2}\right.$$

$$- \frac{y^2\eta_x^{\star}(\Phi_y^{\star})^2\eta_x}{(1 + \eta^{\star})^2} + \frac{y^2\eta_x^{\star}(\Phi_y^{\star})^2\eta}{(1 + \eta^{\star})^3} + \left[y\Phi_y^{\star}\Gamma_x + y\Phi_x^{\star}\Gamma_y\right.$$

$$\left.\left.- \frac{2y^2\eta_x^{\star}\Phi_y^{\star}\Gamma_y}{1 + \eta^{\star}} - \frac{y^2(\Phi_y^{\star})^2\eta_x}{1 + \eta^{\star}} + \frac{y^2(\Phi_y^{\star})^2\eta_x^{\star}\eta}{(1 + \eta^{\star})^2}\right]_x\right\}\,dy.$$

The next step is to express Γ as a function of η. To this end we formulate the boundary-value problem (6.3.3)–(6.3.5) as the integral equation

$$\Gamma = \mathcal{G}_1(F_1(\eta, \Gamma), F_2(\eta, \Gamma), F_3(\eta, \Gamma), \eta) + \mathcal{G}_2(F^{\dagger}(u^{\dagger})), \qquad (6.3.7)$$

where

$$\mathcal{G}_1(P_1, P_2, P_3, p) = \mathcal{F}^{-1}\left[\int_0^1 \{G(y, \tilde{y})(-i\mu\hat{P}_1 - i\lambda\hat{P}_2) + G_{\tilde{y}}(y, \tilde{y})\hat{P}_3\}\,d\tilde{y}\right]$$

$$- \mathcal{F}^{-1}[i\mu G(y, 1)\hat{p}],$$

$$\mathcal{G}_2(P) = \mathcal{F}^{-1}\left[\int_0^1 G(y, \tilde{y})\hat{P}\,d\tilde{y}\right]$$

and

$$G(y, \tilde{y}) = \begin{cases} \dfrac{\cosh(qy)\cosh(q(1 - \tilde{y}))}{q\sinh q}, & 0 \le y \le \tilde{y} \le 1, \\[2ex] \dfrac{\cosh(q\tilde{y})\cosh(q(1 - y))}{q\sinh q}, & 0 \le \tilde{y} \le y \le 1. \end{cases}$$

We study equation (6.3.7) using Lemma 6.3.1 (to decompose terms of the form $G(y, 1)\hat{p}$ and $G(y, y)\hat{P}$ into a leading-order part and a higher-order remainder) and Proposition 6.3.2 (which records the mapping properties of the integral operators appearing on its right-hand side).

Lemma 6.3.1.

(ii) *Suppose that $\mu > 0$. The Green's function G admits the decomposition*

$$G(y,\tilde{y}) = G(y,\tilde{y};\underline{\mu}) + R(y,\tilde{y}),$$

where

$$G(y,\tilde{y};\underline{\mu}) = \begin{cases} \dfrac{\cosh(\underline{\mu}y)\cosh(\underline{\mu}(1-\tilde{y}))}{\underline{\mu}\sinh\underline{\mu}}, & 0 \le y \le \tilde{y} \le 1, \\[4mm] \dfrac{\cosh(\underline{\mu}\tilde{y})\cosh(\underline{\mu}(1-y))}{\underline{\mu}\sinh\underline{\mu}}, & 0 \le \tilde{y} \le y \le 1, \end{cases}$$

and R satisfies the estimates

$$|\partial_y^{m_1}\partial_{\tilde{y}}^{m_2}R(y,\tilde{y})| \le c\left(|\mu - \underline{\mu}| + \lambda^2\right), \qquad m_1, m_2 = 0, 1, 2, \dots,$$

uniformly over $y, \tilde{y}$ in $[0,1]$ and q in each fixed interval $[q_{\min}, q_{\max}]$ containing μ.

(ii) *The Green's function G admits the expansion*

$$G(y,\tilde{y}) = G(y,\tilde{y};0) + R(y,\tilde{y}),$$

where

$$G(y,\tilde{y};0) = \frac{1}{q^2}$$

and R satisfies the estimate

$$|R(y,\tilde{y})| \le c$$

uniformly over $y, \tilde{y}$ in $[0,1]$ and q in each fixed interval $[0, q_{\max}]$.

Proposition 6.3.2.

(i) *The mapping $(P_1, P_2, P_3, p) \mapsto \mathcal{G}_1(P_1, P_2, P_3, p)$ defines a linear function*

$$\begin{cases} H_e^1(\Sigma) \times H_o^1(\Sigma) \times H_o^1(\Sigma) \times H_e^2(\mathbb{R}) \to H_o^2(\Sigma), & \lambda > 0, \\ H_e^1(\Sigma) \times H_o^1(\Sigma) \times H_o^1(\Sigma) \times H_e^2(\mathbb{R}) \to H_{\star,o}^2(\Sigma), & \lambda = 0, \end{cases}$$

which satisfies the estimate

$$\|\nabla\mathcal{G}_1\|_1 + \lambda\|\mathcal{G}_1\|_1 + \lambda^2\|\mathcal{G}_1\|_0$$
$$\le c(\|P_1\|_1 + \|P_2\|_1 + \|P_3\|_1 + \lambda(\|P_2\|_0 + \|P_2\|_0 + \|P_3\|_0)$$
$$+ (1+\lambda^2)\|p\|_0 + \|p_{xx}\|_0).$$

(ii) *Suppose that $\lambda > 0$. The mapping $P \mapsto \mathcal{G}_2(P)$ defines a linear function $L_o^2(\Sigma) \to H_o^2(\Sigma)$, which satisfies the estimate*

$$\|\nabla\mathcal{G}_2\|_1 + \lambda\|\mathcal{G}_2\|_1 + \lambda^2\|\mathcal{G}_2\|_0 \le c(1+\lambda^{-1})\|P\|_0.$$

(iii) *Suppose that $\lambda = 0$ and $P = -P_{1x} - P_{3y}$, where $P_1 \in H_e^1(\Sigma)$ and $P_3 \in \mathring{H}_o^1(\Sigma)$. The mapping $(P_1, P_3) \mapsto \mathcal{G}_2(P)$ defines a linear function $H_e^1(\Sigma) \times H_o^1(\Sigma) \to H_{*,0}^2(\Sigma)$, which satisfies the estimate*

$$\|\nabla \mathcal{G}_2\|_1 \leq c(\|P_1\|_1 + \|P_3\|_1).$$

We now fix $\lambda^\star > 0$ and henceforth suppose that $\lambda \leq \lambda^\star$.

Theorem 6.3.3.

(i) *For each sufficiently small value of ε and each $\eta \in H_e^2(\mathbb{R})$, $F^\dagger \in L_o^2(\Sigma)$, equation (6.3.7) has a unique solution $\tilde{\Gamma} = \tilde{\Gamma}(\eta, u^\dagger) \in H_o^2(\Sigma)$, which satisfies the estimate*

$$\|\nabla \tilde{\Gamma}(\eta, u^\dagger)\|_1 + \lambda \|\tilde{\Gamma}(\eta, u^\dagger)\|_1 + \lambda^2 \|\tilde{\Gamma}(\eta, u^\dagger)\|_0 \leq c\|\eta\|_2 + c_\lambda \|F^\dagger\|_0.$$

(ii) *Suppose that $\lambda = 0$. For each sufficiently small value of ε and each $\eta \in H_e^2(\mathbb{R})$, $\xi_1^\dagger \in H_e^1(\Sigma)$, $\xi_3^\dagger \in \mathring{H}_o^1(\Sigma)$, equation (6.3.7) has a unique solution $\tilde{\Gamma} = \tilde{\Gamma}(\eta, \xi_1^\dagger, \xi_3^\dagger) \in H_{*,0}^2(\Sigma)$, which satisfies the estimate*

$$\|\nabla \tilde{\Gamma}\|_1 \leq c(\|\eta\|_2 + \|\xi_1^\dagger\|_1 + \|\xi_3^\dagger\|_1).$$

Proof. It follows from the estimates

$$\|F_1\|_1, \; \|F_3\|_1 \leq c\varepsilon^r(\|\eta\|_2 + \|\nabla \Gamma\|_1),$$

$$\lambda \|F_1\|_0, \; \lambda \|F_3\|_0 \leq c\varepsilon^r(\lambda \|\eta\|_1 + \lambda \|\nabla \Gamma\|_0),$$

$$\|F_2\|_1 \leq c\varepsilon^r(\lambda \|\eta\|_1 + \lambda \|\Gamma\|_1),$$

$$\lambda \|F_2\|_0 \leq c\varepsilon^r(\lambda^2 \|\eta\|_0 + \lambda^2 \|\Gamma\|_0)$$

and Proposition 6.3.2 (i) that

$$\|\nabla \mathcal{G}_1(\eta, \Gamma)\|_1 + \lambda \|\mathcal{G}_1(\eta, \Gamma)\|_1 + \lambda^2 \|\mathcal{G}_1(\eta, \Gamma)\|_0$$
$$\leq c(\varepsilon^r(\|\nabla \Gamma\|_1 + \lambda \|\Gamma\|_1 + \lambda^2 \|\Phi\|_0) + \lambda \|\eta\|_1),$$

in which $\mathcal{G}_1(\eta, \Gamma)$ is used an abbreviation for $\mathcal{G}_1(F_1(\eta, \Gamma), F_2(\eta, \Gamma), F_3(\eta, \Gamma), \eta)$. Using this estimate together with Proposition 6.3.2(ii) or (iii) as appropriate, one finds that (6.3.7) admits a unique solution that satisfies the stated estimate. □

Substituting $\Phi = \tilde{\Gamma}(\eta, u^\dagger)$ into (6.3.6), we obtain a single equation for η, namely

$$g_\varepsilon(D, \lambda)\eta = \tilde{\mathcal{N}}(\eta, u^\dagger), \tag{6.3.8}$$

where

$$\tilde{\mathcal{N}}(\eta, u^\dagger) = \mathcal{N}(\eta, \tilde{\Gamma}(\eta, u^\dagger), u^\dagger),$$

and the next step is to write (6.3.8) as

$$g_\varepsilon(\mu, \lambda)\hat{\eta}_1 = \chi \mathcal{F}[\tilde{\mathcal{N}}(\eta_1 + \eta_2, u^\dagger)] \tag{6.3.9}$$

and

$$g_\varepsilon(\mu, \lambda)\hat{\eta}_2 = (1 - \chi)\mathcal{F}[\tilde{\mathcal{N}}(\eta_1 + \eta_2, u^\dagger)], \qquad (6.3.10)$$

where $\hat{\eta}_1 = \chi\hat{\eta}$, $\hat{\eta}_2 = (1 - \chi)\hat{\eta}$ and χ is defined by equation (6.2.26) (see the discussion above that equation).

Lemma 6.3.4.

(i) *Suppose that $\lambda > 0$. For each sufficiently small value of ε and each $\eta_1 \in \chi(D)L_e^2(\mathbb{R})$, $u^\dagger \in X_r^0$, equation (6.3.10) has a unique solution $\eta_2 = \check{\eta}_2(\eta_1, u^\dagger) \in H_e^2(\mathbb{R})$, which satisfies the estimate*

$$\|\check{\eta}_2\|_2 \leq c\varepsilon^r\|\eta_1\|_0 + c_\lambda\|u^\dagger\|_{X^0}.$$

(ii) *Suppose that $\lambda = 0$. For each sufficiently small value of ε and each $\eta_1 \in \chi(D)L_e^2(\mathbb{R})$, $\omega^\dagger \in L_e^2(\mathbb{R})$, $\zeta_1^\dagger \in H_e^1(\Sigma)$, $\zeta_3^\dagger \in \mathring{H}_o^1(\Sigma)$, equation (6.3.10) has a unique solution $\eta_2 = \check{\eta}_2(\eta_1, u^\dagger) \in H_e^2(\mathbb{R})$, which satisfies the estimate*

$$\|\check{\eta}_2\|_2 \leq c(\varepsilon^r\|\eta_1\|_0 + \|\omega^\dagger\|_0 + \|\zeta_1^\dagger\|_1 + \|\zeta_3^\dagger\|_1).$$

Proof. Write (6.3.10) as

$$g_\varepsilon(D, \lambda)(\eta_2) - (1 - \chi(D))\tilde{\mathcal{N}}(\eta_2, 0) = (1 - \chi(D))\big(\tilde{\mathcal{N}}(\eta_1, 0) + \tilde{\mathcal{N}}(0, u^\dagger)\big)$$

and use the fact that the linear operator

$$g_\varepsilon(D, \lambda) - (1 - \chi(D))\tilde{\mathcal{N}}(\cdot, 0) : (1 - \chi(D))H_e^2(\mathbb{R}) \to (1 - \chi(D))L_e^2(\mathbb{R})$$

is invertible (because

$$\|\tilde{\mathcal{N}}(\eta, 0)\|_0 \leq c\varepsilon^r\|\eta\|_2$$

and $g_\varepsilon(\mu, \lambda) \geq c(1 + q^2 + \varepsilon^2)$ for $\mu \in \mathrm{supp}(1 - \chi)$) and the estimates

$$\|\tilde{\mathcal{N}}(\eta_1, 0)\|_0 \leq c\varepsilon^r\|\eta_1\|_0,$$

$$\|\tilde{\mathcal{N}}(0, u^\dagger)\|_0 \leq \begin{cases} c_\lambda\|u^\dagger\|_{X^0}, & \lambda > 0, \\ c(\|\omega^\dagger\|_0 + \|\zeta_1^\dagger\|_1 + \|\zeta_3^\dagger\|_1), & \lambda = 0. \end{cases}$$

$\square$

Substituting $\check{\eta}_2 = \check{\eta}_2(\eta_1, u^\dagger)$ into (6.3.9), we obtain a single equation for η_1, namely

$$g_\varepsilon(\mu, \lambda)\hat{\eta}_1 = \chi\mathcal{F}[\check{\mathcal{N}}(\eta_1, u^\dagger)], \qquad (6.3.11)$$

where

$$\check{\mathcal{N}}(\eta_1, u^\dagger) = \tilde{\mathcal{N}}(\eta_1 + \check{\eta}_2(\eta_1, u^\dagger), u^\dagger)$$

$$= \mathcal{N}(\eta_1 + \check{\eta}_2(\eta_1, u^\dagger), \check{\Gamma}(\eta_1, u^\dagger), u^\dagger)$$

and

$$\check{\Gamma}(\eta_1, u^\dagger) = \tilde{\Gamma}(\eta_1 + \check{\eta}_2(\eta_1, u^\dagger), u^\dagger).$$

We henceforth write

$$\check{\mathcal{N}}(\eta_1, u^\dagger) = \check{\mathcal{N}}(\eta_1, 0) + \underbrace{\check{\mathcal{N}}^\dagger(u^\dagger)}_{:= \check{\mathcal{N}}(0, u^\dagger)}$$

and use the corresponding decomposition for the functions $\check{\eta}_2(\eta_1, u^\dagger)$, $\check{\Gamma}(\eta_1, u^\dagger)$ and $\tilde{\Gamma}(\eta, u^\dagger)$, $\tilde{\mathcal{N}}(\eta, u^\dagger)$, which depend upon $u^\dagger$; note the estimates

$$\|\tilde{\mathcal{N}}^\dagger(u^\dagger)\|_0, \ \|\check{\mathcal{N}}^\dagger(u^\dagger)\|_0, \ \|\check{\eta}_2^\dagger(u^\dagger)\|_2$$

$$\leq \begin{cases} c_\lambda \|u^\dagger\|_{X^0}, & \lambda > 0, \\ c(\|\omega^\dagger\|_0 + \|\xi_1^\dagger\|_1 + \|\xi_3^\dagger\|_1), & \lambda = 0, \end{cases}$$

$$\|\nabla\tilde{\Gamma}^\dagger(u^\dagger)\|_1, \ \lambda\|\tilde{\Gamma}^\dagger(u^\dagger)\|_1, \ \|\nabla\check{\Gamma}^\dagger(u^\dagger)\|_1, \ \lambda\|\check{\Gamma}^\dagger(u^\dagger)\|_0$$

$$\leq \begin{cases} c_\lambda \|u^\dagger\|_{X^0}, & \lambda > 0, \\ c(\|\omega^\dagger\|_0 + \|\xi_1^\dagger\|_1 + \|\xi_3^\dagger\|_1), & \lambda = 0. \end{cases}$$

The following lemma shows that equation (6.3.11) is solvable for 'intermediate' values of λ.

Lemma 6.3.5. *There exists $k_{\max} > 0$ with the property that for each $\lambda \in [\varepsilon^r k_{\max}, \lambda^\star]$ and each $u \in X^0$, equation (6.3.11) has a unique solution $\eta_1 \in \chi(D)L_e^2(\mathbb{R})$, which satisfies the estimate*

$$\|\eta_1\|_0 \leq c_\lambda \|u^\dagger\|_{X^0}.$$

Proof. (i) Suppose that $(\beta_0, \alpha_0) \in C_s$. The estimate

$$\|\partial_x^m \check{\mathcal{N}}(\eta_1, 0)\|_0 \leq c\varepsilon^2 \sum_{k=0}^m \varepsilon^k \|\partial_x^{m-k}\eta_1\|_0, \quad m = 0, 1, 2, \ldots,$$

yields

$$\|(\mu^2 + \varepsilon^4 k^2)\mathcal{F}[\check{\mathcal{N}}(\eta_1, 0)]\|_0 \leq c(\varepsilon^6 k^2 \|\eta_1\|_0 + \Delta^{-1}\varepsilon^4 \|\eta_1\|_0 + \Delta\|\eta_{xxx}\|_0), \tag{6.3.12}$$

where Δ is an arbitrary positive number, and because

$$(\mu^2 + \varepsilon^4 k^2)g_\varepsilon(\mu, \varepsilon^2 k) \geq c(\varepsilon^2\mu^2 + \varepsilon^6 k_{\max}^2 + \mu^4 + \varepsilon^4 k_{\max}^2\mu^2 + \varepsilon^4 k_{\max}^2)$$

for $\mu \in \mathrm{supp}\,\chi$, one finds upon choosing Δ sufficiently small and $k_{\max}$ sufficiently large that the linear operator

$$(D^2 + \varepsilon^2 k^2)\big(g_\varepsilon(D, \varepsilon^2 k) - \chi(D)\mathcal{N}(\cdot, 0)\big) : \chi(D)L_e^2(\mathbb{R}) \to \chi(D)L_e^2(\mathbb{R})$$

is invertible.

(ii) Suppose that $(\beta_0, \alpha_0) \in C_w$. Using the estimates

$$\|\check{\mathcal{N}}(\eta_1, 0)\|_0 \le c\varepsilon^2 \|\eta_1\|_0 \qquad (6.3.13)$$

and

$$g_\varepsilon(\mu, \varepsilon k) \ge c\varepsilon^2 (1 + k_{\max}^2)$$

for $\mu \in \text{supp}\,\chi$, one finds upon choosing $k_{\max}$ sufficiently large that the linear operator

$$g_\varepsilon(D, \varepsilon k) - \chi(D)\check{\mathcal{N}}(\cdot, 0) : \chi(D)L_e^2(\mathbb{R}) \to \chi(D)L_e^2(\mathbb{R})$$

is invertible. $\qquad\square$

The next step is to derive an estimate for $\check{\mathcal{N}}(\eta_1, 0)$, which is more precise than (6.3.12) or (6.3.13); the results are recorded in Theorem 6.3.6. Full details of the necessary calculations for weak surface tension are given by Groves, Sun & Wahlén [15]; the corresponding estimate for strong surface tension is obtained in a similar fashion. Note however the different nature of the estimates in the two cases: in part (i) differentiation with respect to x introduces higher powers of ε, while in part (ii) the estimates are conveniently given in terms of the approximation $\zeta_\delta^\star = \chi_0(\varepsilon D)\zeta_{\text{NLS}}^\star$ to $\zeta_{\text{NLS}}^\star$.

Theorem 6.3.6.

(i) *Suppose that $(\beta_0, \alpha_0) \in C_s$ and $\lambda = \varepsilon^2 k$ for $k \le k_{\max}$ and write*

$$\eta_1(x) = \varepsilon^2 \zeta(\varepsilon x). \qquad (6.3.14)$$

The function $\check{\mathcal{N}}(\eta_1, 0)$ is given by the formula

$$\check{\mathcal{N}}(\eta_1, 0) = -3\varepsilon^4 \zeta_{\text{KdV}}^\star(\varepsilon x)\zeta(\varepsilon x) + \varepsilon^{3/2} T_{\text{KP}},$$

where the symbol T_{KP} denotes a quantity that satisfies

$$\|\partial_x^m \partial_y^n \chi(D) T_{\text{KP}}\|_0 \le c\varepsilon^{5/2+m} \|\zeta\|_{m+1}, \qquad m = 0, 1, 2, \ldots, n = 0, 1,$$

and $\partial_x T_{\text{KP}} = T_{\text{KP}}$.

(ii) *Suppose that $(\beta_0, \alpha_0) \in C_w$ and $\lambda = \varepsilon k$ for $k \le k_{\max}$ and write*

$$\eta_1(x) = \frac{1}{2}\varepsilon\zeta(\varepsilon x)e^{i\mu_0 x} + \frac{1}{2}\varepsilon\overline{\zeta(\varepsilon x)}e^{-i\mu_0 x}. \qquad (6.3.15)$$

The function $\check{\mathcal{N}}(\eta_1, 0)$ is given by the formula

$$\check{\mathcal{N}}(\eta_1, 0) = \varepsilon^3 (\zeta_\delta^\star(\varepsilon x))^2 \left((A_3 + A_5)\zeta(\varepsilon x) + A_3\overline{\zeta(\varepsilon x)}\right)\frac{e^{i\mu_0 x}}{2}$$

$$+ \varepsilon^3 (\zeta_\delta^\star(\varepsilon x))^2 \left(A_3\zeta(\varepsilon x) + (A_3 + A_5)\overline{\zeta(\varepsilon x)}\right)\frac{e^{-i\mu_0 x}}{2}$$

$$- 4\varepsilon^3 A_4 \zeta_\delta^\star(\varepsilon x)\vartheta_{\delta x}(\varepsilon x)\left(\frac{1}{2}e^{i\mu_0 x} + \frac{1}{2}e^{-i\mu_0 x}\right) + \varepsilon^2 T_{\text{DS}},$$

where

$$\zeta_\delta^\star = \chi_0(\varepsilon D)\zeta_{\mathrm{NLS}}^\star, \qquad \vartheta_\delta = A_4 \mathcal{F}^{-1}\left[\frac{i\mu}{(1-\alpha_0^{-1})\mu^2 + k^2}\mathcal{F}[2e(\zeta_\delta^\star \zeta)]\right]$$

and the symbol T_{DS} denotes a quantity that satisfies the estimate

$$\|\chi(D)T_{\mathrm{DS}}\|_0 \le c\varepsilon^{3/2}\|\zeta\|_1.$$

It remains to derive the reduced equation in its final form, thus completing the proof of Theorems 6.2.3(i) and 6.2.4(i). Write $\lambda = \varepsilon^r k$ for $k \le k_{\max}$. For $k > 0$ we define $\check{u}_{\varepsilon,k}^{\mathrm{KP}} : \chi_0(\varepsilon D)L_e^2(\mathbb{R}) \times X^0 \to Y^1$, $\check{u}_{\varepsilon,k}^{\mathrm{DS}} : \chi_0(\varepsilon D)L_c^2(\mathbb{R}) \times X^0 \to Y^1$ by

$$\check{u}_{\varepsilon,k}(\zeta, u^\dagger) = (\eta_1 + \check{\eta}_2(\eta_1, u^\dagger), \check{\omega}(\eta_1, u^\dagger), \check{\Gamma}(\eta_1, u^\dagger), \check{\xi}(\eta_1, u^\dagger)),$$

and $\check{u}_{\varepsilon,0}^{\mathrm{KP}} : \chi_0(\varepsilon D)L_e^2(\mathbb{R}) \times Z_r^0 \to Y_\star^1$, $\check{u}_{\varepsilon,0}^{\mathrm{DS}} : \chi_0(\varepsilon D)L_c^2(\mathbb{R}) \times Z_r^0 \to Y_\star^1$ by

$$\check{u}_{\varepsilon,0}(\zeta, \eta^\dagger, \omega^\dagger, \Gamma^\dagger, \xi_1^\dagger, \xi_3^\dagger) = (\eta_1 + \check{\eta}_2(\eta_1, u^\dagger), \check{\omega}(\eta_1, u^\dagger), \check{\Gamma}(\eta_1, u^\dagger), \check{\xi}(\eta_1, u^\dagger))$$

with $\xi^\dagger = -(\xi_1^\dagger)_x - (\xi_3^\dagger)_y$; here η_1 is given by (6.3.14) or (6.3.15), while $\check{\omega}, \check{\xi}$ are given by equations (6.3.1), (6.3.2) with $\Gamma = \check{\Gamma}(\eta_1, u^\dagger)$, $\eta = \eta_1 + \check{\eta}_2(\eta_1, u^\dagger)$. The above theory shows that $\eta_1 \in \chi(D)L_e^2(\mathbb{R})$ solves the reduced equation

$$g_\varepsilon(\mu, \varepsilon^r k)\hat{\eta}_1 = \chi\mathcal{F}[\check{N}(\eta_1, u^\dagger)]$$

if and only if $u = \check{u}_{\varepsilon,k}$ solves the resolvent equations (6.2.20)–(6.2.25); it follows that $\eta_1 \in L_e^2(\mathbb{R})$ solves the equation

$$g_\varepsilon(\mu, \varepsilon^r k)\hat{\eta}_1 = \chi\mathcal{F}[\check{N}(\chi(D)\eta_1, u^\dagger)] \qquad (6.3.16)$$

if and only if $\eta \in \chi(D)L_e^2(\mathbb{R})$ and $u = \check{u}_{\varepsilon,k}$ solves (6.2.20)–(6.2.25). In Sections 6.3.2 and 6.3.3 we compute equation (6.3.16) in terms of the variable ζ, dealing separately with strong and weak surface tension.

6.3.2 Derivation of the Reduced Equation – Strong Surface Tension

We begin with the case $k > 0$. Since all solutions of (6.3.16) have support in the interval $[-\delta, \delta]$, its solution set in $L_e^2(\mathbb{R})$ coincides with its solution set in $H_e^s(\mathbb{R})$ for any $s \ge 0$, and we henceforth work in $H_e^4(\mathbb{R})$ by formulating it as the fourth-order pseudodifferential equation

$$\left(\varepsilon^2\mu^2 + (\beta_0 - \tfrac{1}{3})\mu^4 + \lambda^2\right)\hat{\eta}_1 = q^2\frac{\tilde{g}_\varepsilon(\mu, \varepsilon k)}{g_\varepsilon(\mu, \varepsilon k)}\chi_0\mathcal{F}[\check{N}(\chi(D)\eta_1, u^\dagger)]$$
$$- \left(\varepsilon^2\lambda^2 + (\beta_0 - \tfrac{1}{3})\mu^2\lambda^2\right)\hat{\eta}_1,$$

where

$$\tilde{g}_\varepsilon(\mu, \lambda) = \varepsilon^2 + (\beta_0 - \tfrac{1}{3})\mu^2 + \frac{\lambda^2}{q^2}$$

(so that $g_\varepsilon(\mu,\lambda) = \tilde{g}_\varepsilon(\mu,\lambda) = O(q^4 + q^2\lambda^2)$ as $(\mu,\lambda) \to (0,0)$). In terms of the variable

$$\zeta(x) = \varepsilon^2 \eta(\varepsilon x)$$

this equation reads

$$\left(\tilde{\mu}^2 + (\beta_0 - \tfrac{1}{3})\tilde{\mu}^4 + k^2\right)\hat{\zeta}(\tilde{\mu})$$

$$= (\tilde{\mu}^2 + \varepsilon^2 k^2)\frac{\tilde{g}_\varepsilon(\varepsilon\mu,\varepsilon^2 k)}{g_\varepsilon(\varepsilon\mu,\varepsilon^2 k)}\chi_0(\varepsilon\tilde{\mu})\varepsilon^{-3}\mathcal{F}[\check{\mathcal{N}}(\chi(D)\eta_1,0)](\varepsilon\tilde{\mu})$$

$$- \varepsilon^2\left(k^2 + (\beta_0 - \tfrac{1}{3})\tilde{\mu}^2 k^2\right)\chi_0(\varepsilon\tilde{\mu})\hat{\zeta}(\tilde{\mu}) + \hat{\zeta}_{\varepsilon,k}(u^\dagger)(\tilde{\mu}),$$

where

$$\hat{\zeta}_{\varepsilon,k}(u^\dagger)(\tilde{\mu}) = (\tilde{\mu}^2 + \varepsilon^2 k^2)\frac{\tilde{g}_\varepsilon(\varepsilon\mu,\varepsilon^2 k)}{g_\varepsilon(\mu,\varepsilon^2 k)}\chi_0(\varepsilon\tilde{\mu})\varepsilon^{-3}\mathcal{F}[\check{\mathcal{N}}^\dagger(u^\dagger)](\tilde{\varepsilon}\tilde{\mu}).$$

Using the estimate

$$\|(\chi_0(\varepsilon\tilde{\mu}) - 1)\hat{f}(\tilde{\mu})\|_0^2 = \int_{|\tilde{\mu}| \geq \delta/\varepsilon} |\hat{f}(\tilde{\mu})|^2\,d\tilde{\mu}$$

$$\leq \delta^{-2}\varepsilon^2 \int_{|\tilde{\mu}| \geq \delta/\varepsilon} |\tilde{\mu}|^2 |\hat{f}(\tilde{\mu})|^2\,d\tilde{\mu}$$

$$\leq \delta^{-2}\varepsilon^2 \|f_x\|_0^2, \qquad\qquad (6.3.17)$$

we find from Theorem 6.3.6 (i) that

$$\check{\mathcal{N}}(\chi(D)\eta_1,0) = \varepsilon^4 E(\varepsilon x) + \varepsilon^{3/2} T_{KP},$$

where $E = -3\zeta^\star_{KdV}\zeta$. Furthermore

$$\left\|\mathcal{F}^{-1}\left[(\tilde{\mu}^2 + \varepsilon^2 k^2)\frac{\tilde{g}_\varepsilon(\varepsilon\mu,\varepsilon^2 k)}{g_\varepsilon(\varepsilon\mu,\varepsilon^2 k)}\chi_0(\varepsilon\tilde{\mu})\varepsilon^{-3/2}\mathcal{F}[T_{KP}](\varepsilon\tilde{\mu})\right]\right\|_0$$

$$\leq c(\|\chi_0(D)T_{KP}\|_0 + \varepsilon^{-4}\|(\chi_0(D)T_{KP})_{xx}\|_0)$$

$$\leq c\varepsilon^{1/2}\|\zeta\|_3,$$

where we have used the estimate

$$\left|\frac{\tilde{g}_\varepsilon(\mu,\lambda)}{g_\varepsilon(\mu,\lambda)} - 1\right| \leq cq^2, \quad |\mu| \leq \delta,\ \lambda \leq \lambda^\star, \qquad\qquad (6.3.18)$$

which shows in particular that

$$\left|\frac{\tilde{g}_\varepsilon(\varepsilon\tilde{\mu},\varepsilon^2 k)}{g_\varepsilon(\varepsilon\tilde{\mu},\varepsilon^2 k)}\chi_0(\varepsilon\tilde{\mu})\right| \leq c.$$

Next we write

$$\mathcal{F}^{-1}\left[\frac{\tilde{g}_\varepsilon(\varepsilon\tilde{\mu},\varepsilon^2k^2)}{g_\varepsilon(\varepsilon\tilde{\mu},\varepsilon^2k)}(\tilde{\mu}^2+\varepsilon^2k^4)\chi_0(\varepsilon\tilde{\mu})\hat{E}(\tilde{\mu})\right]$$

$$=\mathcal{F}^{-1}\left[\left(\frac{\tilde{g}_\varepsilon(\varepsilon\tilde{\mu},\varepsilon^2k)}{g_\varepsilon(\varepsilon\tilde{\mu},\varepsilon^2k)}-1\right)(\tilde{\mu}^2+\varepsilon^2k^4)\chi_0(\varepsilon\tilde{\mu})\hat{E}(\tilde{\mu})\right]$$

$$+\mathcal{F}^{-1}[(\tilde{\mu}^2+\varepsilon^2k^4)(\chi_0(\varepsilon\tilde{\mu})-1)\hat{E}(\tilde{\mu})]+\mathcal{F}^{-1}[(\tilde{\mu}^2+\varepsilon^2k^4)\hat{E}(\tilde{\mu})]$$

and note that

$$\left\|\mathcal{F}^{-1}\left[\underbrace{\left(\frac{\tilde{g}_\varepsilon(\varepsilon\tilde{\mu},\varepsilon^2k)}{g_\varepsilon(\varepsilon\tilde{\mu},\varepsilon^2k)}-1\right)}_{\leq c(\varepsilon^2\tilde{\mu}^2+\varepsilon^4k^2)}(\tilde{\mu}^2+\varepsilon^2k^4)\chi_0(\varepsilon\tilde{\mu})\hat{E}(\tilde{\mu})\right]\right\|_0$$

$$\leq c(\varepsilon\|E_{xxx}\|_0+\varepsilon^2\|E\|_0)$$

$$\leq c\varepsilon\|\zeta\|_3,$$

(using (6.3.18)),

$$\|\mathcal{F}^{-1}[(\chi_0(\varepsilon\tilde{\mu})-1)\tilde{\mu}^2\hat{E}(\tilde{\mu})]\|_0\leq c\varepsilon\|E'''\|_0\leq c\varepsilon\|\zeta\|_3,$$

$$\|\mathcal{F}^{-1}[(\chi_0(\varepsilon\tilde{\mu})-1)\hat{E}(\tilde{\mu})]\|_0\leq c\varepsilon\|E'\|_0\leq c\varepsilon\|\zeta\|_0$$

(using (6.3.17)), and

$$\mathcal{F}^{-1}[(\tilde{\mu}^2+\varepsilon^2k^4)\hat{E}(\tilde{\mu})]=E_{xx}+O(\varepsilon^2\|\zeta\|_0)=(3\zeta_{\mathrm{KdV}}^\star\zeta)_{xx}+O(\varepsilon^2\|\zeta\|_0).$$

Finally

$$\|\mathcal{F}^{-1}[(\varepsilon^2\lambda^2+(\beta_0-\tfrac{1}{3})\mu^2\lambda^2)\hat{\eta}_1]\|_0\leq c\varepsilon^2\|\zeta\|_2.$$

Altogether the above calculations show that equation (6.3.21) can be written as

$$\underbrace{-\zeta_{xx}+(\beta_0-\tfrac{1}{3})\zeta_{xxxx}+k^2\zeta-3(\zeta_{\mathrm{KdV}}^\star\zeta)_{xx}-\mathcal{R}_{\varepsilon,k}^{\mathrm{KP}}(\zeta)}_{:=\mathcal{B}_{\varepsilon,k}^{\mathrm{KP}}(\zeta)}=\zeta_{\varepsilon,k}^\dagger(u^\dagger),\qquad(6.3.19)$$

where $\|\mathcal{R}_{\varepsilon,k}^{\mathrm{KP}}(\zeta)\|_0\leq c\varepsilon^{1/2}\|\zeta\|_3$.

In the case $k=0$ we work in $H_{\mathrm{e}}^2(\mathbb{R})$ by formulating (6.3.16) as the second-order pseudodifferential equation

$$\tilde{g}_\varepsilon(\mu,0)\hat{\eta}_1=\frac{\tilde{g}_\varepsilon(\mu,0)}{g_\varepsilon(\mu,0)}\chi\mathcal{F}[\check{\mathcal{N}}^\dagger(\chi(D)\eta_1,u^\dagger)],$$

where

$$\tilde{g}_\varepsilon(\mu,0)=\varepsilon^2+(\beta_0-\tfrac{1}{3})\mu^2$$

is the second-order Taylor polynomial of $g_\varepsilon(\mu,0)$ at $\mu = 0$. Repeating the above calculations, we find that this equation reads

$$\underbrace{-\zeta + (\beta_0 - \tfrac{1}{3})\zeta_{xx} - 3\zeta_{\mathrm{KdV}}\zeta - \mathcal{R}^{\mathrm{KP}}_{\varepsilon,0}(\zeta)}_{:= \mathcal{C}^{\mathrm{KP}}_\varepsilon(\zeta)} = \zeta^\dagger_\varepsilon(\zeta), \qquad (6.3.20)$$

where

$$\mathcal{R}^{\mathrm{KP}}_{\varepsilon,0}(\zeta) = \mathcal{F}^{-1}\left[\frac{\tilde{g}_\varepsilon(\varepsilon\mu,0)}{g_\varepsilon(\varepsilon\mu,0)}\chi_0(\varepsilon\tilde{\mu})\varepsilon^{-3}\mathcal{F}[\check{\mathcal{N}}(\chi(D)\eta_1,0)](\varepsilon\tilde{\mu})\right] - E,$$

$$\mathcal{F}[\zeta^\dagger_{\varepsilon,k}(u^\dagger)](\tilde{\mu}) = \frac{\tilde{g}_\varepsilon(\varepsilon\mu,0)}{g_\varepsilon(\mu,0)}\chi_0(\varepsilon\tilde{\mu})\varepsilon^{-3}\mathcal{F}[\check{\mathcal{N}}^\dagger(u^\dagger)](\tilde{\varepsilon}\tilde{\mu})$$

with $\|\mathcal{R}^{\mathrm{KP}}_{\varepsilon,0}(\zeta)\|_0 \le c\varepsilon^{1/2}\|\zeta\|_1$.

6.3.3 Derivation of the Reduced Equation – Weak Surface Tension

Since all solutions of (6.3.16) have support in the interval $[-\mu_0 - \delta, -\mu_0 + \delta]$ $\cup[\mu_0 - \delta, \mu_0 + \delta]$, its solution set in $L^2_\mathrm{e}(\mathbb{R})$ coincides with its solution set in $H^s_\mathrm{e}(\mathbb{R})$ for any $s \ge 0$, and we henceforth work in $H^2_\mathrm{e}(\mathbb{R})$ by formulating it as the second-order pseudodifferential equation

$$\tilde{g}_\varepsilon(\mu,\varepsilon k)\hat{\eta}^+_1 = \frac{\tilde{g}_\varepsilon(\mu,\varepsilon k)}{g_\varepsilon(\mu,\varepsilon k)}\chi_+\mathcal{F}[\check{\mathcal{N}}^\dagger(\chi(D)\eta_1,u^\dagger)],$$

where $\chi_+ = \chi_0(\mu - \mu_0)$, $\eta^+_1 = \chi_+\eta_1$ (so that $\eta_1 = \eta^+_1 + \overline{\eta^+_1}$) and

$$\tilde{g}_\varepsilon(\mu,\lambda) = \varepsilon^2 + A_1(\mu - \mu_0)^2 + A_2\lambda^2$$

is the second-order Taylor polynomial of g_ε at the point $(\mu_0,0)$. Writing

$$\eta_1(x) = \frac{1}{2}\varepsilon\zeta(\varepsilon x)e^{i\mu_0 x} + \frac{1}{2}\varepsilon\overline{\zeta(\varepsilon x)}e^{-i\mu_0 x}$$

and $\mu = \mu_0 + \varepsilon\tilde{\mu}$ and taking the inverse Fourier transform with respect to $\tilde{\mu}$, we find that

$$A^{-1}_2\zeta - A^{-1}_2 A_1\zeta_{xx} + k^2\zeta$$
$$= \mathcal{F}^{-1}\left[2\frac{\tilde{g}_\varepsilon(\mu_0 + \varepsilon\tilde{\mu},\varepsilon k)}{g_\varepsilon(\mu_0 + \varepsilon\tilde{\mu},\varepsilon k)}\chi_0(\varepsilon\tilde{\mu})\varepsilon^{-2}A^{-1}_2\mathcal{F}[\check{\mathcal{N}}(\chi(D)\eta_1,0)](\mu_0 + \varepsilon\tilde{\mu})\right.$$
$$\left. + \hat{\zeta}^\dagger_{\varepsilon,k}(u^\dagger)(\tilde{\mu})\right], \qquad (6.3.21)$$

where

$$\mathcal{F}[\zeta^\dagger_{\varepsilon,k}(u^\dagger)](\tilde{\mu}) = 2A^{-1}_2\frac{\tilde{g}_\varepsilon(\mu_0 + \varepsilon\tilde{\mu},\varepsilon k)}{g_\varepsilon(\mu_0 + \varepsilon\tilde{\mu},\varepsilon k)}\chi_0(\varepsilon\tilde{\mu})\varepsilon^{-2}\mathcal{F}[\check{\mathcal{N}}^\dagger(u^\dagger)](\mu_0 + \varepsilon\tilde{\mu}).$$

Observe that

$$\check{\mathcal{N}}(\chi(D)\eta_1,0)$$

$$= \varepsilon^3(\zeta_\delta^\star(\varepsilon x))^2 \left((A_3 + A_5)(\chi_0(\varepsilon D)\zeta)(\varepsilon x) + A_3 \overline{(\chi_0(\varepsilon D)\zeta)(\varepsilon x)} \right) \frac{e^{i\mu_0 x}}{2}$$

$$+ \varepsilon^3(\zeta_\delta^\star(\varepsilon x))^2 \left(A_3(\chi_0(\varepsilon D)\zeta)(\varepsilon x) + (A_3 + A_5)\overline{(\chi_0(\varepsilon D)\zeta)(\varepsilon x)} \right) \frac{e^{-i\mu_0 x}}{2}$$

$$+ \varepsilon^3 A_4 \zeta_\delta^\star(\varepsilon x)(\chi_0(\varepsilon D)\vartheta_\delta)_x(\varepsilon x) \left(\frac{1}{2} e^{i\mu_0 x} + \frac{1}{2} e^{-i\mu_0 x} \right) + \varepsilon^2 T_{DS}$$

(see Theorem 6.3.6(ii)). Using the estimate (6.3.17), we find that

$$\| ((\chi_0(\varepsilon D) - 1)\zeta)(\varepsilon x)e^{i\mu_0 x} \|_0^2 = \varepsilon^{-1} \| (\chi_0(\varepsilon \tilde{\mu}) - 1)\hat{\zeta}(\tilde{\mu}) \|_0^2 \le \delta^{-2}\varepsilon \| \zeta \|_1$$

and

$$\| ((\chi_0(\varepsilon D) - 1)\vartheta_\delta)_x(\varepsilon x)e^{i\mu_0 x} \|_0^2 = \varepsilon^{-1} A_4 \left\| \frac{(\chi_0(\varepsilon \tilde{\mu}) - 1)\tilde{\mu}^2}{(1 - \alpha_0^{-1})\tilde{\mu}^2 + k^2} \mathcal{F}[2\mathrm{e}(\zeta_\delta^\star \zeta)](\tilde{\mu}) \right\|_0$$

$$\le c\varepsilon^{-1} \| (\chi_0(\varepsilon \tilde{\mu}) - 1)\mathcal{F}[2\mathrm{e}(\zeta_\delta^\star \zeta)](\tilde{\mu}) \|_0$$

$$\le c\varepsilon \| \zeta \|_1,$$

so that

$$\check{\mathcal{N}}(\chi(D)\eta_1,0) = \frac{\varepsilon^3}{2} E(\varepsilon x)e^{i\mu_0 x} + \frac{\varepsilon^3}{2} F(\varepsilon x)e^{-i\mu_0 x} + \varepsilon^2 T_{DS},$$

where

$$E = (A_3 + A_5)(\zeta_\delta^\star)^2 \zeta + A_3(\zeta_\delta^\star)^2 \overline{\zeta} - 4A_4 \zeta_\delta^\star \vartheta_{\delta x},$$

$$F = A_3(\zeta_\delta^\star)^2 \chi_0(\varepsilon D)\zeta + (A_3 + A_5)(\zeta_\delta^\star)^2 \chi_0(\varepsilon D)\overline{\zeta} - 4A_4 \zeta_\delta^\star (\chi_0(\varepsilon D)\vartheta_\delta)_x.$$

Furthermore

$$\chi_0(\varepsilon \tilde{\mu})\mathcal{F}[\varepsilon F(\varepsilon x)e^{-i\mu_0 x}](\mu_0 + \varepsilon \tilde{\mu}) = \chi_0(\varepsilon \tilde{\mu})\hat{F}(\tilde{\mu} + 2\varepsilon^{-1}\mu_0) = 0$$

because $\mathrm{supp}\,\hat{F} \subseteq [-3\varepsilon^{-1}\delta, 3\varepsilon^{-1}\delta]$.

Using the estimate

$$\left| \frac{\tilde{g}_\varepsilon(\mu,\lambda)}{g_\varepsilon(\mu,\lambda)} - 1 \right| \le c|(\mu - \mu_0, \lambda)|^2, \qquad |\mu - \mu_0| \le \delta, \ \lambda \le \lambda^\star, \qquad (6.3.22)$$

we find in particular that

$$\left| \frac{\tilde{g}_\varepsilon(\mu_0 + \varepsilon \tilde{\mu}, \varepsilon k)}{g_\varepsilon(\mu_0 + \varepsilon \tilde{\mu}, \varepsilon k)} \chi_0(\varepsilon \tilde{\mu}) \right| \le c.$$

It follows that

$$\left\| \mathcal{F}^{-1} \left[\frac{\tilde{g}_\varepsilon(\mu_0 + \varepsilon\tilde{\mu}, \varepsilon k)}{g_\varepsilon(\mu_0 + \varepsilon\tilde{\mu}, \varepsilon k)} \chi_0(\varepsilon\tilde{\mu}) \varepsilon^{-2} \mathcal{F}[T_{\mathrm{DS}}](\mu_0 + \varepsilon\tilde{\mu}) \right] \right\|_0$$

$$\leq c\varepsilon^{-1/2} \|\chi_+(D)T_{\mathrm{DS}}\|_0 \leq c\varepsilon\|\zeta\|_1.$$

Finally, we write

$$\mathcal{F}^{-1}\left[2\frac{\tilde{g}_\varepsilon(\mu_0 + \varepsilon\tilde{\mu}, \varepsilon k)}{g_\varepsilon(\mu_0 + \varepsilon\tilde{\mu}, \varepsilon k)} \chi_0(\varepsilon\tilde{\mu}) \varepsilon^{-2} \mathcal{F}\left[\frac{\varepsilon^3}{2} E(\varepsilon x) \mathrm{e}^{\mathrm{i}\mu_0 x}\right](\mu_0 + \varepsilon\tilde{\mu})\right]$$

$$= \mathcal{F}^{-1}\left[\left(\frac{\tilde{g}_\varepsilon(\mu_0 + \varepsilon\tilde{\mu}, \varepsilon k)}{g_\varepsilon(\mu_0 + \varepsilon\tilde{\mu}, \varepsilon k)} - 1\right)\chi_0(\varepsilon\tilde{\mu})\hat{E}(\tilde{\mu})\right] + \mathcal{F}^{-1}[(\chi_0(\varepsilon\tilde{\mu}) - 1)\hat{E}(\tilde{\mu})] + E(x)$$

and note that

$$\left\| \mathcal{F}^{-1}\left[\left(\frac{\tilde{g}_\varepsilon(\mu_0 + \varepsilon\tilde{\mu}, \varepsilon k)}{g_\varepsilon(\mu_0 + \varepsilon\tilde{\mu}, \varepsilon k)} - 1\right)\chi_0(\varepsilon\tilde{\mu})\hat{E}(\tilde{\mu})\right] \right\|_0 \leq c(\varepsilon\delta\|E'\|_0 + \varepsilon^2 k^2\|E\|_0) \leq c\varepsilon\|\zeta\|_1,$$

$$\|\mathcal{F}^{-1}[(\chi_0(\varepsilon\tilde{\mu}) - 1)\hat{E}(\tilde{\mu})]\|_0 \leq c\varepsilon\|E\|_1 \leq c\varepsilon\|\zeta\|_1$$

(using respectively (6.3.22)) and (6.3.17)).

According to the above calculations equation (6.3.21) can be written as

$$A_2^{-1}\zeta - A_1 A_2^{-1}\zeta_{xx} + k^2\zeta$$
$$= A_2^{-1}(A_3 + A_5)(\zeta_\delta^\star)^2\zeta + A_2^{-1}A_3(\zeta_\delta^\star)^2\bar{\zeta} - 4A_2^{-1}A_4\zeta_\delta^\star\vartheta_{\delta x} + \zeta_{\varepsilon,k}^\dagger(u^\dagger) + \tilde{\mathcal{R}}_{\varepsilon,k}^{\mathrm{DS}}(\zeta)$$
$$= A_2^{-1}(A_3 + A_5)(\zeta_{\mathrm{NLS}}^\star)^2\zeta + A_2^{-1}A_3(\zeta_{\mathrm{NLS}}^\star)^2\bar{\zeta} - 4A_2^{-1}A_4\zeta_{\mathrm{NLS}}^\star\vartheta_x + \zeta_{\varepsilon,k}^\dagger(u^\dagger) + \mathcal{R}_{\varepsilon,k}^{\mathrm{DS}}(\zeta),$$

where

$$\vartheta = A_4\mathcal{F}^{-1}\left[\frac{\mathrm{i}\mu}{(1 - \alpha_0^{-1})\mu^2 + k^2} \mathcal{F}[2\mathrm{e}(\zeta^\star\zeta)]\right]$$

and $\|\tilde{\mathcal{R}}_{\varepsilon,k}^{\mathrm{DS}}(\zeta)\|_0$, $\|\mathcal{R}_{\varepsilon,k}^{\mathrm{DS}}(\zeta)\|_0 \leq c\varepsilon\|\zeta\|_1$. Finally, we recast the equation in terms of real coordinates $\zeta_1 = \mathrm{e}\zeta$, $\zeta_2 = \mathrm{Im}\,\zeta$ in the form

$$\mathcal{B}_{\varepsilon,k}^{\mathrm{DS}}\begin{pmatrix}\zeta_1\\\zeta_2\\\psi\end{pmatrix} = \begin{pmatrix}\zeta_1^\dagger\\\zeta_2^\dagger\\0\end{pmatrix}, \qquad \mathcal{C}_\varepsilon^{\mathrm{DS}}\begin{pmatrix}\zeta_1\\\zeta_2\end{pmatrix} = \begin{pmatrix}\zeta_1^\dagger\\\zeta_2^\dagger\end{pmatrix}$$

for, respectively, $k > 0$ and $k = 0$, where

$$\mathcal{B}_{\varepsilon,k}^{\mathrm{DS}}(\zeta_1, \zeta_2, \psi) =$$

$$\begin{pmatrix} A_2^{-1}\zeta_1 - A_2^{-1}A_1\zeta_{1xx} + k^2\zeta_1 - A_2^{-1}(2A_3 + A_5)(\zeta_{\mathrm{NLS}}^\star)^2\zeta_1 + 4A_2^{-1}A_4\zeta_{\mathrm{NLS}}^\star\psi_x - \mathrm{Re}\,\mathcal{R}_{\varepsilon,k}^{\mathrm{DS}}(\zeta) \\ A_2^{-1}\zeta_2 - A_2^{-1}A_1\zeta_{2xx} + k^2\zeta_2 - A_2^{-1}A_5(\zeta_{\mathrm{NLS}}^\star)^2\zeta_2 - \mathrm{Im}\,\mathcal{R}_{\varepsilon,k}^{\mathrm{DS}}(\zeta) \\ -(1 - \alpha_0^{-1})\psi_{xx} + k^2\psi - 2A_4(\zeta_{\mathrm{NLS}}^\star\zeta_1)_x \end{pmatrix}$$

$$(6.3.23)$$

and

$$\mathcal{C}_\varepsilon^{\mathrm{DS}}(\zeta_1,\zeta_2) =$$

$$\begin{pmatrix} A_2^{-1}\zeta_1 - A_2^{-1}A_1\zeta_{1xx} + k^2\zeta_1 - A_2^{-1}(2A_3+A_5)(\zeta_{\mathrm{NLS}}^\star)^2\zeta_1 + 4A_2^{-1}A_4\zeta_{\mathrm{NLS}}^\star\vartheta_x - \mathrm{Re}\,\mathcal{R}_{\varepsilon,0}^{\mathrm{DS}}(\zeta) \\ A_2^{-1}\zeta_2 - A_2^{-1}A_1\zeta_{2xx} + k^2\zeta_2 - A_2^{-1}A_5(\zeta_{\mathrm{NLS}}^\star)^2\zeta_2 - \mathrm{Im}\,\mathcal{R}_{\varepsilon,0}^{\mathrm{DS}}(\zeta) \end{pmatrix}.$$

$$(6.3.24)$$

6.4 Spectral Theory for the Reduced Equation

In this section we examine the spectra of the operators $\mathcal{B}_{\varepsilon,k}^{\mathrm{KP}} : \mathcal{D}_\mathcal{B}^{\mathrm{KP}} \subseteq W^{\mathrm{KP}} \to W^{\mathrm{KP}}$ defined by (6.3.19) with

$$W^{\mathrm{KP}} = L_{\mathrm{e}}^2(\mathbb{R}), \qquad \mathcal{D}_\mathcal{B}^{\mathrm{KP}} = H_{\mathrm{e}}^4(\mathbb{R})$$

and $\mathcal{B}_{\varepsilon,k}^{\mathrm{DS}} : \mathcal{D}_\mathcal{B}^{\mathrm{DS}} \subseteq W^{\mathrm{DS}} \to W^{\mathrm{DS}}$ defined by (6.3.23) with

$$W^{\mathrm{DS}} = L_{\mathrm{e}}^2(\mathbb{R}) \times L_{\mathrm{o}}^2(\mathbb{R}) \times L_{\mathrm{o}}^2(\mathbb{R}), \qquad \mathcal{D}_\mathcal{B}^{\mathrm{DS}} = H_{\mathrm{e}}^2(\mathbb{R}) \times H_{\mathrm{o}}^2(\mathbb{R}) \times H_{\mathrm{o}}^2(\mathbb{R})$$

and complete the proofs of Theorem 6.2.3 and 6.2.4. We first present an auxiliary result that is needed in the spectral theory below.

Proposition 6.4.1.

(i) *The formula*

$$\mathcal{C}(\zeta) = \zeta - (\beta_0 - \tfrac{1}{3})\zeta_{xx} - 3\,\mathrm{sech}^2\left(\frac{x}{2(\beta_0 - \tfrac{1}{3})^{1/2}}\right)\zeta$$

defines a self-adjoint operator $\mathcal{C} : H_{\mathrm{e}}^2(\mathbb{R}) \subseteq L_{\mathrm{e}}^2(\mathbb{R}) \to L_{\mathrm{e}}^2(\mathbb{R})$ *whose spectrum consists of essential spectrum* $[1,\infty)$ *and simple eigenvalues at* $-\tfrac{5}{4}$ *(with corresponding eigenvector* $\mathrm{sech}^3\left(\tfrac{1}{2}(\beta_0 - \tfrac{1}{3})^{-1/2}x\right)$*) and* $\tfrac{3}{4}$*.*

(ii) *The formulae*

$$\mathcal{C}_1(\zeta_1) = A_2^{-1}\left(\zeta_1 - A_1\zeta_{1xx} - 6\,\mathrm{sech}^2(A_1^{-1/2}x)\zeta_1\right),$$

$$\mathcal{C}_2(\zeta_2) = A_2^{-1}\left(\zeta_2 - A_1\zeta_{2xx} - 2\,\mathrm{sech}^2(A_1^{-1/2}x)\zeta_2\right)$$

define self-adjoint operators $\mathcal{C}_1 : H_{\mathrm{e}}^2(\mathbb{R}) \subseteq L_{\mathrm{e}}^2(\mathbb{R}) \to L_{\mathrm{e}}^2(\mathbb{R})$*,* $\mathcal{C}_1 : H_{\mathrm{o}}^2(\mathbb{R}) \subseteq L_{\mathrm{o}}^2(\mathbb{R}) \to L_{\mathrm{o}}^2(\mathbb{R})$ *whose spectrum consists of essential spectrum* $[A_2^{-1},\infty)$ *and, in the case of* $\mathcal{C}_1$*, a simple eigenvalue at* $-3A_2^{-1}$ *(with corresponding eigenvector* $\mathrm{sech}(A_1^{-1/2}x)$*).*

Proof. The spectrum of the operator $1 - \partial_x^2 - 3\,\mathrm{sech}^2(\tfrac{1}{2}x) : H^2(\mathbb{R}) \subseteq L^2(\mathbb{R}) \to L^2(\mathbb{R})$ consists of essential spectrum $[1,\infty)$ and simple eigenvalues at $-\tfrac{5}{4}$, 0 and $\tfrac{3}{4}$ (with corresponding eigenvectors $\mathrm{sech}^3(\tfrac{1}{2}x)$, $-\tanh(\tfrac{1}{2}x)\,\mathrm{sech}^2(\tfrac{1}{2}x)$ and $(3 - 2\cosh x)\,\mathrm{sech}^3(\tfrac{1}{2}x)$ (see Titchmarsh [21], §4.18]). Similarly, the spectra of the operators $1 - \partial_x^2 - 6\operatorname{sech}^2(x) : H^2(\mathbb{R}) \subseteq L^2(\mathbb{R}) \to L^2(\mathbb{R})$ and $1 - \partial_x^2 - 2\operatorname{sech}^2(x) : H^2(\mathbb{R}) \subseteq L^2(\mathbb{R}) \to L^2(\mathbb{R})$ consists of essential spectrum $[1, \infty)$ and respectively two simple eigenvalues at -3 and 0 (with corresponding eigenvectors $\operatorname{sech}^2(x)$ and $\operatorname{sech}'(x)$) and a simple eigenvalue at 0 (with corresponding eigenvector $\operatorname{sech}(x)$) (see Drazin & Johnson [22, pp. 45–48]).

The assertions follow by a scaling argument and restricting to even or odd functions. □

Define
$$\tilde{\mathcal{B}}_{\varepsilon,k} = \mathcal{B}_{\varepsilon,k} - k^2 I.$$

The spectrum of the operator $\tilde{\mathcal{B}}_{0,k}$ (which does not depend upon k) can be determined precisely using Proposition 6.4.1.

Lemma 6.4.2. *The spectrum of the operator $\tilde{\mathcal{B}}_{0,k}^{\mathrm{KP}}$ consists of essential spectrum $[0, \infty)$ and a simple negative eigenvalue $-k_0^2$.*

Proof. Observe that the spectra of $\tilde{\mathcal{B}}_{0,k}^{\mathrm{KP}} = -\partial_x^2 \mathcal{C}$ and $-\partial_x \mathcal{C} \partial_x$ coincide: on the one hand both operators are compact perturbations of the constant-coefficient operator
$$\mathcal{D}_{\mathcal{B}}^{\mathrm{KP}} \subseteq W \to W, \qquad \zeta \mapsto -\zeta_{xx} + (\beta_0 - \tfrac{1}{3})\zeta_{xxxx},$$
whose essential spectrum is clearly $[0, \infty)$, so that $\sigma_{\mathrm{ess}}(-\partial_x^2 \mathcal{C}) = \sigma_{\mathrm{ess}}(-\partial_x \mathcal{C} \partial_x) = [0, \infty)$ (see Kato, [23, Chapter IV, Theorem 5.26]); on the other hand a straightforward calculation shows that the operators have the same nonzero eigenvalues (the equation $-\partial_x \mathcal{C} \partial_x u = \lambda u$ implies that $w = u_x$ satisfies $-\partial_x^2 \mathcal{C} w = \lambda w$; the equation $-\partial_x^2 \mathcal{C} w = \lambda w$ implies that $w = u_x$, where $u = -\lambda^{-1} \partial_x \mathcal{C} w$ satisfies $-\partial_x \mathcal{C} \partial_x u = \lambda u$).

Because $-\partial_x \mathcal{C} \partial_x$ is self-adjoint with $\sigma_{\mathrm{ess}}(-\partial_x \mathcal{C} \partial_x) = [0, \infty)$ its discrete spectrum lies in the negative real axis. Note that
$$\langle -\partial_x \mathcal{C} \partial_x \zeta, \zeta \rangle_0 = \langle \mathcal{C} \zeta_x, \zeta_x \rangle_0 \geq 0$$
for $\zeta \in H_e^2(\mathbb{R})$ with
$$\left\langle \zeta_x, \operatorname{sech}^3 \left(\frac{x}{2(\beta_0 - \tfrac{1}{3})^{1/2}} \right) \right\rangle_0 = 0$$

(see Proposition 6.4.1(i)). It follows that any subspace of $H_e^2(\mathbb{R})$ upon which $-\partial_x \mathcal{C} \partial_x$ is strictly negative definite is one-dimensional. Define
$$\zeta_R(x) = \phi_R(x) \int_0^x \operatorname{sech}^3 \left(\frac{s}{2(\beta_0 - \tfrac{1}{3})^{1/2}} \right) ds,$$

where $\phi(R) = \chi(x/R)$ and $\chi \in C_0^\infty(\mathbb{R})$ is a cut-off function equal to unity in $[-1, 1]$. The calculation
$$\lim_{R \to \infty} \langle -\partial_x \mathcal{C} \partial_x \zeta_R, \zeta_R \rangle_0 = \lim_{R \to \infty} \langle \mathcal{C} \partial_x \zeta_R, \partial_x \zeta_R \rangle_0 = -\tfrac{5}{4}$$

shows that $\inf \sigma (-\partial_x \mathcal{C} \partial_x) < 0$, so that the spectral subspace of $H_e^4(\mathbb{R})$ corresponding to the part of the spectrum of $-\partial_x \mathcal{C} \partial_x$ in $(-\infty, -\varepsilon)$ is nontrivial and hence one-dimensional for every sufficiently small value of $\varepsilon > 0$. We conclude that $-\partial_x \mathcal{C} \partial_x$ has precisely one simple negative eigenvalue $-\omega_0^2$. □

Lemma 6.4.3. *The spectrum of the operator* $\tilde{\mathcal{B}}_{0,k}^{\mathrm{DS}}$ *consists of essential spectrum* $[0, \infty)$ *and a simple negative eigenvalue* $-k_0^2$.

Proof. First note that $\tilde{\mathcal{B}}_{0,k}^{\mathrm{DS}}$ is a compact perturbation of the constant-coefficient operator

$$\mathcal{D}_{\mathcal{B}}^{\mathrm{DS}} \subseteq W^{\mathrm{DS}} \rightarrow W^{\mathrm{DS}}, \qquad \begin{pmatrix} \zeta_1 \\ \zeta_2 \\ \psi \end{pmatrix} \mapsto \begin{pmatrix} A_2^{-1}\zeta_1 - A_2^{-1}A_2\zeta_{1xx} \\ A_2^{-1}\zeta_2 - A_2^{-1}A_1\zeta_{2xx} \\ -(1-\alpha_0^{-1})\psi_{xx} \end{pmatrix},$$

whose essential spectrum is clearly $[0, \infty)$; it follows that $\sigma_{\mathrm{ess}}(\tilde{\mathcal{B}}_{0,k}^{\mathrm{DS}}) = [0, \infty)$ (see Kato, [23, Chapter IV, Theorem 5.26]). Equipping W^{DS} with the inner product

$$\langle\!\langle (\zeta_1, \zeta_2, \psi), (\tilde{\zeta}_1, \tilde{\zeta}_2, \tilde{\psi}) \rangle\!\rangle = \langle \zeta_1, \tilde{\zeta}_1 \rangle_0 + \langle \zeta_2, \tilde{\zeta}_2 \rangle_0 + 2A_2^{-1}\langle \psi, \tilde{\psi} \rangle_0,$$

one finds that $\tilde{\mathcal{B}}_{0,k}^{\mathrm{DS}}$ is self-adjoint, so that its discrete spectrum lies in the negative real axis. Write $W^1 = L_e^2(\mathbb{R}) \times L_e^2(\mathbb{R})$, $\mathcal{D}_B^1 = H_e^2(\mathbb{R}) \times H_e^2(\mathbb{R})$ and $W^2 = L_o^2(\mathbb{R})$, $\mathcal{D}_B^2 = H_o^2(\mathbb{R})$ and observe that (with a slight abuse of notation)

$$\tilde{\mathcal{B}}_{0,k}^{\mathrm{DS}}(\zeta_1, \zeta_2, \psi) = (\tilde{\mathcal{B}}_{0,k,1}(\zeta_1, \psi), \tilde{\mathcal{B}}_{0,k,2}(\zeta_2)),$$

where $\tilde{\mathcal{B}}_{0,k,1} : \mathcal{D}_B^1 \subseteq W^1 \rightarrow W^1$ and $\tilde{\mathcal{B}}_{0,k,2} : \mathcal{D}_B^2 \subseteq W^2 \rightarrow W^2$ are given by

$$\tilde{\mathcal{B}}_{0,k,1}\begin{pmatrix} \zeta_1 \\ \psi \end{pmatrix} = \begin{pmatrix} A_2^{-1}\zeta_1 - A_2^{-1}A_1\zeta_{1xx} - A_2^{-1}(2A_3 + A_5)(\zeta_{\mathrm{NLS}}^\star)^2\zeta_1 + 4A_2^{-1}A_4\zeta_{\mathrm{NLS}}^\star \psi_x \\ -(1-\alpha_0^{-1})\psi_{xx} - 2A_4(\zeta_{\mathrm{NLS}}^\star\zeta_1)_x \end{pmatrix}$$

and $\tilde{\mathcal{B}}_{0,k,2}(\zeta_2) = \mathcal{C}_2(\zeta_2)$; the eigenvalues of $\tilde{\mathcal{B}}_{0,k}$ are therefore precisely the eigenvalues of $\tilde{\mathcal{B}}_{0,k,1}$ and $\tilde{\mathcal{B}}_{0,k,2}$.

According to Proposition 6.4.1 (ii) the operator $\tilde{\mathcal{B}}_{0,k,2}$ has no negative eigenvalues. Turning to the spectrum of $\tilde{\mathcal{B}}_{0,k,1}$, one finds by an explicit calculation that

$$\langle \tilde{\mathcal{B}}_{0,k,1}(\zeta_1, \psi), (\zeta_1, \psi) \rangle_{W^1}$$

$$= \langle \mathcal{C}_1(\zeta_1), \zeta_1 \rangle_0 + 2A_2^{-1}(1-\alpha_0^{-1})\int_{\mathbb{R}} \left(\psi_x + \frac{2A_4}{1-\alpha_0^{-1}}\zeta_{\mathrm{NLS}}^\star\zeta_1 \right)^2 dx,$$

which quantity is positive for $(\zeta_1, \psi) \in W_+^1$, where

$$W_+^1 = \{(\zeta_1, \psi) \in W^1 : \langle (\zeta_1, \psi), ((\mathrm{sech}^2(A_1^{-1/2}x), 0) \rangle_{W^1} = 0\}$$

(see Proposition 6.4.1 (i)). It follows that any subspace of W^1 upon which $\tilde{\mathcal{B}}_{0,k,1}$ is strictly negative definite is one-dimensional. The calculation

$$\langle \tilde{\mathcal{B}}_{0,k,1}((\operatorname{sech}(A_1^{-1/2}x),0),(\operatorname{sech}(A_1^{-1/2}x),0)\rangle_{W^1} = -\frac{16\sqrt{A_1}A_3}{3A_2A_5} < 0$$

shows that $\inf \sigma(\tilde{\mathcal{B}}_{0,k,1}) < 0$, so that the spectral subspace of W^1 corresponding to the part of the spectrum of $\tilde{\mathcal{B}}_{0,k,1}$ in $(-\infty,-\varepsilon)$ is nontrivial and hence one-dimensional for every sufficiently small value of $\varepsilon > 0$. We conclude that $\tilde{\mathcal{B}}_{0,k,1}$ has precisely one simple negative eigenvalue $-k_0^2$. □

Noting that

$$\|\tilde{\mathcal{B}}_{\varepsilon,k} - \tilde{\mathcal{B}}_{0,k}\|_{\mathcal{L}(\mathcal{D}_B,W)} = \|\tilde{\mathcal{R}}_{\varepsilon,k}\|_{\mathcal{L}(\mathcal{D}_B,W)} \leq c\varepsilon^{1/r}, \tag{6.4.1}$$

we now use a perturbation argument to obtain a qualitative description of a portion of the spectrum of $\tilde{\mathcal{B}}_{\varepsilon,k}$.

Lemma 6.4.4. *Let m and M be positive real numbers with $m < k_0^2 < M$ and γ be an ellipse in the complex plane with major axis $[-M,-m]$. For each $k \in [0,k_{\max}]$ the portion of the spectrum of $\tilde{\mathcal{B}}_{\varepsilon,k}$ within γ consists of precisely one simple real eigenvalue $\lambda_{\varepsilon,k}$ with $\lambda_{0,k} = -k_0^2$. In particular, $\tilde{\mathcal{B}}_{\varepsilon,k}$ is closed and $\tilde{\mathcal{B}}_{\varepsilon,k} - \lambda I : \mathcal{D}_B \to W$ is Fredholm with index 0 for $k \in [0,k_{\max}]$ and $\lambda \in [-M,-m]$.*

Proof. The contour γ defines a separation of $\sigma(\tilde{\mathcal{B}}_{0,k})$: the portion of $\sigma(\tilde{\mathcal{B}}_{0,k})$ in the interior of γ consists of precisely one simple eigenvalue $-k_0^2$ and the remainder of $\sigma(\tilde{\mathcal{B}}_{0,k})$ lies in the exterior of γ. In view of inequality (6.4.1) and the fact that $\tilde{\mathcal{B}}_{0,k}$ does not depend upon k, we may apply a standard argument in spectral perturbation theory (see Kato [23, Theorem 3.16]), which asserts that γ defines the same separation of $\tilde{\mathcal{B}}_{\varepsilon,k}$. Since $\tilde{\mathcal{B}}_{\varepsilon,k}$ is a real operator its eigenvalues arise in complex-conjugate pairs; its simple eigenvalue $\lambda_{\varepsilon,k}$ in the interior of γ is therefore real.

This argument implies in particular that $\tilde{\mathcal{B}}_{\varepsilon,k}$ is closed and that $\tilde{\mathcal{B}}_{\varepsilon,k} - \lambda I : \mathcal{D}_B \to W$ is Fredholm with index 0 for $\lambda \in [-M,-m]$ (because $[-M,-m] \setminus \{\lambda_{\varepsilon,k}\} \subseteq \rho(\tilde{\mathcal{B}}_{\varepsilon,k})$ and $\lambda_{\varepsilon,k}$ is a simple eigenvalue of $\tilde{\mathcal{B}}_{\varepsilon,k}$). □

Choose $k_{\max} > k_0$ and $k_{\min} \in (0,k_0)$ and apply Lemma 6.4.4, selecting m, M such that $M > k_{\max}^2$ and $m < k_{\min}^2$, so that the point $-k^2$ lies in the interior of γ for all $k \in [k_{\min},k_{\max}]$ (see Figure 6.6). The next step is to establish that the linear mapping $\mathcal{R}_{\varepsilon,k} : \mathcal{D}_B \to W$ is a continuously differentiable function of $k > 0$, which satisfies the estimate

$$\|\mathcal{R}_{\varepsilon,k}^{\mathrm{KP}}(\zeta)'\|_0 \leq c\varepsilon^{1/2}\|\zeta\|_3, \qquad \|\mathcal{R}_{\varepsilon,k}^{\mathrm{DS}}(\zeta)'\|_0 \leq c\varepsilon\|\zeta\|_1$$

uniformly over $k \in [k_{\min},k_{\max}]$. This property of $\mathcal{R}_{\varepsilon,k}$ is established by systematically examining each step in its derivation, checking the continuous

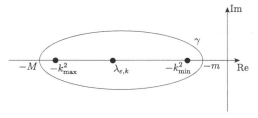

Figure 6.6. The portion of $\sigma(\tilde{\mathcal{B}}_{\varepsilon,k})$ within γ consists of precisely one simple real eigenvalue $\lambda_{\varepsilon,k} \in [-k_{\max}^2, -k_{\min}^2]$ with $\lambda_{0,k} = -k_0^2$.

differentiability of the operators that arise in this process and estimating them with the techniques used to prove Theorem 6.3.6.

Corollary. *The linear mapping* $\tilde{\mathcal{B}}_{\varepsilon,k} : \mathcal{D}_{\mathcal{B}} \to W$ *satisfies the estimate*

$$\|\tilde{\mathcal{B}}'_{\varepsilon,k}\|_{\mathcal{L}(\mathcal{D}_{\mathcal{B}},W)} \le c\varepsilon^{1/r}$$

uniformly over $k \in [k_{\min}, k_{\max}]$.

The following result follows from the relation $\mathcal{B}_{\varepsilon,k} = \tilde{\mathcal{B}}_{\varepsilon,k} + k^2 I$.

Lemma 6.4.5. *Fix* $k \in [k_{\min}, k_{\max}]$.

(i) *The operator* $\mathcal{B}_{\varepsilon,k} : \mathcal{D}_{\mathcal{B}} \subseteq W \to W$ *is invertible provided that* $\lambda_{\varepsilon,k} + k^2 \ne 0$.
(ii) *The operator* $\mathcal{B}_{\varepsilon,k} : \mathcal{D}_{\mathcal{B}} \subseteq W \to W$ *has a simple zero eigenvalue whenever* $\lambda_{\varepsilon,k} + k^2 = 0$.

Next we show that the equation

$$\lambda_{\varepsilon,k} + k^2 = 0 \tag{6.4.2}$$

has precisely one solution k_ε in the interval $[k_{\min}, k_{\max}]$. The solution set is known explicitly for $\varepsilon = 0$, and we approach the problem for $\varepsilon > 0$ using a perturbation argument and the contraction-mapping principle. The necessary estimates for $\lambda_{\varepsilon,k}$, which are stated in the next lemma, are deduced from the corresponding estimates

$$\|\tilde{\mathcal{B}}_{\varepsilon,k} - \tilde{\mathcal{B}}_{0,k}\|_{\mathcal{L}(\mathcal{D}_{\mathcal{B}},W)} \le c\varepsilon^{1/r}, \qquad \|\tilde{\mathcal{B}}'_{\varepsilon,k}\|_{\mathcal{L}(\mathcal{D}_{\mathcal{B}},W)} \le c\varepsilon^{1/r}$$

for $\tilde{\mathcal{B}}_{\varepsilon,k}$ using standard spectral-theoretical techniques (e.g., see Groves, Haragus & Sun [14, Lemma 3.15]).

Lemma 6.4.6. *The eigenvalue* $\lambda_{\varepsilon,k}$ *of* $\tilde{\mathcal{B}}_{\varepsilon,k} : \mathcal{D}_{\mathcal{B}} \subseteq V \to V$ *is a differentiable function of* k *and satisfies the estimates*

$$|\lambda_{\varepsilon,k} - \lambda_{0,k}| \le c^\star \varepsilon^{1/r}, \quad |\lambda'_{\varepsilon,k}| \le c^\star \varepsilon^{1/r}$$

uniformly over $k \in [k_{\min}, k_{\max}]$.

Theorem 6.4.7. *Equation (6.4.2) has precisely one solution k_ε in the interval $[k_{\min}, k_{\max}]$. This solution lies in the set*

$$\mathcal{S} = \left\{ k : |k - k_0| \le \frac{c^\star \varepsilon^{1/r}}{k_0} \right\}.$$

Finally, we examine the reduced equation

$$\mathcal{C}_\varepsilon(\zeta) = \zeta^\dagger, \tag{6.4.3}$$

where $\mathcal{C}_\varepsilon^{\mathrm{KP}} : \mathcal{D}_\mathcal{C}^{\mathrm{KP}} \subseteq V^{\mathrm{KP}} \to V^{\mathrm{KP}}$ is given by

$$V^{\mathrm{KP}} = L_e^2(\mathbb{R}), \quad \mathcal{D}_\mathcal{C}^{\mathrm{KP}} = H_e^2(\mathbb{R}),$$

$$\mathcal{C}_\varepsilon^{\mathrm{KP}}(\zeta) = -\mathcal{C}(\zeta) - \mathcal{R}_{\varepsilon,0}^{\mathrm{KP}}(\zeta)$$

(see (6.3.20)) and $\mathcal{C}_\varepsilon^{\mathrm{DS}} : \mathcal{D}_\mathcal{C}^{\mathrm{DS}} \subseteq V^{\mathrm{DS}} \to V^{\mathrm{DS}}$ is given by

$$V^{\mathrm{DS}} = L_e^2(\mathbb{R}) \times L_o^2(\mathbb{R}), \quad \mathcal{D}_\mathcal{C}^{\mathrm{DS}} = H_e^2(\mathbb{R}) \times H_o^2(\mathbb{R}),$$

$$\mathcal{C}_\varepsilon^{\mathrm{DS}} \begin{pmatrix} \zeta_1 \\ \zeta_2 \end{pmatrix} = \begin{pmatrix} \mathcal{C}_1 \zeta_1 - \operatorname{Re}\mathcal{R}_{\varepsilon,0}^{\mathrm{DS}}(\zeta) \\ \mathcal{C}_2 \zeta_2 - \operatorname{Im}\mathcal{R}_{\varepsilon,0}^{\mathrm{DS}}(\zeta) \end{pmatrix}$$

(see (6.3.24)). Observe that $\mathcal{C}_0 : \mathcal{D}_\mathcal{C} \to V$ is invertible (Proposition 6.4.1) and $\|\mathcal{R}_{\varepsilon,0}\|_{\mathcal{L}(\mathcal{D}_\mathcal{C}, V)} \le c\varepsilon^{1/r}$.

Lemma 6.4.8. *The reduced equation (6.4.3) has a unique solution $\zeta \in \mathcal{D}_\mathcal{C}$ for each $\zeta^\dagger \in V$.*

Appendix: Asymptotic Expansions of the Line Solitary Waves

An asymptotic expansion of the Amick-Kirchgässner line solitary waves for strong surface tension is obtained by applying the change of variable (6.2.2) to the asymptotic expressions given in their paper. One finds that

$$\eta^\star(x) = \eta_l^\star(x) + \eta_r^\star(\varepsilon x), \qquad \Phi^\star(x,y) = \Phi_l^\star(x) + \Phi_r^\star(\varepsilon x, y),$$

where

$$\eta_l^\star(x) = \varepsilon^2 \zeta_{\mathrm{KdV}}^\star(\varepsilon x), \qquad \Phi_l^\star(x) = \varepsilon(\partial_x^{-1}\zeta_{\mathrm{KdV}}^\star)(\varepsilon x), \tag{A.1}$$

$\zeta_{\mathrm{KdV}}^\star$ is defined by equation (6.1.7) and $\eta_r^\star = \mathcal{O}(\varepsilon^4)$, $\nabla\Gamma_r^\star = \mathcal{O}(\varepsilon^3)$, the symbol $\mathcal{O}(\varepsilon^a)$ denoting a smooth quantity whose derivatives are all $O(\varepsilon^a e^{-\rho\varepsilon|x|})$ for some $\rho > 0$ (uniformly over $y \in [0,1]$).

An asymptotic expansion of the Iooss-Kirchgässner line solitary waves for weak surface tension may be computed in the framework of their existence theory (see Dias & Iooss [12], noting the omission of certain terms in equation (3.41)). One finds that

$$\eta^\star(x) = \eta_1^\star(x) + \eta_2^\star(x) + \eta_r^\star(x), \qquad \Phi^\star(x,y) = \Phi_1^\star(x,y) + \Phi_2^\star(x,y) + \Phi_r^\star(x,y),$$

where

$$\eta_1^\star(x) = \varepsilon \zeta_{\mathrm{NLS}}^\star(\varepsilon x) \cos(\mu_0 x), \tag{A.2}$$

$$\eta_2^\star(x) = C_0 \varepsilon^2 \zeta_{\mathrm{NLS}}^\star(\varepsilon x) \xi_{\mathrm{NLS}}^\star(\varepsilon x) \sin(\mu_0 x) - C_1 \varepsilon^2 (\zeta_{\mathrm{NLS}}^\star)^2(\varepsilon x) \cos(2\mu_0 x)$$
$$- C_2 \varepsilon^2 (\zeta_{\mathrm{NLS}}^\star)^2(\varepsilon x) + C_3 \varepsilon^2 (\zeta_{\mathrm{NLS}}^\star)'(\varepsilon x) \sin(\mu_0 x), \tag{A.3}$$

$$\Phi_1^\star(x,y) = \varepsilon \zeta_{\mathrm{NLS}}^\star(\varepsilon x) \sin(\mu_0 x) \frac{\cosh(\mu_0 y)}{\sinh(\mu_0)}, \tag{A.4}$$

$$\Phi_2^\star(x,y) = -C_0 \varepsilon^2 \xi_{\mathrm{NLS}}^\star(\varepsilon x) \cos(\mu_0 x) \frac{\cosh(\mu_0 y)}{\sinh \mu_0}$$
$$+ \varepsilon^2 (\zeta_{\mathrm{NLS}}^\star)^2(\varepsilon x) \sin(2\mu_0 x) \frac{\mu_0 y \sinh(\mu_0 y)}{2\sinh(\mu_0)}$$
$$- C_4 \varepsilon \partial_x^{-1} (\zeta_{\mathrm{NLS}}^{\star 2})(\varepsilon x) - C_5 \varepsilon^2 (\zeta_{\mathrm{NLS}}^\star)^2(\varepsilon x) \sin(2\mu_0 x) \frac{\cosh(2\mu_0 y)}{\sinh(2\mu_0)}, \tag{A.5}$$

$\zeta_{\mathrm{KdV}}^\star$ is defined by equation (6.1.7), $\xi_{\mathrm{NLS}}^\star(x) = x\zeta_{\mathrm{NLS}}^\star(x)$ and $\eta_r^\star$, $\nabla \Gamma_r^\star = \mathcal{O}(\varepsilon^3)$. The positive coefficients C_1, C_2, C_4 and C_5 are given by

$$C_1 = \frac{\mu_0^2(\cosh(2\mu_0) + 2)}{4\sinh^2(\mu_0)g_0(2\mu_0, 0)}, \qquad C_2 = \frac{\mu_0(\sinh(2\mu_0) + \mu_0)}{4\sinh^2(\mu_0)(\alpha_0 - 1)},$$

$$C_4 = \frac{\mu_0(\alpha_0 \sinh(2\mu_0) + \mu_0)}{4\sinh^2(\mu_0)(\alpha_0 - 1)}, \qquad C_5 = \frac{\mu_0^2(\cosh(2\mu_0) + 2)}{4\sinh^2(\mu_0)g_0(2\mu_0, 0)} + \frac{\mu_0 \sinh(2\mu_0)}{4\sinh^2(\mu_0)},$$

while the values (and signs) of the coefficients C_0 and C_3 are unimportant.

References

[1] ABLOWITZ, M. J. & CLARKSON, P. A. 1991 *Solitons, Nonlinear Evolution Equations and Inverse Scattering*. LMS Lecture Note Series **149**. Cambridge: Cambridge University Press.

[2] TAJIRI, M. & MURAKAMI, Y. 1990 The periodic solution resonance: solutions of the Kadomtsev-Petviashvili equation with positive dispersion. *Phys. Lett. A* **143**, 217–220.

[3] HARAGUS, M. & KIRCHGÄSSNER, K. 1995 Breaking the dimension of a steady wave: some examples. In *Nonlinear Dynamics and Pattern Formation in the Natural Environment* (eds. Doelman, A. & van Harten, A.). *Pitman Res. Notes Math. Ser.* **335**, 119–129.

[4] DJORDJEVIC, V. D. & REDEKOPP, L. G. 1977 On two-dimensional packets of capillary-gravity waves. *J. Fluid Mech.* **79**, 703–714.

[5] ABLOWITZ, M. J. & SEGUR, H. 1979 On the evolution of packets of water waves. *J. Fluid Mech.* **92**, 691–715.

[6] SULEM, C. & SULEM, P. L. 1999 *The Nonlinear Schrödinger Equation*. Applied Mathematical Sciences **139**. New York: Springer-Verlag.

[7] GROVES, M. D., SUN, S.-M. & WAHLÉN, E. 2014 Periodic solitons for the elliptic-elliptic focussing Davey-Stewartson equations. Preprint.

[8] AMICK, C. J. & KIRCHGÄSSNER, K. 1989 A theory of solitary water waves in the presence of surface tension. *Arch. Rat. Mech. Anal.* **105**, 1–49.

[9] KIRCHGÄSSNER, K. 1988 Nonlinearly resonant surface waves and homoclinic bifurcation. *Adv. Appl. Mech.* **26**, 135–181.

[10] SACHS, R. L. 1991 On the existence of small amplitude solitary waves with strong surface-tension. *J. Diff. Eqns.* **90**, 31–51.

[11] IOOSS, G. & KIRCHGÄSSNER, K. 1990 Bifurcation d'ondes solitaires en présence d'une faible tension superficielle. *C. R. Acad. Sci. Paris, Sér. 1* **311**, 265–268.

[12] DIAS, F. & IOOSS, G. 1993 Capillary-gravity solitary waves with damped oscillations. *Physica D* **65**, 399–423.

[13] IOOSS, G. & PÉROUÈME, M. C. 1993 Perturbed homoclinic solutions in reversible 1:1 resonance vector fields. *J. Diff. Eqns.* **102**, 62–88.

[14] GROVES, M. D., HARAGUS, M. & SUN, S.-M. 2002 A dimension-breaking phenomenon in the theory of gravity-capillary water waves. *Phil. Trans. Roy. Soc. Lond. A* **360**, 2189–2243.

[15] GROVES, M. D., SUN, S.-M. & WAHLÉN, E. 2015 The dimension-breaking route to three-dimensional solitary gravity-capillary water waves. *Arch. Rat. Mech. Anal.*, published online.

[16] GROVES, M. D. & MIELKE, A. 2001 A spatial dynamics approach to three-dimensional gravity-capillary steady water waves. *Proc. Roy. Soc. Edin. A* **131**, 83–136.

[17] GROVES, M. D. & HARAGUS, M. 2003 A bifurcation theory for three-dimensional oblique travelling gravity-capillary water waves. *J. Nonlinear Sci.* **13**, 397–447.

[18] DEVANEY, R. L. 1976 Reversible diffeomorphisms and flows. *Trans. Amer. Math. Soc.* **218**, 89–113.

[19] IOOSS, G. 1999 Gravity and capillary-gravity periodic travelling waves for two superposed fluid layers, one being of infinite depth. *J. Math. Fluid Mech.* **1**, 24–63.

[20] BAGRI, G. & GROVES, M. D. 2014 A spatial dynamics theory for doubly periodic travelling gravity-capillary surface waves on water of infinite depth. *J. Diff. Dyn. Eqns*, published online.

[21] TITCHMARSH, E. C. 1962 *Eigenfunction Expansions Associated with Second-Order Differential Equations. Part I*, 2nd edn. Oxford: Clarendon Press.

[22] DRAZIN, P. G. & JOHNSON, R. S. (1989) *Solitons: An Introduction.* Cambridge: Cambridge University Press.

[23] KATO, T. 1976 *Perturbation Theory for Linear Operators,* 2nd edn. New York: Springer-Verlag.

7

Validity and Non-Validity of the Nonlinear Schrödinger Equation as a Model for Water Waves

Guido Schneider

Abstract

In 1968 V.E. Zakharov [1] derived the Nonlinear Schrödinger (NLS) equation for the description of slow spatial and temporal modulations of spatially and temporarily oscillating surface water waves. It took more than 40 years before this formal approximation had been justified by rigorous error estimates in case of no surface tension in infinite and finite depth. Shortly after, in case of positive surface tension a counter example was established showing that the NLS equation can fail to make correct predictions. It is the purpose of the present paper to give an overview about validity and non-validity results for the NLS approximation of the water wave problem. We explain which results can be expected in case of positive surface tension and which cannot. Finally, we discuss the same question for the NLS approximation describing modulations of periodic traveling surface water waves.

7.1 The Water Wave Problem

The 2D water wave problem consists in finding the irrotational flow of an inviscid incompressible fluid in an infinitely long canal of finite or infinite depth with a free surface under the influence of gravity and surface tension. It turns out that under these assumptions the dynamics of the problem is completely determined by the evolution of the free surface $\Gamma(t)$ (Figure 7.1).

The coordinates are denoted with x_1 in the horizontal and with x_2 in the vertical direction. In case of finite depth h the fluid is contained in the unbounded domain $\Omega(t)$ between the impermeable bottom $\{(x_1, -h) : x_1 \in \mathbb{R}\}$ and the free unknown top surface $\Gamma(t)$. The velocity field $u = (u_1, u_2)$ satisfies Euler's equations in $\Omega(t)$. The assumption of the irrotationality of the flow, i.e., $\nabla \times u = 0$, is preserved by Euler's equations. Hence, the assumptions of irrotationality and incompressibility leads to an elliptic system

$$\nabla \times u = 0, \qquad \nabla \cdot u = 0 \qquad (7.1.1)$$

121

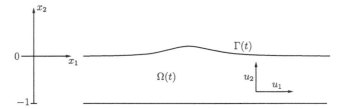

Figure 7.1. The Water Wave Problem.

in $\Omega(t)$. The impermeability of the bottom gives the lower boundary condition

$$u_2|_{x_2=-h} = 0. \tag{7.1.2}$$

System (7.1.1) can be solved uniquely if another boundary condition such as the value of $u_1|_{\Gamma(t)}$ at the top surface is assumed. Hence, from these assumptions we have a relation between the velocity components $u_1|_{\Gamma(t)}$ and $u_2|_{\Gamma(t)}$ at the top surface.

On the free surface $\Gamma(t) = \{x \in \mathbb{R}^2 : x_2 = \eta(x_1,t)\}$ we have the kinematic boundary condition

$$\partial_t \eta = u_2|_{\Gamma(t)} - u_1|_{\Gamma(t)}(\partial_{x_1}\eta), \tag{7.1.3}$$

and the balance of forces

$$\partial_t u_1|_{\Gamma(t)} = -\frac{1}{2}\partial_{x_1}((u_1|_{\Gamma(t)})^2 + (u_2|_{\Gamma(t)})^2) - g\partial_{x_1}\eta + \sigma\partial_{x_1}^2\left[\frac{\partial_{x_1}\eta}{\sqrt{1+(\partial_{x_1}\eta)^2}}\right],$$
$$\tag{7.1.4}$$

with g being the gravitational constant and σ the surface tension parameter. Without loss of generality we will set $g = 1$ and the depth of the fluid at rest, i.e., $\eta = 0$, to one or infinity, i.e. $h = 1$ or $h = \infty$, in the following. Using equation (7.1.1) the water wave problem is completely described by the evolution of $\eta = \eta(x_1,t)$ and $w = w(x_1,t) = u_1(x_1,\eta(x_1,t),t)$.

Equations (7.1.1)–(7.1.4) are called the Eulerian formulation of the water wave problem. There are various other formulations of the water wave problem which differ by the chosen parametrization of the top surface $\Gamma(t)$. For the Eulerian formulation local existence and uniqueness results have been shown for instance in [2–4] and for the Lagrangian formulation, where more general $\Gamma(t) = \{x \in \mathbb{R}^2 : x = x(\alpha,t), \alpha \in \mathbb{R}\}$, such results have been established for instance in [5–13]. For a third formulation of the water wave problem in which the top surface is parametrized by arc length a local existence and uniqueness theorem can be found in [14]. There exist local existence and uniqueness theorems for the water wave problem in 2D and 3D, for finite and infinite depth, with and without surface tension, different regularities of the initial conditions and for the different parameterizations of the top surface

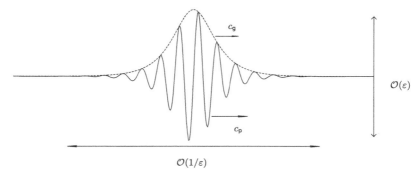

Figure 7.2. The envelope, advancing with the group velocity c_g, of the oscillating wave packet, advancing with the phase velocity $c_p = \omega_0/k_0$, is described by the amplitude A, which solves the NLS equation (7.2.1).

which have been chosen to formulate the problem. Recently, global and almost global existence and uniqueness results for small initial conditions have been established in a number of papers, cf. [15–17].

7.2 The NLS Approximation

In 1968 V.E. Zakharov [1] derived the Nonlinear Schrödinger (NLS) equation

$$\partial_T A = i\nu_1 \partial_X^2 A + i\nu_2 A|A|^2, \qquad (7.2.1)$$

with $T \in \mathbb{R}$, $X \in \mathbb{R}$, $A(X,T) \in \mathbb{C}$, and coefficients $\nu_j = \nu_j(k_0) \in \mathbb{R}$ from the equations of the 2D water wave problem in case of infinite depth and no surface tension in order to describe slow spatial and temporal modulations of a spatially and temporarily oscillating wave packet $e^{i(k_0 x_1 - \omega_0 t)}$, with a basic spatial wave number $k_0 \neq 0$ and a basic temporal wave number $\omega_0 \neq 0$. They are related via the linear dispersion relation of the water wave problem (7.1.1)–(7.1.4), namely

$$\omega^2 - (k + \sigma k^3)\tanh(k) = 0, \quad \text{and} \quad \omega^2 - (|k| + \sigma |k|^3) = 0 \qquad (7.2.2)$$

in case $h = 1$, respectively, $h = \infty$.

The ansatz to derive the NLS equation is given by

$$\begin{pmatrix} \eta \\ w \end{pmatrix} = \varepsilon \Psi_{NLS} + \mathcal{O}(\varepsilon^2)$$

where

$$\varepsilon \Psi_{NLS} = \varepsilon A(\varepsilon(x_1 - c_g t), \varepsilon^2 t) e^{i(k_0 x_1 - \omega_0 t)} \varphi(k_0) + c.c. \qquad (7.2.3)$$

Here $0 < \varepsilon \ll 1$ is a small perturbation parameter, $\varphi(k_0) \in \mathbb{C}^2$, c_g the group velocity of the wave packet and $\omega_0 > 0$ the basic temporal wave number

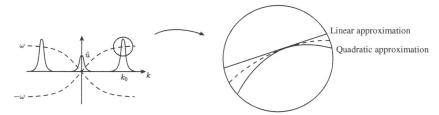

Figure 7.3. The left panel shows the two curves of eigenvalues $k \mapsto \pm\omega(k)$. The solution $\widehat{u} = (\widehat{\eta}, \widehat{w})$ is strongly concentrated at integer multiples of the basic wave number k_0. Hence, for the dynamics the form of $k \mapsto \omega(k)$ at the wave number k_0 plays a prominent role. The linear approximation gives the group velocity c_g and the quadratic approximation is related to the coefficient ν_1 appearing in the NLS equation (7.2.1).

associated to the basic spatial wave number $k_0 > 0$. $T = \varepsilon^2 t$ is the slow time scale and $X = \varepsilon(x_1 - c_g t)$ is the slow spatial scale, i.e., the time scale of the modulations is $\mathcal{O}(1/\varepsilon^2)$ and the spatial scale of the modulations is $\mathcal{O}(1/\varepsilon)$. The complex-valued amplitude $A = A(X, T)$ solves in lowest order the NLS equation (7.2.1). See Figure 7.2.

The NLS equation appears as a universal amplitude equation for various dispersive wave systems. The reason can be seen by considering the NLS ansatz $\varepsilon A(\varepsilon x_1) e^{ik_0 x_1}$ in Fourier space, which is given by

$$\frac{1}{2\pi} \int_{\mathbb{R}} \varepsilon A(\varepsilon x_1) e^{ik_0 x_1} e^{-ikx_1} dx_1 = \widehat{A}\left(\frac{k - k_0}{\varepsilon}\right).$$

Hence the ansatz is strongly concentrated around the basic spatial wave number k_0 and so for the evolution the form of the curves of eigenvalues $k \mapsto \omega(k)$ at k_0 plays a fundamental role. The linear approximation of $\omega(k)$ gives the group velocity c_g. The quadratic approximation is related to the coefficient ν_1 in the NLS equation (7.2.1). See Figure 7.3.

7.3 Approximation Results in Case $\sigma = 0$

It took more than 40 years for a first approximation result. The first approximation result, i.e., error estimates that the NLS approximation makes correct predictions for the water wave problem with infinite depth and no surface tension can be found in [18]. The given proof uses special properties of this problem and cannot be generalized to other quasilinear dispersive wave systems in an obvious way. The problem for general dispersive wave systems with quasilinear quadratic terms remained unsolved so far in general.

7.3.1 The Result for Finite Depth

The classical approach for the justification of the NLS approximation that is based on a near identity change of variables can be transferred to the Lagrangian formulation of the water wave problem with finite depth and without surface tension [19]. The result is as follows.

Theorem 7.3.1. *Fix* $s_A - 4 \geq s \geq 6$. *Then for all* $k_0 > 0$ *and for all* $C_1, T_0 > 0$ *there exist* $T_1 > 0$, $C_2 > 0$, $\varepsilon_0 > 0$ *such that for all solutions* $A \in C([0, T_0], H^{s_A}(\mathbb{R}, \mathbb{C}))$ *of the NLS equation (7.2.1) with*

$$\sup_{T \in [0, T_0]} \|A(\cdot, T)\|_{H^{s_A}(\mathbb{R}, \mathbb{C})} \leq C_1$$

the following holds. For all $\varepsilon \in (0, \varepsilon_0)$ *there exists a solution*

$$(\eta, w) \in C([0, T_1/\varepsilon^2], (H^s(\mathbb{R}, \mathbb{R}))^2)$$

of the 2D water wave problem (7.1.1)-(7.1.4) with $h = 1$, *which satisfies*

$$\sup_{t \in [0, T_1/\varepsilon^2]} \left\| \begin{pmatrix} \eta \\ w \end{pmatrix} (\cdot, t) - \varepsilon \Psi_{NLS}(\cdot, t) \right\|_{(H^s(\mathbb{R}, \mathbb{R}))^2} \leq C_2 \varepsilon^{3/2}.$$

In the following we explain the overall structure of the proof and its major difficulties. We write the water wave problem as

$$\partial_t v = \Lambda v + B(v, v) + g(v), \tag{7.3.1}$$

where Λv stands for the linear, $B(v, v)$ for the quadratic terms, and $g(v)$ for the cubic and higher order terms. W.l.o.g. we assume B to be a symmetric bilinear mapping.

The NLS approximation is abbreviated by $\varepsilon \Psi_{NLS}$. It is advantageous to work with a modified approximation $\varepsilon \Psi$, which is $\mathcal{O}(\varepsilon^2)$ close to $\varepsilon \Psi_{NLS}$. This approximation is mainly obtained by adding higher order terms to the original approximation such that the residual

$$\text{Res}(v) = -\partial_t v + \Lambda v + B(v, v) + g(v)$$

is small for $v = \varepsilon \Psi$. In detail, for all $\gamma > 0$ there exists a formal approximation $\varepsilon \Psi$ close to $\varepsilon \Psi_{NLS}$ such that

$$\text{Res}(\varepsilon \Psi) = \mathcal{O}(\varepsilon^\gamma). \tag{7.3.2}$$

This residual first has been estimated in [20]. The difference

$$\varepsilon^\beta R = v - \varepsilon \Psi$$

between the correct solution v and the approximation $\varepsilon \Psi$ satisfies

$$\partial_t R = \Lambda R + 2\varepsilon^\alpha B(\Psi, R) + \varepsilon^\beta B(R, R) + \cdots + \varepsilon^{-\beta} \text{Res}(\varepsilon \Psi). \tag{7.3.3}$$

For a $\beta > 1$ we have to prove that R is of order $\mathcal{O}(1)$ for all $t \in [0, T_1/\varepsilon^2]$. In order to do so a few things have to be checked. Since Λ generates a uniformly bounded semigroup we would be done aside from possible complicated functional analytic details, if a) $\alpha \geq 2$, b) $\beta > 2$, and c) $\varepsilon^{-\beta} \text{Res}(\varepsilon \Psi) = \mathcal{O}(\varepsilon^2)$. The result then would follow by a rescaling of time, $T = \varepsilon^2 t$, and an application of Gronwall's inequality, e.g., [21]. In fact, we can choose γ and $\beta = \gamma + 2$ arbitrarily large. However, we have $\alpha = 1$ and so the major difficulty is to control the influence of the term $2\varepsilon B(\Psi, R)$ on the dynamics. This term turns out to be oscillatory and so it can be controlled by averaging or normal form methods. In fact except of the wave numbers $k = 0, \pm k_0$ this term can be eliminated with a near identity change of variables

$$R = w + \varepsilon M(\Psi, w) \tag{7.3.4}$$

with M a bilinear mapping. This change of variables will be called normal form transform in the following. For general dispersive wave systems this idea goes back to [22]. We find

$$\partial_t w = \Lambda w + 2\varepsilon B(\Psi, R) + \varepsilon \Lambda M(\Psi, w) - \varepsilon M(\Lambda \Psi, w) - \varepsilon M(\Psi, \Lambda w) + \mathcal{O}(\varepsilon^2).$$

In order to eliminate the terms of order $\mathcal{O}(\varepsilon)$ we have to find a M such that

$$2B(\Psi, R) + \Lambda M(\Psi, w) - M(\Lambda \Psi, w) - M(\Psi, \Lambda w) = 0. \tag{7.3.5}$$

However, for quasilinear systems, such as the water wave problem, a loss of regularity comes with the normal form transform, i.e., B and finally M will loose regularity. Therefore, the transformation cannot be inverted with Neumann's series. However, for the Langrangian formulation of the water wave problem that was used in [19] the right-hand side of the water wave problem and finally M only loose half a derivative. Since the transformation can be inverted with the help of energy estimates, after the transformation the right-hand side of the Lagrangian formulation looses one derivative, which allows us to solve the transformed system with the Cauchy-Kowalevskaya theorem. In Section 7.3.2 we provide more details about the normal form transform.

Remark Modifying the NLS ansatz with a cut-off function in Fourier space makes the used NLS approximation analytic in a strip symmetric around the real axis in the complex plane. In order to gain the missing derivative the strip is made smaller with a velocity of order $\mathcal{O}(\varepsilon^2)$. This leads to an artificial smoothing of one derivative. The velocity is sufficiently fast to control the terms on the right-hand side of the water wave problem after the normal form transform. However, it leads to a restriction on the length of the approximation interval $[0, T_1/\varepsilon^2]$. Its length has the correct order $\mathcal{O}(1/\varepsilon^2)$, however, $T_1 = T_0$ would be optimal in Theorem 7.3.1.

7.3.2 The Normal Form Transform

In Fourier space after a diagonalization the water wave problem, respectively, a general dispersive wave system, is of the form

$$\partial_t \widehat{v}_j(k,t) = i\omega_j(k)\widehat{v}_j(k,t)$$
$$+ \int \sum_{j_1 j_2} b_{jj_1 j_2}(k, k-l, l)\widehat{v}_{j_1}(k-l,t)\widehat{v}_{j_2}(l,t)dl + \cdots,$$

with curves of eigenvalues $k \mapsto \omega_j(k)$, kernels $b_{jj_1 j_2}(k, k-l, l)$, and j, j_1, and j_2 in some index set. The equations for the error functions

$$\varepsilon^\beta \widehat{R}_j(k,t) = \widehat{v}_j(k,t) + \varepsilon \widehat{\Psi}(k,t)$$

are then of the form

$$\partial_t \widehat{R}_j(k,t) = i\omega_j(k)\widehat{R}_j(k,t)$$
$$+2\varepsilon \int \sum_{j_1 j_2} b_{jj_1 j_2}(k, k-l, l)\widehat{\Psi}_{j_1}(k-l,t)\widehat{R}_{j_2}(l,t)dl + \mathcal{O}(\varepsilon^2),$$

The function M of the near identity change of variables (7.3.4) to eliminate the terms of order $\mathcal{O}(\varepsilon)$ is chosen to have the same structure as the function B, namely

$$M_{jj_1 j_2}(\widehat{\Psi}, \widehat{R}) = 2\int m_{jj_1 j_2}(k, k-l, l)\widehat{\Psi}_{j_1}(k-l,t)\widehat{R}_{j_2}(l,t)dl.$$

Condition (7.3.5) transfers into

$$m_{jj_1 j_2}(k, k-l, l) = -b_{jj_1 j_2}(k, k-l, l)/(\omega_j(k) - \omega_{j_1}(k-l) - \omega_{j_2}(l)).$$

Since $\widehat{\Psi}$ is strongly concentrated at $\pm k_0$ the kernel functions $m_{jj_1 j_2}$ can be set to zero outside small neighborhoods of $k - l \approx k_0$ and $k - l \approx -k_0$, in fact the $m_{jj_1 j_2}(k, k-l, l)$ can be expressed in terms of $b_{jj_1 j_2}(k, k-l, l)$ if

$$\omega_j(k) - \omega_{j_1}(\pm k_0) - \omega_{j_2}(k \mp k_0) \neq 0.$$

For the water wave problem without surface tension this non-resonance condition is satisfied except for the wave numbers $k = 0$ and $k = \pm k_0$. We will come back to this difficulty below.

Next we will explain the loss of regularity associated to this transformation. For the Lagrangian formulation of the water wave problem without surface tension we have the following asymptotics. In the worst case

$$\omega_j(k) - \omega_{j_1}(\pm k_0) - \omega_{j_2}(k \mp k_0) = \mathcal{O}(1)$$

for $|k| \to \infty$, but

$$b_{jj_1 j_2}(k, \pm k_0, k \mp k_0) = \mathcal{O}(|k|^{1/2})$$

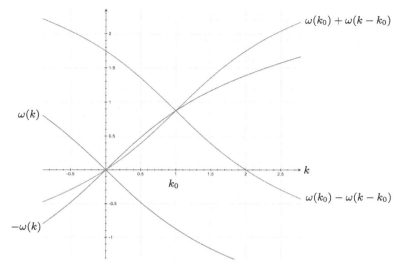

Figure 7.4. The curves of eigenvalues $k \mapsto \pm\omega(k)$ and the curves $k \mapsto \omega(k_0) \pm \omega(k - k_0)$ in case of finite depth. Intersection points correspond to wave numbers of quadratic resonances. In case of no surface tension $\sigma = 0$ only two intersections occur, namely at $k = 0$ and $k = k_0$.

such that

$$m_{jj_1j_2}(k, \pm k_0, k \mp k_0) = \mathcal{O}(|k|^{1/2}).$$

Hence in (7.3.4) the term M is not a small perturbation of the identity in contrast to the case of a semilinear B, where $b_{jj_1j_2}(k, \pm k_0, k \mp k_0) = \mathcal{O}(1)$ for $|k| \to \infty$. As a consequence, the normal form cannot be inverted with Neumann's series and secondly higher order terms are created on the right-hand side of the water wave problem that loose finally one derivative, which is half a derivative more than the linear part.

The loss of regularity only appears for the diagonal terms $m_{jj_1j}(k, \pm k_0, k \mp k_0)$. The associated one-dimensional subproblems can be inverted with the help of energy estimates. The "off-diagonal" terms do not loose regularity and so the remaining part can be inverted with the help of Neumann's series.

After the transformation the right-hand side of the Lagrangian formulation looses one derivative that allows us to solve the transformed system with the Cauchy-Kowalevskaya theorem (Figure 7.4).

Remark   Normal form transforms play a significant role in global existence results for dispersive wave systems [23], especially recently in long time and global existence results for the water wave problem [17]. The loss of regularity also appears there. However, in this approach spaces H^{s_2} and $W^{s_\infty,\infty}$ with $s_\infty < s_2$ occur. The transformation is only necessary in the estimates for the

$W^{s\infty,\infty}$ norm, where the nonlinear terms already loose regularity before the transformation and so the additional loss of regularity is no problem. The global existence and uniqueness result gives estimates for solutions in a small ball around 0 and not estimates for solutions in a small ball around Ψ_{NLS}. Hence, the statement in [18] is a nontrivial extension of [16]. Moreover, the L^1 norm of the NLS approximation is of order $\mathcal{O}(1)$ for $\varepsilon \to 0$ and dispersive L^1-L^∞ decay estimates do not gain powers of ε on the relevant time scale.

Remark   It is of no advantage to define the transformation $R \mapsto w$ explicitly through $w = R + \varepsilon \widetilde{M}(\Psi, R)$ with a symmetric bilinear mapping $\widetilde{M}$. In the end after the transformation there is a loss of one derivative for the transformed Lagrangian formulation of the water wave problem, too.

7.3.3 The Resonance at the Wave Numbers $k = 0$ and $k = k_0$

As already explained the denominator $\omega_j(k) - \omega_{j_1}(\pm k_0) - \omega_{j_2}(k \mp k_0)$ appearing in the normal form transform possesses two zeroes, namely at $k = 0$ and $k = k_0$ due to $\omega_j(0) = 0$. The resonance at the wave number $k = 0$ is trivial in the sense that the nonlinear terms vanish at this wave number, too. However, the resonance at the wave number $k = k_0$ is nontrivial. The problem can be overcome by a wave number dependent scaling of the error function R, namely

$$\varepsilon^\beta \vartheta(k)\hat{R} = u - \varepsilon \Psi,$$

where $\vartheta(k) = \min(\varepsilon + |k|/\delta, 1)$ with $\delta > 0$ sufficiently small, but independent of $0 < \varepsilon \ll 1$ (Figure 7.5).

By this scaling non-resonant terms of order $\mathcal{O}(1)$ occur. After their elimination a second normal form transform is necessary to eliminate the remaining non-resonant terms of order $\mathcal{O}(\varepsilon)$. The remaining resonant terms of order $\mathcal{O}(\varepsilon)$ are of long wave form similar to $\varepsilon \partial_x(B(\varepsilon x)R(x))$ and can be controlled by some energy estimates. This has been carried out for a Boussinesq equation as a model problem in [24]. These ideas of [24] are the basis of the analysis used in [19].

Other famous systems fall into this class. In [25] it has been explained that the proofs of the approximation theorems given for the partial differential

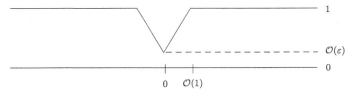

Figure 7.5. The weight function ϑ.

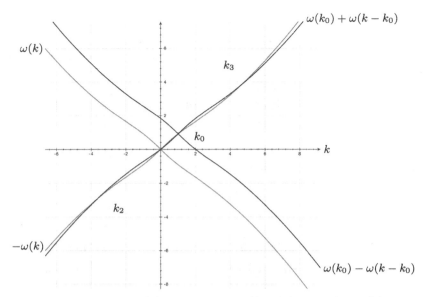

Figure 7.6. The curves of eigenvalues $k \mapsto \pm\omega(k)$ and the curves $k \mapsto \omega(k_0) \pm \omega(k - k_0)$ in case of finite depth and small surface tension. Intersection points correspond to wave numbers of quadratic resonances. Beside the two intersections points at $k = 0$ and $k = k_0$ there are two additional intersections points at k_2 and k_3 satisfying $k_3 - k_0 + k_2 = 0$.

equation (PDE) systems can be transferred almost line for line to the FPU system by looking at the Fourier transformed FPU system.

7.4 The Water Wave Problem with Surface Tension

Next we come to the water wave problem with surface tension. It is again of the form (7.3.1) and the error R satisfies again an equation of the form (7.3.3). However, the right-hand side now looses $3/2$ derivatives and the normal form transform looses one derivative such that the previous idea with the Cauchy-Kowalevskaya theorem does not apply. So far the justification of the NLS approximation of the water wave problem with surface tension in general is an open problem.

From the point of controlling the terms of order $\mathcal{O}(\varepsilon)$ the water wave problem with surface tension with $\sigma \in (0, 1/3)$ in case of finite depth $h = 1$ and $\sigma > 0$ in case of infinite depth is more interesting than the case $\sigma = 0$. The reason is the occurrence of additional quadratic resonances for these values of σ. The resonance structure for finite depth $h = 1$ and $\sigma > 1/3$ is the same as for $\sigma = 0$.

Even for semilinear toy problems with this resonance structure there are still many open problems. For the water wave problem with surface tension there is one rigorous negative result showing that the NLS approximation can fail to make correct predictions [26]. In the following we explain the basic ideas of this counter example.

7.4.1 The Counter Example

In case of small surface tension a counter example can be constructed that shows that the NLS equation can fail to approximate the water wave problem correctly. In [26] the water wave problem with small surface tension under spatially periodic boundary conditions has been considered. In this case the NLS equation degenerates into an ordinary differential equation (ODE).

The construction of the counter example uses the quadratic resonances and is based on the fact that it is possible to put the resonant wave numbers on integer multiples of a basic wave number k_*. The dynamics of the resonant modes can approximately be described by the associated three wave interaction (TWI) system. The TWI system possesses one unstable invariant subspace, which for the water wave problem is associated to the wave number with largest modulus.

The other two modes increase on an $\mathcal{O}(\varepsilon^{-1}|\ln(\varepsilon)|)$ time scale to the same size $\mathcal{O}(\varepsilon)$ as the mode associated to the largest wave number. If the basic wave number k_0 of the NLS approximation equals the unstable wave number k_3 a counter example can be constructed by proving an approximation theorem for the TWI approximation on an $\mathcal{O}(\varepsilon^{-1}|\ln(\varepsilon)|)$ time scale. Since the predictions of the TWI approximation differs from the predictions of the NLS approximation the NLS approximation cannot make correct predictions on an $\mathcal{O}(\varepsilon^{-2})$ time scale. In the following we provide some more details.

7.4.2 Analysis of the Resonances

The TWI approximation is given by the multiple scaling ansatz

$$\binom{\eta}{w} \approx \varepsilon \psi_{twi}(x,t) = \varepsilon A_1(\varepsilon t)e^{i(k_1 x - \omega_1 t)}\varphi_1 + \varepsilon A_2(\varepsilon t)e^{i(k_2 x - \omega_2 t)}\varphi_2$$

$$+\varepsilon A_3(\varepsilon t)e^{i(k_3 x - \omega_3 t)}\varphi_3 + c.c., \qquad (7.4.1)$$

with unit vectors $\varphi_j \in \mathbb{C}^2$, which depend only on k_j, ω_j and can be computed explicitly, with $0 < \varepsilon \ll 1$ a small perturbation parameter, where the the spatial and temporal wave numbers are related via the linear dispersion relation (7.2.2) of the water wave problem and satisfy the resonance condition

$$k_1 + k_2 + k_3 = 0, \qquad \omega_1 + \omega_2 + \omega_3 = 0, \qquad (7.4.2)$$

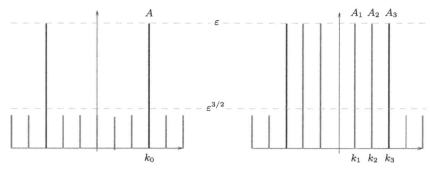

Figure 7.7. Left panel: qualitative form of the initial mode distribution. If the NLS approximation would be valid the solutions of the water wave problem should have the same qualitative mode distribution on a time interval of length $\mathcal{O}(1/\varepsilon^2)$. Only the NLS modes are of order $\mathcal{O}(\varepsilon)$. The other modes are much smaller. Right panel: Mode distribution at a time $t = \mathcal{O}(\varepsilon^{-1}|\ln(\varepsilon)|)$ predicted by the TWI approximation. Since the TWI approximation is valid according to the subsequent Theorem 7.4.1 the NLS approximation cannot make correct predictions.

The amplitudes A_j satisfy

$$
\begin{aligned}
\partial_T A_1 &= i\gamma_1 \overline{A_2 A_3}, \\
\partial_T A_2 &= i\gamma_2 \overline{A_1 A_3}, \\
\partial_T A_3 &= i\gamma_3 \overline{A_1 A_2},
\end{aligned}
\tag{7.4.3}
$$

with $A_j = A_j(T) \in \mathbb{C}$ and coefficients $\gamma_j \in \mathbb{C}$. This system possesses three invariant subspaces subspaces $M_1 = \{(A_1, 0, 0) : A_1 \in \mathbb{C}\}$, $M_2 = \{(0, A_2, 0) : A_2 \in \mathbb{C}\}$, and $M_3 = \{(0, 0, A_3) : A_3 \in \mathbb{C}\}$. The distance to the invariant subspace M_3 can be measured by $E_3 = \varrho_1 |A_1|^2 + \varrho_2 |A_2|^2$ (Figure 7.7). We find

$$
\frac{d}{dt} E_3 = -i(\varrho_1 \gamma_1 + \varrho_2 \gamma_2)(A_2 A_3 A_1 - \overline{A_2 A_3 A_1}).
$$

Hence we have $\frac{d}{dt} E_3 = 0$ for positive ρ_1, ρ_2 if $\gamma_1 \gamma_2 < 0$. For $\gamma_1 \gamma_2 > 0$ instability of M_3 occurs. The explicit formulas [26, Appendix 2] for the coefficients γ_j show the instability of the subspace associated to the wave number with largest modulus. Since we have conservation of energy in the water wave problem the energy surface of the associated TWI system is an ellipsoid. A sketch of the dynamics in the energy surface can be found in Figure 7.8.

7.4.3 The Extended TWI Approximation Result

An approximation result that the dynamics of the TWI system really occurs in the water wave problem we have to prove has been established in [27]. In case of an $\mathcal{O}(\varepsilon^{-1})$ time scale beside functional analytic difficulties a

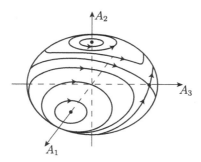

Figure 7.8. The phase portrait of the TWI system in the energy surface which is an ellipsoid since due to conservation of energy not all γ_j have the same sign. The axes are invariant subspaces associated to the wave numbers k_j. There are one unstable and two unstable subspaces.

TWI approximation result is easily be obtained with the help of Gronwall's inequality. Solutions of order $\mathcal{O}(\varepsilon)$ have to estimated on an $\mathcal{O}(\varepsilon^{-1})$ time scale.

In [26] the result of [27] has been improved in two directions. First, periodic boundary conditions have to be considered and secondly and more essential the approximation time has to be extended beyond the natural time scale $\mathcal{O}(\varepsilon^{-1})$ of the TWI approximation to a time scale $\mathcal{O}(\varepsilon^{-1}|\ln(\varepsilon)|)$. This is the time scale w.r.t. the original t variable, which is needed to come from a small $\mathcal{O}(\varepsilon^{1/2})$ neighborhood of the unstable fixed point to an $\mathcal{O}(1)$ distance to this unstable fixed point in the TWI system.

In [26] an extended approximation $\varepsilon \psi_{twi,ext}$ has been considered that is formally close to the original TWI approximation $\varepsilon \psi_{twi}$.

Theorem 7.4.1. *For the extended TWI approximation $\varepsilon \psi_{twi,ext}$ there exist $C_0, \varepsilon_0 > 0$ such that for all $\varepsilon \in (0, \varepsilon_0)$ there are solutions of the water wave problem (7.1.1)–(7.1.4) satisfying*

$$\sup_{t \in [0, \widetilde{2t}]} \left\| \begin{pmatrix} \eta \\ w \end{pmatrix} (x,t) - \varepsilon \psi_{twi,ext}(x,t) \right\|_{(C_b^{s-2})^2} \leq C_0 \varepsilon^{3/2},$$

with $\widetilde{t} = \widetilde{T}/\varepsilon$, where $\widetilde{T} := -\ln(2) - \frac{1}{2}\ln(\varepsilon)$.

Idea of the proof. An energy E which measures the magnitude of the error in the H^s norm satisfies an inequality of the form

$$\partial_t E \leq C_1 \varepsilon E + C_2(E) \varepsilon^\beta E^{3/2} + C_3 \varepsilon^{2\delta-1}$$

where the term $C_3 \varepsilon^{2\delta-1}$ comes from the residual, i.e., from the terms that do not cancel after inserting the extended approximation $\varepsilon \psi_{twi,ext}$ into the equations of the water wave problem. Since from the first term on the right-hand side we have a linear growth like $e^{C_1 \varepsilon t}$, an extension of the time scale can only be obtained by making the residual smaller. This is the reason for the occurrence of the extended TWI approximation. On the other hand we then need bounds for the extended TWI system beyond the natural time scale of the TWI system. This problem is solved in [26] by using conserved quantities of the classical TWI system (7.4.3).   □

Finally, the extended TWI system should show qualitatively the same behavior as the TWI system, although the solutions of the classical TWI system do not approximate the solutions of the extended TWI system on the extended time scale. For this problem estimates are derived that show that solutions leave a small neighborhood of the origin along the unstable manifold.

Remark There are essentially two consistent descriptions of oscillating wave packets in dispersive wave systems by amplitude equations, namely the NLS limit and the N wave interaction limit. The justification of these amplitude equations via approximation theorems goes very similarly. For some model equations as original dispersive wave systems this has been explained in [28]. If the spatially periodic boundary conditions in the previous counter example is given up the question remained unsolved if an approximation property holds or not for the NLS approximation. However, due to the different scaling, $X = \varepsilon^2 x$ instead of $X = \varepsilon(x - c_g t)$, the four wave interaction (FWI) approximation is expected to fail in general to approximate the water wave problem for $\sigma \in (0, 1/3)$ in case $h = 1$ and for $\sigma > 0$ in case of infinite depth, also in this case. The construction of this counter example for an amplitude equation making wrong predictions on the natural time scale of the approximation without imposing spatially periodic boundary conditions on the original system is the subject of ongoing research.

7.5 What can be Expected in General for Capillary-Gravity Waves

The previous counter example only works if the subspace in the TWI system associated to the basic wave number of the NLS approximation is unstable and if the involved resonances are integer multiples of a basic wave number k_*. So what happens in the other cases if the spatial periodicity is given up? There exist a number of approximation results for semilinear dispersive wave systems such as

$$\partial_t^2 u = \partial_x^2 u + \partial_x^2 (u^2) + \partial_t^2 \partial_x^2 u + \mu \partial_x^6 u, \qquad (7.5.1)$$

with a resonance structure similar to the one of the water wave problem with surface tension with $\sigma \in (0, 1/3)$ in case of finite depth $h = 1$ and $\sigma > 0$ in case of infinite depth.

In case that the subspace in the TWI system associated to the basic wave number of the NLS approximation is stable, which is called stable resonance in the following, an approximation result has been established in [24]. This idea will be presented in more detail in Section 7.5.1. The case of an unstable resonance remains open if the periodic boundary conditions are dropped. In this case the different group velocities at the resonant wave numbers have to be taken into account. It is expected that the approximation property holds for localized NLS solutions, and that it does not hold for non-localized NLS solutions. See the discussion in [31, Section 4.2.]. The second approach that we will present here is based on the idea that for analytic solutions of the NLS equation the resonant modes are initially so small that they cannot grow to a reasonable size before the relevant $\mathcal{O}(1/\varepsilon^2)$ time scale. This has been carried out in detail in [30].

7.5.1 Stable and Unstable Resonances

In [29, Theorem 3.8] a NLS approximation theorem has been shown in case that the subspace $\{A_2 = A_3 = 0\}$ associated to the wave number $k_1 = k_0$ is stable in the TWI system. The proof is based on a mixture of normal form transforms for the non-resonant wave numbers and energy estimates for the resonant wave numbers. In [24] the ideas of [31] and [29] are brought together to handle a Boussinesq equation such as (9.3.4).

Theorem 7.5.1. *Consider* (9.3.4) *with* $\mu > 0$. *Choose a wave number* $k_0 > 0$ *for which the NLS subspaces in the TWI systems associated to the resonances are stable. Moreover, let* $\theta_A - \theta \geq 8$ *and* $\theta \geq 1$. *Let* $A \in C([0, T_0], H^{\theta_A}(\mathbb{R}, \mathbb{C}))$ *be a solution of the NLS equation* (7.2.1). *Then there exist* $\varepsilon_0 > 0$ *and* $C > 0$ *such that for all* $\varepsilon \in (0, \varepsilon_0)$ *there are solutions* u *of* (9.3.4) *with*

$$\sup_{t \in [0, T_0/\varepsilon^2]} \|u(\cdot, t) - \varepsilon \psi_{\mathrm{NLS}}(\varepsilon, \cdot, t)\|_{H^\theta} \leq C\varepsilon^{3/2}.$$

The situation on the whole real line for an unstable resonance is still open. As already said in this case the different group velocities $\omega'(k_j)$ at the resonant wave numbers k_j no longer can be neglected. For a thorough discussion see [29, §4.2]. A recent attempt to understand this situation can be found in [32]. However, the concept of space-time resonances used in [32] would give weaker results as in [24] in case of stable resonances since stronger decay rates for $|x| \to \infty$ of the initial conditions have to be taken.

7.5.2 Resonances Bounded Away from Integer Multiples of k_0

The second approach that we will present can be found in [30]. It is based on the fact that for analytic solutions of the NLS equation the resonant modes are

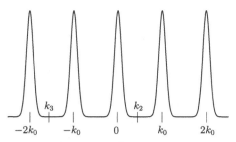

Figure 7.9.   The inverse of the weight $\vartheta^{\omega,a}$, which is proportional to the mode distribution of the NLS approximation in Fourier space. The resonant wave numbers k_2 and k_3 are bounded away from integer multiples of k_0.

initially so small that they cannot grow to a reasonable size before the relevant $\mathcal{O}(1/\varepsilon^2)$ time scale. The analyticity of the solutions of the NLS equation corresponds in Fourier space to an exponential decay of the Fourier modes for $|k| \to \infty$, i.e., in Fourier space solutions of the NLS equation are in a space

$$W^{\omega,a} = \{u : \mathbb{R} \to \mathbb{C} : \|u\|_{W^{\omega,a}} = \|\widehat{u}\rho^{\omega,a}\|_{L^1} < \infty\}$$

where $\rho^{\omega,a}(k) = e^{a|k|}$. If we assume for the solutions of the NLS equation that $A(T) \in W^{\omega,a}$ for fixed T then the solutions of the original systems will have a Fourier mode distribution that is bounded from above by

$$1/\vartheta^{\omega,a}(k) = \sup_{m \in \mathbb{Z}} |e^{-\frac{a}{\varepsilon}|k - mk_0|}|,$$

i.e., at integer multiples of the basic wave number k_0 there are small peaks of width of order $\mathcal{O}(\varepsilon)$. See Figure 7.9. As a consequence the modes associated to the resonant wave numbers are exponentially small w.r.t. ε initially if the set of resonant wave numbers and integer multiples of the basic wave number k_0 have a positive distance, i.e., these modes are initially of order $\mathcal{O}(e^{-r_0/\varepsilon})$ for a $r_0 > 0$ independent of $0 < \varepsilon \ll 1$. Due to the resonance these modes will grow with some rate $\mathcal{O}(e^{r_1 \varepsilon t})$ for a $r_1 > 0$ independent of $0 < \varepsilon \ll 1$. Hence, these modes are less than $\mathcal{O}(\varepsilon^2)$ for $t \in [0, \frac{r_0}{2r_1}\varepsilon^{-2}]$.

The idea can be transferred into a proof by introducing spaces

$$\mathcal{M}^{\omega,a} = \{u : \mathbb{R} \to \mathbb{C} : \|u\|_{\mathcal{M}^{\omega,a}} = \|\widehat{u}\vartheta^{\omega,a}\|_{L^1} < \infty\}$$

and by measuring the error w.r.t. the $\mathcal{M}^{\omega,a}$ norm, but with α depending on time. We choose

$$\alpha(t) = \beta_0/\varepsilon - \beta_1 \varepsilon t,$$

with constants β_0 and β_1, which can be chosen independently of $0 < \varepsilon \ll 1$. W.r.t. this time dependent norm the resonant modes are damped with some exponential rate.

Theorem 7.5.2. *Assume that the resonant wave numbers are bounded away from integer multiples of the basic wave number k_0. Let $A_1 \in C([0, T_0], W^{\omega, 2\alpha_0})$ be a solution of the NLS equation (7.2.1) for an $\alpha_0 > 0$. Then there exist $\varepsilon_0 > 0$, $T_1 \in (0, T_0]$, and a $C > 0$ such that for all $\varepsilon \in (0, \varepsilon_0)$ we have solutions u of (9.3.4), which satisfy*

$$\sup_{t \in [0, T_1/\varepsilon^2]} \|u(\cdot, t) - \varepsilon \psi_{NLS}(\cdot, t)\|_{\mathcal{M}^{\omega, \alpha(t)}} \leq C\varepsilon^2,$$

where $\alpha(t) = \alpha_0/\varepsilon - \alpha_0 \varepsilon t/T_1$ and as consequence

$$\sup_{t \in [0, T_1/\varepsilon^2]} \sup_{x \in \mathbb{R}} |u(x, t) - \varepsilon \psi_{NLS}(x, t)| \leq C\varepsilon^2.$$

Remark The approximation result is not optimal in the sense that error estimates can only be proved on the correct time scale, namely for $t \in [0, T_1/\varepsilon^2]$, but not necessarily for all $t \in [0, T_0/\varepsilon^2]$. Hence, we can only guarantee that parts of the NLS dynamics can be seen in the original system.

7.6 Modulations of Periodic Wave Trains

Finally, we comment on the validity of amplitude equations, such as the KdV equation, Whitham's system, and the NLS equation for the description of small modulations in time and space of periodic wave trains. In [33] for big parameter regions short wave instabilities for periodic traveling surface water waves have been found. For periodic waves of order $\mathcal{O}(1)$ these instabilities are of order $\mathcal{O}(1)$ for $\varepsilon \to 0$, where $0 < \varepsilon \ll 1$ is the small perturbation parameter that occurs in the derivation of the amplitude equations. This allows to prove that the NLS, KdV, and Whitham approximation in general make wrong predictions in this situation. The error will grow linearly as $\mathcal{O}(e^{rt})$ for a $r > 0$, which for solutions in Sobolev spaces cannot be kept small on the time scales under consideration, namely $\mathcal{O}(1/\varepsilon^2)$ for NLS, $\mathcal{O}(1/\varepsilon^3)$ for KdV, and $\mathcal{O}(1/\varepsilon)$ for Whitham.

Due to the spatial scaling, according to the Paley-Wiener theorem, for analytic initial conditions the unstable modes can be initially of order $\mathcal{O}(e^{-1/\varepsilon})$. Hence, for the NLS and KdV approximation even for analytic initial conditions the non-validity can be expected. For the Whitham approximation the initial size $\mathcal{O}(e^{-1/\varepsilon})$ of the unstable modes permits the previous construction of a counter example via linear unstable modes. Similar situations as for the Whitham approximation appear for the resonant TWI and FWI approximation in case of analytic initial conditions. These last considerations have to be taken into account in a precise formulation of possible counter examples.

Acknowledgment

I am grateful for the stimulating atmosphere at the Newton Institute in Cambridge during the visit of the programme "Theory of water waves" in summer 2014 and to Katie Oliveras and Bernard Deconinck for providing information about the linear stability and instability of periodic wave trains.

References

[1] V. E. Zakharov. Stability of periodic waves of finite amplitude on the surface of a deep fluid. *Sov. Phys. J. Appl. Mech. Tech. Phys*, 4:190–194, 1968.
[2] M. Shinbrot. The initial value problem for surface waves under gravity. I. The simplest case. *Indiana Univ. Math. J.*, 25(3):281–300, 1976.
[3] T. Kano and T. Nishida. Sur les ondes de surface de l'eau avec une justification mathématique des équations des ondes en eau peu profonde. *J. Math. Kyoto Univ.*, 19(2):335–370, 1979.
[4] D. Lannes. Well-posedness of the water-waves equations. *J. Amer. Math. Soc.*, 18(3):605–654 (electronic), 2005.
[5] V. I. Nalimov. The Cauchy-Poisson problem. *Dinamika Splošn. Sredy*, 254 (Vyp. 18 Dinamika Zidkost. so Svobod. Granicami):104–210, 1974.
[6] H. Yosihara. Gravity waves on the free surface of an incompressible perfect fluid of finite despth. *Publ. Res. Inst. Math. Sci.*, 18:49–96, 1982.
[7] H. Yosihara. Capillary-gravity waves for an incompressible ideal fluid. *J. Math. Kyoto Univ.*, 23:649–694, 1983.
[8] W. Craig. An existence theory for water waves and the Boussinesq and Korteweg-de Vries scaling limits. *Comm. Partial Differential Equations*, 10(8):787–1003, 1985.
[9] S. Wu. Well-posedness in Sobolev spaces of the full water wave problem in 2-D. *Invent. Math.*, 130(1):39–72, 1997.
[10] S. Wu. Well-posedness in Sobolev spaces of the full water wave problem in 3-D. *J. Amer. Math. Soc.*, 12(2):445–495, 1999.
[11] G. Schneider and C. E. Wayne. The long wave limit for the water wave problem I. The case of zero surface tension. *Comm. Pure. Appl. Math.*, 53(12):1475–1535, 2000.
[12] T. Iguchi. Well-posedness of the initial value problem for capillary-gravity waves. *Funkcial. Ekvac.*, 44(2):219–241, 2001.
[13] G. Schneider and C. E. Wayne. The rigorous approximation of long-wavelength capillary-gravity waves. *Arch. Rat. Mech. Anal.*, 162:247–285, 2002.
[14] D. M. Ambrose and N. Masmoudi. The zero surface tension limit of two-dimensional water waves. *Comm. Pure Appl. Math.*, 58(10):1287–1315, 2005.
[15] S. Wu. Almost global wellposedness of the 2-D full water wave problem. *Invent. Math.*, 177(1):45–135, 2009.
[16] S. Wu. Global wellposedness of the 3-D full water wave problem. *Invent. Math.*, 184(1):125–220, 2011.
[17] P. Germain, N. Masmoudi, and J. Shatah. Global solutions for the gravity water waves equation in dimension 3. *Ann. of Math. (2)*, 175(2):691–754, 2012.

[18] N. Totz and S. Wu. A rigorous justification of the modulation approximation to the 2D full water wave problem. *Commun. Math. Phys.*, 310(3):817–883, 2012.

[19] W.-P. Düll, G. Schneider, and C. E. Wayne. Justification of the Nonlinear Schrödinger equation for the evolution of gravity driven 2D surface water waves in a canal of finite depth. *Arch. Rat. Mech. Anal.*, published online 2015.

[20] W. Craig, C. Sulem, and P.-L. Sulem. Nonlinear modulation of gravity waves: a rigorous approach. *Nonlinearity*, 5(2):497–522, 1992.

[21] P. Kirrmann, G. Schneider, and A. Mielke. The validity of modulation equations for extended systems with cubic nonlinearities. *Proc. Roy. Soc. Edinburgh Sect. A*, 122(1-2):85–91, 1992.

[22] L. A. Kalyakin. Asymptotic decay of a one-dimensional wavepacket in a nonlinear dispersive medium. *Math. USSR Sbornik*, 60(2):457–483, 1988.

[23] J. Shatah. Normal forms and quadratic nonlinear Klein-Gordon equations. *Comm. Pure Appl. Math.*, 38(5):685–696, 1985.

[24] W.-P. Düll and G. Schneider. Justification of the nonlinear Schrödinger equation for a resonant Boussinesq model. *Indiana Univ. Math. J.*, 55(6):1813–1834, 2006.

[25] G. Schneider. Bounds for the nonlinear Schrödinger approximation of the Fermi-Pasta-Ulam system. *Appl. Anal.*, 89(9):1523–1539, 2010.

[26] G. Schneider, D. A. Sunny, and D. Zimmermann. The NLS approximation makes wrong predictions for the water wave problem in case of small surface tension and spatially periodic boundary conditions. *J. Dyn. Diff. Eq.*, published online 2014.

[27] G. Schneider and C. E. Wayne. Estimates for the three-wave interaction of surface water waves. *European J. Appl. Math.*, 14(5):547–570, 2003.

[28] G. Schneider and O. Zink. Justification of the equations for the resonant four wave interaction. In *EQUADIFF 2003*, pages 213–218. World Sci. Publ., Hackensack, NJ, 2005.

[29] G. Schneider. Justification and failure of the nonlinear Schrödinger equation in case of non-trivial quadratic resonances. *J. Diff. Eq.*, 216(2):354–386, 2005.

[30] W.-P. Düll, A. Hermann, G. Schneider, and D. Zimmermann. Justification of the 2D NLS equation for a fourth order nonlinear wave equation – quadratic resonances do not matter much in case of analytic initial conditions -. *Preprint*, Universität Stuttgart:19p., 2014.

[31] G. Schneider. Approximation of the Korteweg-de Vries equation by the nonlinear Schrödinger equation. *J. Diff. Eq.*, 147:333–354, 1998.

[32] N. Masmoudi and K. Nakanishi. Multifrequency NLS scaling for a model equation of gravity-capillary waves. *Commun. Pure Appl. Math.*, 66(8):1202–1240, 2013.

[33] B. Deconinck and K. Oliveras. The instability of periodic surface gravity waves. *J. Fluid Mech.*, 675:141–167, 2011.

8

Vortex Sheet Formulations and Initial Value Problems: Analysis and Computing

David M. Ambrose

Abstract

We place the irrotational water wave problem in the larger context of vortex sheets. We describe the evolution equations for vortex sheets in 2D or in 3D. The numerical method of Hou, Lowengrub, and Shelley (HLS) for the solution of the initial value problem for the vortex sheet with surface tension in 2D is discussed; furthermore, we indicate how the HLS formulation of the problem is useful for a proof of well-posedness. We then show how one may take the zero surface tension limit in the water wave case. We close with a brief discussion of the extension of the HLS ideas to 3D, for both analysis and computing.

8.1 Introduction

The irrotational water wave is a special case of the irrotational vortex sheet. For the vortex sheet problem, two fluids whose motions are described by the incompressible, irrotational Euler equations meet at an interface. This interface, the vortex sheet, is free to move, and moves according to the velocities of the two fluids restricted to the interface. Each fluid has its own non-negative, constant density. Different geometries are possible, but to be definite, at present we consider the case in which the fluids are two-dimensional and such that each fluid region has one component, which is of infinite vertical extent and horizontally periodic. Thus, we may say that we have an upper fluid and a lower fluid. In the water wave case, the density of the upper fluid is equal to zero.

Without surface tension, if each of the two fluids has positive density, then the vortex sheet is known to have an ill-posed initial value problem; this has been demonstrated by several authors. We note that when discussing ill-posedness of a problem, to be precise, one should mention the function spaces under consideration; for example, Caflisch and Orellana have shown that the vortex sheet initial value problem is ill-posed in Sobolev spaces [1]. In

140

analytic function spaces, however, solutions of the vortex sheet problem have been shown to exist by a Cauchy-Kowalewski argument [2].

The ill-posedness of the vortex sheet initial value problem (when the two fluids have positive densities) is caused by the presence of the Kelvin-Helmholtz instability. Another viewpoint on the same issue is that ill-posedness of the intial value problem for the unregularized vortex sheet occurs because the evolution equations are elliptic; by contrast, the water wave problem and the vortex sheet with surface tension are dispersive. Instead of having well-posed initial value problems, elliptic equations tend to possess solutions to boundary value problems. This idea was explored by Duchon and Robert [3], who considered the space-time domain $\mathbb{R} \times [0, \infty)$. They showed that on this domain, if one specifies a sufficiently small position of the vortex sheet at time $t = 0$, and specifies that the solution decays to zero as $t \to \infty$, then the vortex sheet problem has a solution. Furthermore, the Duchon-Robert solutions at $t = 0$ only are of finite smoothness, but at positive times become analytic. Using the reversibility of time in the Euler equations, this can be shown to imply ill-posedness of the vortex sheet problem: small solutions possess infinitely many derivatives at an initial time, and a short time later, they possess only finitely many derivatives, so blow-up of some derivative has occured. If one works out the details, one sees that this gives discontinuous dependence of solutions on the initial conditions. The Duchon-Robert ideas have been further explored by the author and collaborators in [4, 5] for the vortex sheet and for equations of Boussinesq type, and by Beck, Sosoe, and Wong for the Rayleigh-Taylor problem [6] (Duchon and Robert treated the vortex sheet problem in the density-matched case, but allowing the upper density to be larger than the lower density, one arrives at the Rayleigh-Taylor problem). Other treatments of the unregularized vortex sheet problem are by Wu, by Lebeau, and by Kamotski and Lebeau [7–9].

While the vortex sheet initial value problem is ill-posed, the Kelvin-Helmholtz instability vanishes in the water wave case. Thus one may expect to find well-posedness, and indeed, that is the case. Some early existence results for gravity water waves are by Nalimov [10], Yosihara [11], and Craig [12]. These are short-time, small data existence results. A breakthrough was made by Wu in the papers [13, 14], which were the first robust well-posedness theorems for the irrotational water wave problem: these are still short-time theorems, but a wide class of initial data was used. After Wu, there were several other versions of short-time existence theory given in 2D or 3D, such as [15–20].

For the water wave with surface tension, some existence results are [21, 22], and [23]. The water wave case is well-posed with or without surface tension, and this reflects the fact that without surface tension, there is no Kelvin-Helmholtz instability to overcome. All of these three results considered the case in which the free surface is given by a graph (rather than using a more general parameterized surface); they all give existence of solutions for a short time, and the result of [23] is also in the case of small initial data.

In the presence of surface tension, the Kelvin-Helmholtz instability is regularized [24]. That is, the Kelvin-Helmholtz instability can be characterized by the unbounded rate of growth of disturbances; for a perturbation from equilibrium with wavenumber ξ, the disturbance grows with exponential growth rate $|\xi|$. In the presence of surface tension, there is only growth for small values of ξ, so there exists a finite maximum growth rate. One then expects well-posedness in this case, and again, this is what has been found. Iguchi, Tanaka, and Tani studied interfacial capillary-gravity waves, proving existence of solutions for a short time, for sufficiently small initial data [25]. The author's work [26] gave a well-posedness proof for vortex sheets with surface tension for a general class of initial data (similarly to Wu [13], there was no restriction on the size of the data and the free surface is given by a parameterized curve, so the interface may have multi-valued height). Specializing to the water wave case, the result of [26] gives well-posedness for the initial value problem for water waves with surface tension in 2D. In 3D, some related works are [27] and [28].

Also in the presence of surface tension, a breakthrough was made for the computation of solutions of initial value problems by Hou, Lowengrub, and Shelley [29, 30]. For a fully explicit timestepping scheme, an interfacial Euler flow with surface tension would face a $3/2$-order timestepping constraint. In the related problem of interfacial Darcy flow (in which the fluid velocity follows Darcy's Law, giving a velocity field, which is proportional to the pressure gradient), the presence of surface tension gives a third-order stiffness constraint. The works [29, 30] give a reformualtion of the problem in terms of geometrically inspired variables, and use this as the basis of a non-stiff numerical method. The analysis performed by the author in [26] follows this reformulation, and this forms the basis for the present exposition. After [26], and the related works [31] and [16] by the author and Masmoudi, other authors have used the HLS formulation for analysis of problems in 2D free-surface fluid dynamics [18, 32–37]. As will be discussed in Section 8.7, these ideas were then generalized to analysis for 3D free-surface fluid dynamics by the author and Masmoudi [17, 27, 38], and then by other authors [39, 40]. The ideas of the Hou, Lowengrub, and Shelley work have been extended to a numerical method for 3D gravity water waves [41], to 3D capillary flows in the axisymmetric case [42], and to 3D capillary flows in the case of horizontally doubly periodic boundary conditions [43, 44].

There are many other developments for the solvability of initial value problems for free-surface fluid dynamics that we will not have the opportunity to discuss. This includes existence theory in the absence of the irrotationality assumption, such as in the works by Cheng, Coutand, and Shkoller [45, 46], Lindblad [47], Ogawa and Tani [48], Schweizer [49], Shatah and Zeng [50], or Zhang and Zhang [51, 52]. Also, there are now results about the existence of waves for all time, assuming the initial data is sufficiently small, such as those by Alazard and Delort [53, 54], Germain, Masmoudi, and Shatah [55–57], Hunter, Ifrim, and Tataru [58–60], Ionescu and Pusateri [61], and Wu [62, 63]. Other results of interest include singularity formation (or the absence of singularities of certain forms) [64–69], gain of regularity in the presence of surface tension [32, 70–72], and efforts to lower the regularity requirements for existence theory [15, 73].

The following is the plan for the remainder of this article: in Section 8.2, we describe the equations of motion for the 2D irrotational vortex sheet. In Section 8.3 we will describe the Hou, Lowengrub, and Shelley (HLS) formulation of the problem, and in Section 8.4 we describe the HLS numerical method. In Section 8.5 we describe the well-posedness proof for vortex sheets with surface tension, giving a fairly complete argument for well-posedness of a simpler model system. In Section 8.6, the zero surface tension limit of 2D irrotational water waves is discussed. Section 8.7 discusses extensions to three spatial dimensions.

8.2 The Equations of Motion

Before we begin describing the equations of motion for the irrotational vortex sheet, we mention that two good, general references are the books of Saffman [74] and Majda and Bertozzi [75]. We describe now the situation in 2D; a good description of the problem in 3D can be found, for instance, in [76].

We study two two-dimensional, incompressible, irrotational fluids with velocity and pressure given by the Euler equations:

$$\rho_i(\mathbf{v}_{i,t} + \mathbf{v}_i \cdot \nabla \mathbf{v}_i) = -\nabla p_i,$$

$$\text{div}(\mathbf{v}_i) = 0,$$

$$\mathbf{v}_i = \nabla \phi_i.$$

These equations hold for $\vec{x} \in \Omega_i(t)$, where $\Omega_i(t)$ is the region occupied by fluid i at time t. Of course, ρ_i is the density of fluid i, $\mathbf{v}_i$ is the velocity of fluid i, p_i is the pressure for fluid i, and ϕ_i is the velocity potential associated to $\mathbf{v}_i$. We have $\partial \Omega_1(t) = \partial \Omega_2(t)$ for all t, and this boundary, which we take to have a single component, is the vortex sheet. That is, since the fluids are irrotational, if we compute the curl of the velocity, we would get zero in each fluid region. However, the velocity can jump across the interface (we have not yet discussed the boundary conditions), and computing the curl of the velocity in the entire plane thus requires taking derivatives of functions that have jumps. The vorticity is therefore measure-valued and supported on the interface (it is an amplitude times the Dirac mass supported on the interface).

Combining the equation $\mathbf{v}_i = \nabla \phi_i$ with the equation $\text{div}(\mathbf{v}_i) = 0$, we see that the velocity potentials are harmonic. If $\hat{\mathbf{n}}$ is the normal vector to the interface at a point, then we would have the normal velocity at that point given by either $\nabla \phi_1 \cdot \hat{\mathbf{n}}$ or $\nabla \phi_2 \cdot \hat{\mathbf{n}}$. Physical considerations dictate that these be the same; if the normal velocities on the two side of the interface were to be unequal, then either the fluid regions could penetrate each other, or the fluid regions could pull apart and separate from each other; either way, this violates incompressibility. Thus, we must have continuity of the normal derivative across the interface; this is the first of the boundary conditions that should be mentioned. Since the potentials are harmonic and the normal derivative is continuous across the interface, this suggests using a double-layer potential representation; let $(x(\alpha, t), y(\alpha, t))$ be the location of the interface at time t, with the parameter $\alpha \in (-\infty, \infty)$. Letting $\vec{x}$ be in fluid region i at time t, we have

$$\phi_i(\vec{x}, t) = \frac{1}{2\pi} \int_{-\infty}^{\infty} \mu(\alpha', t) \frac{\vec{x} - (x(\alpha', t), y(\alpha', t))}{|\vec{x} - (x(\alpha, t), y(\alpha', t))|^2} \cdot \hat{\mathbf{n}}(\alpha', t) \, ds_{\alpha'}, \qquad (8.2.1)$$

where $ds_{\alpha'}$ indicates that this is an integral with respect to arclength along the interface.

To find the fluid velocity, one takes the gradient of (8.2.1). One can also then take the limit as $\vec{x}$ approaches the interface, from either fluid region. In doing this, one applies the Plemelj formulas [77]. The relevant integral that arises is the Birkhoff-Rott integral, $\mathbf{W} = (W_1, W_2)$, which can be written as

$$\mathbf{W}(\alpha, t) = \frac{1}{2\pi} \text{PV} \int_{-\infty}^{\infty} \gamma(\alpha, t) \frac{(-(y(\alpha, t) - y(\alpha', t)), x(\alpha, t) - x(\alpha', t))}{(x(\alpha, t) - x(\alpha', t))^2 + (y(\alpha, t) - y(\alpha', t))^2} \, d\alpha'.$$
$$(8.2.2)$$

Here, the quantity γ is known as the vortex sheet strength, and is defined by $\gamma = \partial_\alpha \mu$. The Plemelj formulas imply the jump condition

$$[\![\mathbf{v}]\!] \cdot \hat{\mathbf{t}} = \frac{\gamma}{s_\alpha}, \qquad (8.2.3)$$

where $[\![\cdot]\!]$ denotes the jump across the interface, $\hat{\mathbf{t}}$ is a unit tangent vector and s_α is the arclength element. (Thus, the quantity γ / s_α is sometimes known as the true vortex sheet strength.) The relevance of the Birkhoff-Rott integral for the fluid velocity is discussed in more detail in Sections 8.3 and 8.5.3.

In the case that the interface is spatially periodic, the Birkhoff-Rott integral (8.2.2) may be rewritten as an integral over one period. Say that the interface is 2π-periodic, so that $x(\alpha + 2\pi, t) = x(\alpha, t) + 2\pi$ and $y(\alpha + 2\pi, t) = y(\alpha, t)$. Then, if we write the interface with complex notation, $z(\alpha,t) = x(\alpha,t) + iy(\alpha,t)$, using an identity of Mittag-Leffler [78], (8.2.2) becomes

$$(W_1 - iW_2)(\alpha,t) = \frac{1}{4\pi i} \text{PV} \int_0^{2\pi} \gamma(\alpha',t) \cot\left(\frac{1}{2}(z(\alpha,t) - z(\alpha',t))\right) d\alpha'.$$

(8.2.4)

To complete the specification of the equations of motion, it remains to give the evolution equation for γ. To find the evolution equation for γ, one starts by finding an evolution equation for μ and differentiating with respect to α. To find the evolution equation for μ, one starts with the Bernoulli equations satisfied by each potential, ϕ_i, in each fluid region. The Bernoulli equations can be studied up to the boundary, and the jump taken across the boundary, yielding the equation for μ_t. In doing this, the jump condition for the pressure becomes important. The jump condition for the pressure is the Laplace-Young condition, that the jump in pressure is proportional to the curvature, κ, of the interface. For full details, see, for instance, the appendix of [16]. We will give the γ_t equation in Section 8.3.

8.3 The HLS Formulation

Given the curve $(x(\alpha,t),y(\alpha,t))$, we may write a frame of unit normal and tangent vectors as

$$\hat{\mathbf{t}} = \frac{(x_\alpha, y_\alpha)}{|(x_\alpha, y_\alpha)|}, \qquad \hat{\mathbf{n}} = \frac{(-y_\alpha, x_\alpha)}{|(x_\alpha, y_\alpha)|}.$$

We then decompose the velocity of the curve into its normal and tangential components:

$$(x,y)_t = U\hat{\mathbf{n}} + V\hat{\mathbf{t}}. \tag{8.3.1}$$

A central idea of [29, 30] is to describe the motion of the curve (x,y) by using geometric dependent variables rather than the Cartesian coordinates. To this end, we define the tangent angle the curve forms with the horizontal, θ, and the arclength element, s_α, to be

$$\theta = \tan^{-1}\left(\frac{y_\alpha}{x_\alpha}\right), \qquad s_\alpha = \sqrt{x_\alpha^2 + y_\alpha^2}. \tag{8.3.2}$$

The main motivation for studying θ and s_α rather than (x,y) is that the pressure jump is given by curvature, and we have the relationship $\kappa = \frac{\theta_\alpha}{s_\alpha}$, whereas κ is much more nonlinear in terms of the Cartesian description of the surface.

We are able to infer evolution equations for θ and s_α from (8.3.1) and (8.3.2). To do so, we will need a little geometry of curves; specifically, we notice that we can write $\hat{\mathbf{t}} = (\cos(\theta), \sin(\theta))$ and $\hat{\mathbf{n}} = (-\sin(\theta), \cos(\theta))$. We can then see the following:

$$\hat{\mathbf{t}}_\alpha = \theta_\alpha \hat{\mathbf{n}}, \qquad \hat{\mathbf{n}}_\alpha = -\theta_\alpha \hat{\mathbf{t}}. \tag{8.3.3}$$

Differentiating (8.3.1) with respect to α, we make use of (8.3.3), finding the following:

$$(x,y)_{\alpha t} = (U_\alpha + V\theta_\alpha)\hat{\mathbf{n}} + (V_\alpha - \theta_\alpha U)\hat{\mathbf{t}}.$$

Using the definition of θ in (8.3.2), we then differentiate θ with respect to t:

$$\theta_t = \frac{1}{1+(y_\alpha/x_\alpha)^2} \cdot \frac{x_\alpha y_{\alpha t} - y_\alpha x_{\alpha t}}{x_\alpha^2} = \frac{1}{s_\alpha}(x,y)_{\alpha t} \cdot \hat{\mathbf{n}} = \frac{U_\alpha + V\theta_\alpha}{s_\alpha}. \qquad (8.3.4)$$

Similarly, we differentiate s_α with respect to t:

$$s_{\alpha t} = \frac{1}{2\sqrt{x_\alpha^2 + y_\alpha^2}} \cdot (2x_\alpha x_{\alpha t} + 2y_\alpha y_{\alpha t}) = (x,y)_{\alpha t} \cdot \hat{\mathbf{t}} = V_\alpha - \theta_\alpha U. \qquad (8.3.5)$$

In the spatially periodic case, we define the length of one period of the interface to be $L = L(t) = \int_0^{2\pi} s_\alpha(\alpha,t)\, d\alpha$. Taking the derivative of this with respect to time, and using (8.3.5) and periodicity, we see that

$$L_t = -\int_0^{2\pi} \theta_\alpha(\alpha,t)U(\alpha,t)\, d\alpha.$$

The HLS formulation includes the choice of a preferred parameterization, and the most convenient parameterization is a normalized arclength parameterization, $s_\alpha = L/2\pi$. In this case, differentiating with respect to t, we find $s_{\alpha t} = L_t/2\pi$. Using this equation together with (8.3.5), and solving for V_α, we find

$$V_\alpha = -\theta_\alpha U + \frac{1}{2\pi}\int_0^{2\pi} \theta_\alpha U\, d\alpha. \qquad (8.3.6)$$

Thus, in the HLS formulation, the tangential velocity of the interface is chosen to satisfy (8.3.6), to enforce a preferred parameterization. Unlike the tangential velocity, the normal velocity must instead be chosen according to the underlying physics, which are as described earlier in Section 8.2. This means that the normal velocity U must be the normal component of the Birkhoff-Rott integral, $U = \mathbf{W} \cdot \hat{\mathbf{n}}$, with $\mathbf{W} = (W_1, W_2)$ specified by (8.2.4).

With these choices made, the evolution equation for γ becomes

$$\gamma_t = \frac{2\pi\tau}{L}\theta_{\alpha\alpha} + \frac{2\pi(V - \mathbf{W}\cdot\hat{\mathbf{t}})\gamma}{L}$$

$$- 2\mathrm{At}\left(\frac{L}{2\pi}\mathbf{W}_t \cdot \hat{\mathbf{t}} + \frac{\pi^2}{L^2}\gamma\,\gamma_\alpha - (V - \mathbf{W}\cdot\hat{\mathbf{t}})\mathbf{W}_\alpha \cdot \hat{\mathbf{t}} - gy_\alpha\right). \qquad (8.3.7)$$

Here, τ is the non-negative, constant coefficent of surface tension, At is the Atwood number, given by $\mathrm{At} = (\rho_1 - \rho_2)/(\rho_1 + \rho_2)$, and g is the constant acceleration due to gravity.

8.3.1 The Small-Scale Decomposition

Another essential idea in the HLS formulation is that the evolution equations (8.3.4), (8.3.7) can best be understood by extracting the leading-order terms from the right-hand sides. This requires understanding the leading-order behavior of the Birkhoff-Rott integral. To this end, we add and subtract in (8.2.4):

$$W_1 - iW_2 = \frac{1}{4\pi i} \text{PV} \int_0^{2\pi} \gamma(\alpha') \frac{1}{z_\alpha(\alpha')} \cot\left(\frac{1}{2}(\alpha - \alpha')\right) d\alpha'$$

$$+ \frac{1}{4\pi i} \int_0^{2\pi} \gamma(\alpha') \left[\cot\left(\frac{1}{2}(z(\alpha) - z(\alpha'))\right) - \frac{1}{z_\alpha(\alpha')} \cot\left(\frac{1}{2}(\alpha - \alpha')\right) \right] d\alpha'.$$

The integral operator with kernel $\frac{1}{2\pi} \cot\left(\frac{1}{2}(\alpha - \alpha')\right)$ is the (periodic) Hilbert transform; this is a very classical singular integral operator, which we denote by H. Details of the Hilbert transform can be found, for instance, in [79]. If $\mathcal{F}$ denotes the (periodic) Fourier transform, then the symbol of H is given by $\mathcal{F}Hf(k) = -i\text{sgn}(k)$. One useful property that can be seen from the symbol is that, if the mean of f is equal to zero, then $H^2(f) = -f$. If we define the operator $\mathcal{K}$ to be

$$\mathcal{K}[z]f(\alpha) = \frac{1}{4\pi i} \int_0^{2\pi} f(\alpha') \left[\cot\left(\frac{1}{2}(z(\alpha) - z(\alpha'))\right) \right.$$

$$\left. - \frac{1}{z_\alpha(\alpha')} \cot\left(\frac{1}{2}(\alpha - \alpha')\right) \right] d\alpha',$$

then we can write

$$W_1 - iW_2 = \frac{1}{2i} H\left(\frac{\gamma}{z_\alpha}\right) + \mathcal{K}[z]\gamma.$$

One can show that the operator $\mathcal{K}$ is a smoothing operator [26, 80]. Converting back to real notation, this implies that the leading-order part of **W** is proportional to $H(\gamma \hat{\mathbf{n}})$. With these considerations in mind, Hou, Lowengrub, and Shelley were able to introduce their small-scale decomposition (SSD),

$$\theta_t = \frac{2\pi}{L^2}^2 H(\gamma_\alpha) + P, \tag{8.3.8}$$

$$\gamma_t = \frac{2\pi \tau}{L} \theta_{\alpha\alpha} + Q, \tag{8.3.9}$$

where P and Q are collections of smooth terms that are of lower order. For their aims, namely developing a non-stiff numerical method, the SSD works very well. Analytically, however, the author found in [26] that a modification of (8.3.9) was more useful, namely

$$\gamma_t = \frac{2\pi \tau}{L} \theta_{\alpha\alpha} + \frac{2\pi^2 \gamma}{L^2}^2 H(\theta_\alpha) + \tilde{Q}, \tag{8.3.10}$$

where again $\tilde{Q}$ is a collection of lower-order terms.

8.4 The HLS Numerical Method

Starting from the formulation of Section 8.3, Hou, Lowengrub, and Shelley developed a non-stiff numerical method for the solution of the corresponding initial value problems. For a fully explicit timestepping scheme, one would expect a stiffness constraint of order $3/2$ (so that the timestep, Δt, must be taken to be proportional to $(\Delta \alpha)^{3/2}$, for any spatial step of size $\Delta \alpha$) for the vortex sheet with surface tension. In fact, the method also applies for stiffer problems (interfacial Darcy flow with surface tension), in which a stiffness constraint of order 3 would be present in a fully explicit method. For any of these cases, the HLS method reduces the stiffness constraint to first order, so that significantly larger timesteps may be taken as compared to explicit methods, and thus the long-time behavior of the problems can be studied.

The HLS method is a pseudospectral method, in which certain operations are carried out after using the Fast Fourier Transform (FFT), and other operations are instead carried out using the original variables. Two operations that appear many times in the problem are spatial derivatives, ∂_α, and the Hilbert transform, H. These operators are very well-suited to being carried out after taking the FFT, since they are multipliers in Fourier space.

Given the discrete values γ_j, z_j, for the vortex sheet strength and interface location at grid points $j \in \{0, 1, \dots, N\}$, the Birkhoff-Rott integral is computed according to the alternate-point trapezoid rule,

$$W_{1,j} - iW_{2,j} = \frac{2h}{4\pi i} \sum_{j+k \text{ odd}} \gamma_k \cot\left(\frac{z_j - z_k}{2}\right).$$

Here, h is the grid spacing, given by $h = \frac{2\pi}{N}$. This alternate-point quadrature is spectrally accurate, and is an elegant solution to the problem of avoiding the singularity when evaluating a singular integral. Other approaches are possible, such as subtracting off the singularity and then evaluating, or simply skipping the singular point; these approaches, however, are either more complicated or less accurate than the alternating point trapezoid rule. As is typically the case with spectral or pseudospectral methods, filtering is performed to address aliasing errors.

There are various options available for timestepping. An excellent feature of the HLS formulation for the 2D problem, however, is that the problem (8.3.8), (8.3.9) has become semilinear with respect to the spatial variable. This makes the use of an implicit-explicit scheme very straightforward, with the leading-order terms being treated implicitly and the lower-order terms being treated explicitly. A good choice for the timestepping scheme, then, is the fourth-order, implicit-explicit scheme of Ascher, Ruuth, and Wetton [81].

8.5 Well-Posedness with Surface Tension in 2D

We prove well-posedness by the energy method. The following are the steps of the energy method:

(i) Introduce an approximate system of evolution equations. There are two goals in introducing the approximate system: (a) to be able to show existence of solutions to the approximate system, and (b) to be able to estimate the growth of solutions of the approximate system. We will call the approximation parameter ε, so that a solution of the original system will correspond to $\varepsilon = 0$.

(ii) Prove that solutions of the approximate system exist. Typically, this will mean proving that solutions exist on some time interval $[0, T_\varepsilon]$, and T_ε will depend badly on ε (i.e., $T_\varepsilon \to 0$ as $\varepsilon \to 0$).

(iii) Prove the energy estimate: show that solutions of the approximate problem can be bounded independently of ε. This will show that solutions exist on a common time interval, $[0, T]$.

(iv) Pass to the limit as $\varepsilon \to 0$, to get a limiting function.

(v) Show that this limiting function solves the original problem.

(vi) Prove estimates similar to the energy estimate of Step 3 to prove uniqueness and continuous dependence.

8.5.1 Norms and Operators

We consider horizontally periodic Sobolev spaces in one dimension. Thus, for any $\sigma \geq 0$, the space H^σ consists of all functions $f : [0, 2\pi] \to \mathbb{R}$ such that

$$\|f\|_{H^\sigma}^2 = \int_0^{2\pi} f^2(\alpha) + (\partial_\alpha^\sigma f(\alpha))^2 \, d\alpha < \infty,$$

and such that f satisfies periodic boundary conditions.

Since the symbol of the Hilbert transform is $-i\,\mathrm{sgn}(k)$ and the symbol of a spatial derivative is ik, we see that the symbol of $H\partial_\alpha$ is

$$\mathcal{F}(H\partial_\alpha)(k) = |k|.$$

Given a periodic function f, we see that the norm $\|f\|_{H^{1/2}}$ can be written as

$$\|f\|_{H^{1/2}}^2 = \int_0^{2\pi} f^2 + f(H\partial_\alpha f) \, d\alpha,$$

since by the Plancherel theorem we have

$$\int fH\partial_\alpha f \, d\alpha = \sum_{k=-\infty}^{\infty} \hat{f}(k)|k|\overline{\hat{f}(k)} = \sum_{k=-\infty}^{\infty} \left(|k|^{1/2}\hat{f}(k) \right) \overline{\left(|k|^{1/2}\hat{f}(k) \right)}.$$

Notice also that the operator $H\partial_\alpha$ is self-adjoint; thus, we have the following useful formula for time derivatives:

$$\frac{d}{dt}\frac{1}{2}\int_0^{2\pi} fH\partial_\alpha f\, d\alpha = \int_0^{2\pi} f_t H\partial_\alpha f\, d\alpha. \tag{8.5.1}$$

One other tool for Sobolev spaces that will be of use is the following elementary interpolation inequality; the proof may be found, for example, in [26].

Lemma 8.5.1. *Let $m \geq 0$ and $\ell \geq m$ be given. Let $f \in H^\ell$ be given. Then, the following inequality holds:*

$$\|f\|_m \leq c\|f\|_\ell^{m/\ell}\|f\|_0^{1-m/\ell}. \tag{8.5.2}$$

As described earlier, the first step in the energy method is to introduce an approximate evolution equation. To this end, we introduce mollifier operators; for any $\varepsilon > 0$, we have the operator $\mathcal{J}_\varepsilon$, which is an approximate identity operator. There are various versions of mollifiers available, but a simple choice in the spatially periodic case is that the operator $\mathcal{J}_\varepsilon$ represents truncation of the Fourier series, zeroing out modes with wavenumber larger than $\frac{1}{\varepsilon}$. For more on mollifiers, the interested reader might consult [75]. The following are the essential properties we will use, which can be proved in a straightforward way using the Plancherel theorem:

$$\|\mathcal{J}_\varepsilon f\|_{H^s} \leq \|f\|_{H^s}, \tag{8.5.3}$$

$$\|\mathcal{J}_\varepsilon \partial_\alpha^m f\|_{L^2} \leq \frac{c}{\varepsilon^m}\|f\|_{L^2}. \tag{8.5.4}$$

Also, note that H, ∂_α, and $\mathcal{J}_\varepsilon$ all commute, as they are all multipliers in Fourier space.

Finally, we will sometimes need to commute the Hilbert transform with multiplication by a function. We introduce the commutator $[H, \psi]$ to be the operator given by $[H, \psi]f = H(\psi f) - \psi H(f)$. The following lemma contains the estimate we will use; versions of lemmas like this, with proofs, can be found in [80] or [26], for instance.

Lemma 8.5.2. *Let $m \geq 2$ and let $\psi \in H^m$. Then, the operator $[H, \psi]$ is bounded from L^2 to H^{m-1}, and from H^{-1} to H^{m-2}, with the estimate*

$$\|[H, \psi]f\|_{H^{m-1-i}} \leq c\|\psi\|_{H^m}\|f\|_{H^{-i}},$$

for $i \in \{0, 1\}$.

8.5.2 Existence of Solutions for a Simple Model

In previous work, the author has introduced a simple model of the density-matched case of the vortex sheet with surface tension [82, 83]. This model has the same linear, dispersive terms as the full equations of motion, and also would be ill-posed if the surface tension parameter were set equal to zero. The following are the evolution equations for this model:

$$\theta_t = H(\gamma_a), \tag{8.5.5}$$

$$\gamma_t = \tau\theta_{aa} + \mathbb{P}(\gamma^2 H(\theta_a)). \tag{8.5.6}$$

This model is formed from (8.3.8) and (8.3.10) by eliminating the collections of lower-order terms P and $\tilde{Q}$, and by ignoring factors of 2, π, and L. The operator $\mathbb{P}$ is the projection onto functions with zero mean; it is included here so that the mean of γ is a conserved quantity in the model, as it is also for the full system.

We introduce the following approximate system:

$$\theta_t^\varepsilon = \mathcal{J}_\varepsilon^2 H(\gamma_a^\varepsilon), \tag{8.5.7}$$

$$\gamma_t^\varepsilon = \tau\mathcal{J}_\varepsilon^2\theta_{aa}^\varepsilon + \mathbb{P}((\gamma^\varepsilon)^2 \mathcal{J}_\varepsilon^2 H(\theta_a^\varepsilon)). \tag{8.5.8}$$

The placement of the operators $\mathcal{J}_\varepsilon$ may, at this point, appear arbitrary. They are, however, placed very carefully, so that the estimates that follow are possible.

We will now give many of the details of the proof of well-posedness of the initial value problem for the model system. In what follows, we will always have $\tau > 0$, as the estimates would not be possible otherwise. Also, the regularity index for solutions, s, is taken to be sufficiently large. In one sense, this means that whenever we might need s to be large enough for a certain property (such as Sobolev embedding) to hold, we assume that s is large enough for that purpose. On the other hand, this means that there exists $s_0 > 0$ such that we assume $s > s_0$, and a close reading of the following argument would allow one to determine the minimal value of s_0. Our existence theory begins with application of the Picard theorem for our approximate system (8.5.7), (8.5.8), so we state the Picard Theorem.

Theorem 8.5.3 (Picard Theorem). *Let B be a Banach space, and let $O \subseteq B$ be an open set. Let $F : O \to B$ be such that F is locally Lipschitz: for any $X \in O$, there exists $\lambda > 0$ and an open set $U \subseteq O$ such that for all $Y \in U$ and for all $Z \in U$,*

$$\|F(Y) - F(Z)\|_B \le \lambda\|Y - Z\|_B.$$

Then, for all $X_0 \in O$, there exists $T > 0$ and a unique $X \in C^1([-T,T];O)$ such that X solves the initial value problem

$$\frac{dX}{dt} = F(X), \quad X(0) = X_0.$$

For application to the system (8.5.7), (8.5.8), we can simply let the open set $\mathcal{O}$ be given by $\mathcal{O} = H^s \times H^{s-1/2}$.

Lemma 8.5.4. *Let* $(\theta_0, \gamma_0) \in \mathcal{O}$ *be given. For any* $\varepsilon > 0$, *there exists* $T_\varepsilon > 0$ *and* $(\theta^\varepsilon, \gamma^\varepsilon) \in C^1([0, T]; \mathcal{O})$ *such that* $(\theta^\varepsilon, \gamma^\varepsilon)$ *satisfies (8.5.7) and (8.5.8), and satisfies* $(\theta^\varepsilon(\cdot, 0), \gamma^\varepsilon(\cdot, 0)) = (\theta_0, \gamma_0)$.

Proof. We let F_1 and F_2 be defined as

$$F_1(\theta^\varepsilon, \gamma^\varepsilon) = \mathcal{J}_\varepsilon^2 H(\gamma_\alpha^\varepsilon),$$
$$F_2(\theta^\varepsilon, \gamma^\varepsilon) = \tau \mathcal{J}_\varepsilon^2 \theta_{\alpha\alpha}^\varepsilon + \mathbb{P}\left((\gamma^\varepsilon)^2 \mathcal{J}_\varepsilon^2 H(\theta_\alpha^\varepsilon)\right).$$

The conclusion of the lemma will follow from the Picard theorem, if we show that $F = (F_1, F_2)$ is locally Lipschitz. Using (8.5.4), we see that F maps into $H^s \times H^{s-1/2}$.

We define U to be the ball of radius one centered at (θ_0, γ_0) in $\mathcal{O}$. Let $(u, v) \in U$ and $(w, z) \in U$ be given. Clearly, then, we have

$$\|u\|_{H^s} + \|v\|_{H^{s-1/2}} \le K, \quad \|w\|_{H^s} + \|z\|_{H^{s-1/2}} \le K, \tag{8.5.9}$$

where $K = 1 + \|\theta_0\|_{H^s} + \|\gamma_0\|_{H^{s-1/2}}$.

We begin with F_1:

$$\|F_1(u, v) - F_1(w, z)\|_{H^s} = \|\mathcal{J}_\varepsilon^2 \partial_\alpha H(v - z)\|_{H^s}.$$

Using properties (8.5.3) and (8.5.4), we have the following:

$$\|F_1(u, v) - F_2(w, z)\|_{H^s} \le \|\mathcal{J}_\varepsilon \partial_\alpha H(v - z)\|_{H^s} \le \frac{c}{\varepsilon^2} \|H(v - z)\|_{H^{s-1}}.$$

For the Hilbert transform, we have the property $\|Hf\|_{H^s} \le \|f\|_{H^s}$ (which can be seen clearly from the symbol $\hat{H}(k) = -i\,\mathrm{sgn}(k)$ and the Plancherel Theorem), so we can conclude

$$\|F_1(u, v) - F_1(w, z)\|_{H^s} \le \frac{c}{\varepsilon^2} \|v - z\|_{H^{s-1}}, \tag{8.5.10}$$

which is to say F_1 satisfies the appropriate Lipschitz condition.

We now turn to F_2. We write F_2 as $F_2 = F_{2,1} + F_{2,2}$, with

$$F_{2,1}(\theta^\varepsilon, \gamma^\varepsilon) = \tau \mathcal{J}_\varepsilon^2 \theta_{\alpha\alpha}^\varepsilon,$$
$$F_{2,2}(\theta^\varepsilon, \gamma^\varepsilon) = \mathbb{P}\left((\gamma^\varepsilon)^2 \mathcal{J}_\varepsilon^2 H(\theta_\alpha^\varepsilon)\right).$$

The estimate for $F_{2,1}$ is very similar to the above estimate for F_1; to start, we write the following:

$$\|F_{2,1}(u, v) - F_{2,1}(w, z)\|_{H^{s-1/2}} = \tau \|\mathcal{J}_\varepsilon^2 \partial_\alpha^2 (u - w)\|_{H^{s-1/2}}.$$

Once again using properties (8.5.3) and (8.5.4) yields the following:

$$\|F_{2,1}(u, v) - F_{2,1}(w, z)\|_{H^{s-1/2}} \le \tau \|\mathcal{J}_\varepsilon \partial_\alpha^2 (u - w)\|_{H^{s-1/2}} \le \frac{c\tau}{\varepsilon^2} \|u - w\|_{H^{s-1/2}}. \tag{8.5.11}$$

This is the desired Lipschitz estimate for $F_{2,1}$.

To complete the proof, we consider $F_{2,2}$:

$$\|F_{2,2}(u,v) - F_{2,2}(w,z)\|_{H^{s-1/2}} = \|\mathbb{P}\left([v^2 \mathcal{J}_\varepsilon^2 H u_\alpha] - [z^2 \mathcal{J}_\varepsilon^2 H z_\alpha]\right)\|_{H^{s-1/2}}.$$

It is clear (using the Plancherel Theorem) that for any f, we have $\|\mathbb{P}f\|_{H^{s-1/2}} \leq \|f\|_{H^{s-1/2}}$; using this fact yields

$$\|F_{2,2}(u,v) - F_{2,2}(w,z)\|_{H^{s-1/2}} \leq \|v^2 \mathcal{J}_\varepsilon^2 H u_\alpha - z^2 \mathcal{J}_\varepsilon^2 H w_\alpha\|_{H^{s-1/2}}.$$

To continue to estimate this, we add and subtract and use the triangle inequality:

$$\|F_{2,2}(u,v) - F_{2,2}(w,z)\|_{H^{s-1/2}}$$
$$\leq \|(v^2 - z^2)\mathcal{J}_\varepsilon^2 H u_\alpha\|_{H^{s-1/2}} + \|z^2 \mathcal{J}_\varepsilon^2 H(u_\alpha - w_\alpha)\|_{H^{s-1/2}}$$
$$= \|(v-z)(v+z)\mathcal{J}_\varepsilon^2 H u_\alpha\|_{H^{s-1/2}} + \|z^2 H \mathcal{J}_\varepsilon^2 \partial_\alpha (u-w)\|_{H^{s-1/2}}.$$

We can then use the Sobolev algebra property (since $s - \frac{1}{2} > \frac{1}{2}$), the triangle inequality, and the mollifier inequalities (8.5.3) and (8.5.4) to write this as

$$\|F_{2,2}(u,v) - F_{2,2}(w,z)\|_{H^{s-1/2}}$$
$$\leq c\|v-z\|_{H^{s-1/2}} (\|v\|_{H^{s-1/2}} + \|z\|_{H^{s-1/2}}) \left(\frac{1}{\varepsilon}\|u\|_{H^{s-1/2}}\right)$$
$$+ c\|z\|_{H^{s-1/2}}^2 \left(\frac{1}{\varepsilon}\|u-w\|_{H^{s-1/2}}\right).$$

Using the bounds (8.5.9), then, we have

$$\|F_{2,2}(u,v) - F_{2,2}(w,z)\|_{H^{s-1/2}} \leq \frac{3cK^2}{\varepsilon}\|(u,v) - (w,z)\|_{H^s \times H^{s-1/2}}. \qquad (8.5.12)$$

Putting together (8.5.10), (8.5.11), and (8.5.12), we have established the appropriate local Lipschitz estimate, and we have completed the proof of the lemma. ∎

Next, we must show that there exists $T > 0$ and $\varepsilon_0 > 0$ such that for all $\varepsilon \in (0, \varepsilon_0)$, the solutions $(\theta^\varepsilon, \gamma^\varepsilon)$ from Lemma 8.5.4 are in fact elements of $C([0,T]; \mathcal{O})$. To do this, we would like to use the Continuation Theorem:

Theorem 8.5.5 (Continuation Theorem for ODEs). *Let B be a Banach space, and let $\Omega \subseteq B$ be an open set. Let $F : \Omega \to B$ be locally Lipschitz continuous. Let $X_0 \in \Omega$ be given. Let X be the solution of the initial value problem*

$$\frac{dX}{dt} = F(X), \qquad X(0) = X_0,$$

and let $T > 0$ be the maximal time such that $X \in C^1([0,T); \Omega)$. Then either $T = \infty$, or $T < \infty$ with $X(t)$ leaving the set Ω as $t \nearrow T$.

So, we see that in order to apply the continuation theorem uniformly in ε, it is enough to have control of the norm of $(\theta^\varepsilon, \gamma^\varepsilon)$, uniformly in ε. This is why we prove an energy estimate; this is the content of the following lemma.

Lemma 8.5.6. *Let $(\theta_0, \gamma_0) \in \mathcal{O}$ be given. There exists $T > 0$ such that for all $\varepsilon \in (0, 1]$, the initial value problem (8.5.7), (8.5.8) with $(\theta(\cdot, 0), \gamma(\cdot, 0)) = (\theta_0, \gamma_0)$ has a solution $(\theta^\varepsilon, \gamma^\varepsilon) \in C([0, T]; \mathcal{O})$.*

Proof. Let $\varepsilon \in (0, 1]$ be given. By Lemma 8.5.4, we know that there exists $T_\varepsilon > 0$ and $(\theta^\varepsilon, \gamma^\varepsilon) \in C^1([-T_\varepsilon, T_\varepsilon]; \mathcal{O})$, which solves the initial value problem. We now will show that this solution can be continued until some time, T, which is independent of ε.

We define the energy, $E(t)$, to be

$$E(t) = E_0(t) + E_1(t) + E_2(t) + E_3(t),$$

with

$$E_0(t) = \frac{1}{2} \int_0^{2\pi} (\theta^\varepsilon)^2 + (\gamma^\varepsilon)^2 \, d\alpha,$$

$$E_1(t) = \frac{1}{2} \int_0^{2\pi} (\partial_\alpha^s \theta^\varepsilon)^2 \, d\alpha,$$

$$E_2(t) = \frac{1}{2\tau} \int_0^{2\pi} (\partial_\alpha^{s-1} \gamma^\varepsilon)(H \partial_\alpha^s \gamma^\varepsilon) \, d\alpha,$$

$$E_3(t) = \frac{1}{2\tau^2} \int_0^{2\pi} (\gamma^\varepsilon)^2 (\partial_\alpha^{s-1} \gamma^\varepsilon)^2 \, d\alpha.$$

Notice that the energy controls the square of the norm of $(\theta^\varepsilon, \gamma^\varepsilon)$:

$$\|\theta^\varepsilon\|_{H^s}^2 + \|\gamma^\varepsilon\|_{H^{s-1/2}}^2 \leq cE. \tag{8.5.13}$$

We take the time derivative of each of E_0, E_1, E_2, and E_3. We begin with E_0:

$$\frac{dE_0}{dt} = \int_0^{2\pi} \theta^\varepsilon \theta_t^\varepsilon + \gamma^\varepsilon \gamma_t^\varepsilon \, d\alpha.$$

Substituting from (8.5.7) and (8.5.8), this becomes

$$\frac{dE_0}{dt} = \int_0^{2\pi} \theta^\varepsilon \mathcal{J}_\varepsilon^2 H(\gamma_\alpha^\varepsilon) + \tau \gamma^\varepsilon \mathcal{J}_\varepsilon^2 \theta_{\alpha\alpha}^\varepsilon + \gamma^\varepsilon \mathcal{J}_\varepsilon \mathbb{P}\left((\gamma^\varepsilon)^2 \mathcal{J}_\varepsilon H(\theta_\alpha^\varepsilon)\right) d\alpha.$$

Using the mollifier property (8.5.3), we see that this clearly satisfies the following estimate:

$$\frac{dE_0}{dt} \leq c\left(\|\theta^\varepsilon\|_{H^0} \|\gamma^\varepsilon\|_{H^1} + \|\gamma^\varepsilon\|_{H^0} \|\theta^\varepsilon\|_{H^2} + \|\gamma^\varepsilon\|_{H^0} \|\gamma^\varepsilon\|_{L^\infty}^2 \|\theta^\varepsilon\|_{H^1}\right).$$

In light of (8.5.13), and using Sobolev embedding, we conclude that the growth of E_0 is controlled by the energy, as long as $s \geq 2$:

$$\frac{dE_0}{dt} \leq c(E + E^2). \tag{8.5.14}$$

We move on to E_1, calculating the following:

$$\frac{dE_1}{dt} = \int_0^{2\pi} (\partial_\alpha^s \theta^\varepsilon)(\partial_\alpha^s \theta_t^\varepsilon) \, d\alpha.$$

Applying ∂_α^s to (8.5.7) and substituting yields the following:

$$\frac{dE_1}{dt} = \int_0^{2\pi} (\partial_\alpha^s \theta^\varepsilon)(\mathcal{J}_\varepsilon^2 \partial_\alpha^{s+1} H \gamma^\varepsilon) \, d\alpha. \tag{8.5.15}$$

Notice that we have used the fact that the operators H, ∂_α, and $\mathcal{J}_\varepsilon$ all commute. Notice also that the right-hand side of (8.5.15) cannot be bounded by the energy: we can control s derivatives of θ^ε and $s - 1/2$ derivatives of γ^ε. The expression in (8.5.15) thus has $3/2$ more derivatives than we can control. We will need to cancel this term with part of dE_2/dt; we will see this soon, but for now, we rewrite (8.5.15) simply by using the fact that $\mathcal{J}_\varepsilon$ is self-adjoint:

$$\frac{dE_1}{dt} = \int_0^{2\pi} (\mathcal{J}_\varepsilon \partial_\alpha^s \theta^\varepsilon)(\mathcal{J}_\varepsilon \partial_\alpha^{s+1} H \gamma^\varepsilon) \, d\alpha. \tag{8.5.16}$$

Next, we take the time derivative of E_2. Making use of (8.5.1), we get

$$\frac{dE_2}{dt} = \frac{1}{\tau} \int_0^{2\pi} (\partial_\alpha^{s-1} \gamma_t^\varepsilon)(H \partial_\alpha^s \gamma^\varepsilon) \, d\alpha. \tag{8.5.17}$$

We apply ∂_α^{s-1} to (8.5.8), keeping in mind that $\partial_\alpha \mathbb{P} = \partial_\alpha$:

$$\partial_\alpha^{s-1} \gamma_t^\varepsilon = \tau \mathcal{J}_\varepsilon^2 \partial_\alpha^{s+1} \theta^\varepsilon + \partial_\alpha^{s-1} \left((\gamma^\varepsilon)^2 \mathcal{J}_\varepsilon^2 H(\theta_\alpha^\varepsilon) \right). \tag{8.5.18}$$

For the second term on the right-hand side of (8.5.18), it is helpful to use the product rule when applying ∂_α^{s-1} :

$$\partial_\alpha^{s-1} \gamma_t^\varepsilon = \tau \mathcal{J}_\varepsilon^2 \partial_\alpha^{s+1} \theta^\varepsilon + (\gamma^\varepsilon)^2 \mathcal{J}_\varepsilon^2 H(\partial_\alpha^s \theta^\varepsilon)$$

$$+ \sum_{j=0}^{s-2} \binom{s-1}{j} (\partial_\alpha^{s-1-j}(\gamma^\varepsilon)^2) \mathcal{J}_\varepsilon^2 H(\partial_\alpha^{j+1} \theta^\varepsilon). \tag{8.5.19}$$

We substitute (8.5.19) into (8.5.17):

$$\frac{dE_2}{dt} = \int_0^{2\pi} (\mathcal{J}_\varepsilon^2 \partial_\alpha^{s+1} \theta^\varepsilon)(H \partial_\alpha^s \gamma^\varepsilon) \, d\alpha + \frac{1}{\tau} \int_0^{2\pi} (\gamma^\varepsilon)^2 \left(\mathcal{J}_\varepsilon^2 H(\partial_\alpha^s \theta^\varepsilon) \right) (H \partial_\alpha^s \gamma^\varepsilon) \, d\alpha$$

$$+ \frac{1}{\tau} \int_0^{2\pi} \left(\sum_{j=0}^{s-2} \binom{s-1}{j} (\partial_\alpha^{s-1-j}(\gamma^\varepsilon)^2) \left(\mathcal{J}_\varepsilon^2 H(\partial_\alpha^{j+1} \theta^\varepsilon) \right) \right) (H \partial_\alpha^s \gamma^\varepsilon) \, d\alpha.$$

We rewrite this in several ways: for the first integral on the right-hand side, we use the fact that $\mathcal{J}_\varepsilon$ is self-adjoint, and we integrate by parts once. For the third integral, we give the name Φ_1 to the summation:

$$\Phi_1 = \sum_{j=0}^{s-2} \binom{s-1}{j} (\partial_\alpha^{s-1-j}(\gamma^\varepsilon)^2)(\mathcal{J}_\varepsilon^2 H(\partial_\alpha^{j+1}\theta^\varepsilon)). \tag{8.5.20}$$

These considerations yield the following:

$$\frac{dE_2}{dt} = -\int_0^{2\pi} (\mathcal{J}_\varepsilon \partial_\alpha^s \theta^\varepsilon)(\mathcal{J}_\varepsilon \partial_\alpha^{s+1} H\gamma^\varepsilon)\, d\alpha$$
$$+ \frac{1}{\tau}\int_0^{2\pi} (\gamma^\varepsilon)^2(\mathcal{J}_\varepsilon^2 H(\partial_\alpha^s \theta^\varepsilon))(H\partial_\alpha^s \gamma^\varepsilon)\, d\alpha + \frac{1}{\tau}\int_0^{2\pi} \Phi_1(H\partial_\alpha^s \gamma^\varepsilon)\, d\alpha. \tag{8.5.21}$$

Next, we add (8.5.16) and (8.5.21); importantly, the first terms on the right-hand sides of (8.5.16) and (8.5.21) cancel with each other:

$$\frac{dE_1}{dt} + \frac{dE_2}{dt} = \frac{1}{\tau}\int_0^{2\pi} (\gamma^\varepsilon)^2(\mathcal{J}_\varepsilon^2 H(\partial_\alpha^s \theta^\varepsilon))(H\partial_\alpha^s \gamma^\varepsilon)\, d\alpha + \frac{1}{\tau}\int_0^{2\pi} \Phi_1(H\partial_\alpha^s \gamma^\varepsilon)\, d\alpha. \tag{8.5.22}$$

The first term on the right-hand side of (8.5.22) cannot be bounded in terms of the energy; notice that it has $1/2$ more derivatives than we can control. This is the reason we need to include E_3 in the definition of the energy.

We take the time derivative of E_3:

$$\frac{dE_3}{dt} = \frac{1}{\tau^2}\int_0^{2\pi} (\gamma^\varepsilon)^2(\partial_\alpha^{s-1}\gamma_t^\varepsilon)(\partial_\alpha^{s-1}\gamma^\varepsilon)\, d\alpha + \frac{1}{\tau^2}\int_0^{2\pi} \gamma^\varepsilon \gamma_t^\varepsilon(\partial_\alpha^{s-1}\gamma^\varepsilon)^2\, d\alpha.$$

For the first term on the right-hand side, we substitute from (8.5.19):

$$\frac{dE_3}{dt} = \frac{1}{\tau}\int_0^{2\pi} (\gamma^\varepsilon)^2(\mathcal{J}_\varepsilon^2 \partial_\alpha^{s+1}\theta^\varepsilon)(\partial_\alpha^{s-1}\gamma^\varepsilon)\, d\alpha + \frac{1}{\tau^2}\int_0^{2\pi} \Phi_2(\partial_\alpha^{s-1}\gamma^\varepsilon)\, d\alpha, \tag{8.5.23}$$

where the collection of terms Φ_2 is given by

$$\Phi_2 = (\gamma^\varepsilon)^4 \mathcal{J}_\varepsilon^2 H(\partial_\alpha^s \theta^\varepsilon) + (\gamma^\varepsilon)^2 \Phi_1 + \gamma^\varepsilon \gamma_t^\varepsilon(\partial_\alpha^{s-1}\gamma^\varepsilon)$$
$$= (\gamma^\varepsilon)^4 \mathcal{J}_\varepsilon^2 H(\partial_\alpha^s \theta^\varepsilon) + (\gamma^\varepsilon)^2 \Phi_1 + \gamma^\varepsilon(\tau\theta_{\alpha\alpha}^\varepsilon + \mathbb{P}((\gamma^\varepsilon)^2 \mathcal{J}_\varepsilon^2 H(\theta_\alpha^\varepsilon)))(\partial_\alpha^{s-1}\gamma^\varepsilon). \tag{8.5.24}$$

We integrate the first integral on the right-hand side of (8.5.23) by parts, arriving at the following:

$$\frac{dE_3}{dt} = -\frac{1}{\tau}\int_0^{2\pi} (\gamma^\varepsilon)^2(\mathcal{J}_\varepsilon^2 \partial_\alpha^s \theta^\varepsilon)(\partial_\alpha^s \gamma^\varepsilon)\, d\alpha + \frac{1}{\tau^2}\int_0^{2\pi} \tilde{\Phi}_2(\partial_\alpha^{s-1}\gamma^\varepsilon)\, d\alpha, \tag{8.5.25}$$

with $\tilde{\Phi}_2$ given by

$$\tilde{\Phi}_2 = \Phi_2 - 2\tau \gamma^\varepsilon \gamma_\alpha^\varepsilon (\mathcal{J}_\varepsilon^2 \partial_\alpha^s \theta^\varepsilon). \tag{8.5.26}$$

We would like to now add (8.5.22) and (8.5.25), finding another important cancellation. However, the leading-order term on the right-hand side of (8.5.22) has two Hilbert transforms present, while the leading-order term on the right-hand side of (8.5.25) has no Hilbert transforms. We therefore must rewrite (8.5.22) a bit more before proceeding. Since the adjoint of H is $-H$ (this can be seen from the symbol of H), and since H and $\mathcal{J}_\varepsilon$ commute, we have the following:

$$\frac{dE_1}{dt} + \frac{dE_2}{dt} = -\frac{1}{\tau} \int_0^{2\pi} (\mathcal{J}_\varepsilon^2 \partial_\alpha^s \theta^\varepsilon) H\left((\gamma^\varepsilon)^2 (H\partial_\alpha^s \gamma^\varepsilon)\right) d\alpha + \frac{1}{\tau} \int_0^{2\pi} \Phi_1 (H\partial_\alpha^s \gamma^\varepsilon) \, d\alpha.$$

We continue to rewrite, pulling $(\gamma^\varepsilon)^2$ through the Hilbert transform, incurring a commutator, and using the identity $H^2 \partial_\alpha = -\partial_\alpha$:

$$\frac{dE_1}{dt} + \frac{dE_2}{dt} = \frac{1}{\tau} \int_0^{2\pi} (\gamma^\varepsilon)^2 (\mathcal{J}_\varepsilon^2 \partial_\alpha^s \theta^\varepsilon)(\partial_\alpha^s \gamma^\varepsilon) \, d\alpha$$

$$- \frac{1}{\tau} \int_0^{2\pi} (\mathcal{J}_\varepsilon^2 \partial_\alpha^s \theta^\varepsilon) \left[H, (\gamma^\varepsilon)^2\right] (H\partial_\alpha^s \gamma^\varepsilon) \, d\alpha$$

$$+ \frac{1}{\tau} \int_0^{2\pi} \Phi_1 (H\partial_\alpha^s \gamma^\varepsilon) \, d\alpha. \tag{8.5.27}$$

We add (8.5.25) and (8.5.27), now seeing clearly the cancellation of the leading-order terms on the right-hand sides:

$$\frac{dE_1}{dt} + \frac{dE_2}{dt} + \frac{dE_3}{dt} = -\frac{1}{\tau} \int_0^{2\pi} (\mathcal{J}_\varepsilon^2 \partial_\alpha^s \theta^\varepsilon) \left[H, (\gamma^\varepsilon)^2\right] (H\partial_\alpha^s \gamma^\varepsilon) \, d\alpha$$

$$+ \frac{1}{\tau} \int_0^{2\pi} \Phi_1 (H\partial_\alpha^s \gamma^\varepsilon) \, d\alpha + \frac{1}{\tau^2} \int_0^{2\pi} \tilde{\Phi}_2 (\partial_\alpha^{s-1} \gamma^\varepsilon) \, d\alpha. \tag{8.5.28}$$

Just as we bounded $\frac{dE_0}{dt}$, we wish to bound (8.5.28) by some combination of powers of E. Since the energy is quartic and the time-derivative of γ^ε is cubic, it turns out that the highest power we will need is a third power of E (since the energy is quadratic in $(\theta^\varepsilon, \gamma^\varepsilon)$, this corresponds to sixth powers of $(\theta^\varepsilon, \gamma^\varepsilon)$ appearing in the estimates). Thus, we will seek to demonstrate the inequality

$$\frac{dE}{dt} \leq c(E + E^3). \tag{8.5.29}$$

In light of (8.5.13), it is sufficient to bound the integrals on the right-hand side of (8.5.28) by combinations of $\|\theta^\varepsilon\|_{H^s}$ and $\|\gamma^\varepsilon\|_{H^{s-1/2}}$. Thus, it is sufficient to establish the following bounds:

$$\left\| [H,(\gamma^\varepsilon)^2](H\partial_\alpha^s\gamma^\varepsilon) \right\|_{L^2} \le c\|\gamma^\varepsilon\|_{H^{s-1/2}}^3, \tag{8.5.30}$$

$$\|\Phi_1\|_{H^{1/2}} \le c\|\theta^\varepsilon\|_{H^s}\|\gamma^\varepsilon\|_{H^{s-1/2}}^2, \tag{8.5.31}$$

$$\|\tilde{\Phi}_2\|_{L^2} \le c\|\theta^\varepsilon\|_{H^s}\left(\|\gamma^\varepsilon\|_{H^{s-1/2}}^2 + \|\gamma^\varepsilon\|_{H^{s-1/2}}^4\right). \tag{8.5.32}$$

Of these, the estimate (8.5.30) follows immediately from Lemma 8.5.2, if we take, say $s \ge 3$. To bound (8.5.31) and (8.5.32) requires simply using the definitions (8.5.20), (8.5.24), (8.5.26) together with the inequality (8.5.3), the Sobolev algebra property, and the fact that s is sufficiently large. This establishes (8.5.29). A Gronwall-type argument thus completes the proof of the lemma. ∎

Having established that the solutions $(\theta^\varepsilon, \gamma^\varepsilon)$ all exist on a common time interval, we may then pass to the limit as $\varepsilon \to 0$. This can be done in one of two ways: in the case of a compact domain, such as a periodic interval, this compactness can be exploited by use of the Arzela-Ascoli theorem. Otherwise, one may show directly that the sequence $(\theta^\varepsilon, \gamma^\varepsilon)$ is a Cauchy sequence as $\varepsilon \to 0$; this is a more involved argument than the compactness argument, but it works out nonetheless. Since we have chosen to work with a periodic interval, we focus now on the compactness argument.

The energy estimate demonstrates that the sequence $(\theta^\varepsilon, \gamma^\varepsilon)$ is uniformly bounded in $C([0,T]; H^s \times H^{s-1/2})$. This implies $\theta_\alpha^\varepsilon$ and $\gamma_\alpha^\varepsilon$ are uniformly bounded in L^∞, and that θ_t^ε and γ_t^ε (using (8.5.7) and (8.5.8)) are as well, since s is taken to be sufficiently large. Thus, we have an equicontinuous family, and there is a subsequence (which we do not relabel) which converges uniformly to (θ, γ). Using the uniform bound in $H^s \times H^{s-1/2}$ and the interpolation inequality of Lemma 8.5.1, we see that $(\theta^\varepsilon, \gamma^\varepsilon)$ converges to (θ, γ) in $H^{s'} \times H^{s'-1/2}$ for all $s' < s$.

Integrating (8.5.7), (8.5.8) with respect to time, we get

$$\theta^\varepsilon(\cdot,t) = \theta_0 + \int_0^t \mathcal{J}_\varepsilon^2 H(\gamma_\alpha^\varepsilon(\cdot,\tau))\, d\tau, \tag{8.5.33}$$

$$\gamma^\varepsilon(\cdot,t) = \gamma_0 + \int_0^t \tau \mathcal{J}_\varepsilon^2 \theta_{\alpha\alpha}^\varepsilon(\cdot,\tau) + \mathbb{P}((\gamma^\varepsilon(\cdot,\tau))^2 \mathcal{J}_\varepsilon^2 H(\theta_\alpha^\varepsilon(\cdot,\tau)))\, d\tau. \tag{8.5.34}$$

Again, since s is sufficiently large, we are able to pass to the limit in (8.5.33) and (8.5.34), finding that (θ, γ) is indeed a solution of the initial value problem (8.5.5), (8.5.6) with $\theta(\cdot,0) = \theta_0$, $\gamma(\cdot,0) = \gamma_0$.

We exclude the rest of the details, but one can go on to show the highest regularity, $(\theta, \gamma) \in C([0,T]; H^s \times H^{s-1/2})$. This completes the existence theorem for solutions of the initial value problem. Uniqueness and continuous dependence can also be established, and these require a further estimate, for the difference of two solutions. If E_d is a carefully chosen energy functional that measures a norm of $(\theta - \theta', \gamma - \gamma')$, where (θ, γ) and (θ', γ') are two solutions of the initial value problem with different initial data, then it can be shown that

$$\frac{dE_d}{dt} \leq cE_d.$$

Gronwall's inequality then implies both existence and uniqueness (if the initial data is the same, then E_d remains zero at positive times; otherwise, the initial data may be taken sufficiently small so that E_d remains small over the time interval $[0, T]$). Carrying out the details of this argument establishes well-posedness of the initial value problem for the model. Full details are carried out for similar problems in, for instance, [84] or [85], among other places.

8.5.3 Well-Posedness of the Full System

There are, naturally, some issues present in the well-posedness proof of the initial value problem for the full vortex sheet with surface tension that did not arise in the proof for the simple model in Section 8.5.2.

One difference between the full equations and the simple model is that while the model is indeed dispersive, nonlinear, and nonlocal, the nonlocality is quite straightforward: the Hilbert transform is present, but the more complicated Birkhoff-Rott integral does not appear in the simple model. As described earlier, the Birkhoff-Rott integral is approximated with Hilbert transforms; in proving well-posedness of the full system, estimates for the error in making this approximation must be used. Examples of such estimates can be found in Lemma 3.5 in [26].

Before discussing the Birkhoff-Rott integral further, it is necessary to discuss the curve itself. The simple model did not require reconstructing an interface from θ, but in the full equations, this is necessary. To perform such a reconstruction, we first note that the length of the curve is determined by θ and the periodicity assumption. The periodicity assumption is $z(\alpha + 2\pi, t) = z(\alpha, t) + 2\pi$. Using a normalized arclength parameterization, we can write $z_\alpha = \frac{L}{2\pi} e^{i\theta}$. Integrating this from 0 to 2π, we get

$$z(2\pi, t) - z(0, t) = \frac{L}{2\pi} \int_0^{2\pi} \cos(\theta(\alpha, t)) + i\sin(\theta(\alpha, t)) \, d\alpha.$$

We notice two things from this: first, in order for this to be possible, it must be the case that

$$\int_0^{2\pi} \sin(\theta(\alpha, t)) \, d\alpha = 0. \tag{8.5.35}$$

Second, we see that in this case, we can solve for L:

$$L(t) := L[\theta](t) = \frac{4\pi^2}{\int_0^{2\pi} \cos(\theta(\alpha,t))\, d\alpha},$$

as long as

$$\int_0^{2\pi} \cos(\theta(\alpha,t))\, d\alpha \neq 0. \tag{8.5.36}$$

With conditions (8.5.35) and (8.5.36) satisfied, the curve z can be reconstructed, then, up to a single point. Integrating the relationship $z_\alpha = \frac{L}{2\pi}e^{i\theta}$ from 0 to α, we get

$$z(\alpha,t) - z(0,t) = \frac{L[\theta](t)}{2\pi} \int_0^\alpha \exp\{i\theta(\alpha',t)\}\, d\alpha'. \tag{8.5.37}$$

We remark that the choice of $z(0,t)$ does not actually affect the evolution: if we define $z_d(\alpha,t) = z(\alpha,t) - z(0,t)$, then the Birkhoff-Rott integral can be written in terms of z_d rather than z:

$$W_1 - iW_2 = \frac{1}{4\pi i} \mathrm{PV} \int_0^{2\pi} \gamma(\alpha,t) \cot\left(\frac{1}{2}(z_d(\alpha,t) - z_d(\alpha,t))\right)\, d\alpha'.$$

For completeness, one can determine the value of $z(0,t)$ after one knows that a solution exists by integrating $z_t(0,t)$ with respect to time. We may denote z_d as $z_d[\theta]$.

The most significant issue that is not present for the simple model is the non–self-intersection condition. The Birkhoff-Rott integral (8.2.4) does not make sense for a self-intersecting curve; that is, if there exist $\alpha \neq \alpha'$ such that $z(\alpha) = z(\alpha')$, then the integral in (8.2.4) does not converge. Thus, for the full system, we use an initial curve that satisfies a non–self-intersection condition, and we must then ensure that this condition holds at positive times. The most commonly used condition is the chord-arc condition; this was used, for instance, in the papers of Wu on well-posedness of irrotational water waves in 2D and 3D, and subsequently by the author and collaborators, and others [13, 14, 16, 26]:

$$\exists \bar{c} > 0 \text{ s.t. } \forall \alpha, \alpha', \qquad \left| \frac{z_d(\alpha) - z_d(\alpha')}{\alpha - \alpha'} \right| > \bar{c}. \tag{8.5.38}$$

In order to ensure that this condition holds at positive times, given that it holds initially, the Picard Theorem and the Continuation Theorem can be used. For the Picard Theorem, the initial condition is given in a set O, and a solution of the evolution equation is then found in the set O; this solution can then be continued as long as the solution does not leave the set O. The condition (8.5.38), then, can be used as part of the definition of the set O. To ensure that the condition continues to hold at positive times, it is sufficient to have an upper bound for $\|z_{\alpha t}\|_{L^\infty}$; such a bound follows from the energy estimate.

Another issue arises in the case $\rho_1 \neq \rho_2$; thus, this issue is present in the water wave case. This is that the equation (8.3.7) for the evolution of γ is an integral equation, since γ_t is present inside the integral in the term $\mathbf{W}_t \cdot \hat{\mathbf{t}}$ on the right-hand side. This integral equation is known to be solvable [86], and there are good estimates for the corresponding inverse operator [18].

With these considerations, we are able in [26] to prove well-posedness of the initial value problem for vortex sheets with surface tension, with initial data in sufficiently regular Sobolev spaces.

Theorem 8.5.7. *Let $\tau > 0$ be given and let $\rho_1 \geq 0$, $\rho_2 \geq 0$ be given such that $\rho_1 + \rho_2 > 0$. Let $\theta_0 \in H^s$ and $\gamma_0 \in H^{s-1/2}$ be given, such that θ_0 satisfies $\int_0^{2\pi} \sin(\theta_0(\alpha)) \, d\alpha = 0$ and $\int_0^{2\pi} \cos(\theta_0(\alpha)) \, d\alpha \neq 0$. Assume that $z_d[\theta_0]$, defined as in (8.5.37), satisfies (8.5.38). There exists $T > 0$ and $\theta \in C([0,T];H^s)$, $\gamma \in C([0,T];H^{s-1/2})$ such that (θ,γ) are the unique solution of (8.3.4), (8.3.7) with $\theta(\alpha,0) = \theta_0$, $\gamma(\alpha,0) = \gamma_0$. Furthermore, solutions of the initial value problem depend continuously on the initial data.*

8.6 The Zero Surface Tension Limit in 2D

The proof of Theorem 8.5.7 is based on energy estimates, which are very similar to those of Section 8.5.2. As can be seen in Section 8.5.2, the proof of the energy estimate in Lemma 8.5.6 uses in a fundamental way the condition $\tau > 0$. If instead $\tau = 0$, this same argument simply does not work. In fact, the situation is somewhat worse than this, because if $\tau = 0$ and $\rho_1 \rho_2 > 0$ (which is to say, both fluids are present), then the initial value problem is known to be ill-posed [1, 3, 7–9].

In one case, however, the problem is well-posed without surface tension present, and that is the water wave case, $\rho_1 > 0$, $\rho_2 = 0$. In this case, energy estimates can be made which do not rely on the positivity of the surface tension parameter. To show well-posedness of the water wave without surface tension, it is certainly not necessary to consider the presence of surface tension at all. However, one approach to studying the water wave without surface tension is to attempt to take the limit of water waves with surface tension, as the surface tension parameter vanishes; this is what the author and Masmoudi carried out in [16, 17] for 2D and 3D water waves.

Thus, similarly to how we took the limit of solutions $(\theta^\varepsilon, \gamma^\varepsilon)$ as $\varepsilon \to 0$ in Section 8.5.2 to find solutions of the problem with $\varepsilon = 0$, we may attempt to take the limit of solutions as $\tau \to 0$ to find solutions when $\tau = 0$. In this case, we rephrase the result of Theorem 8.5.7 to say that we have solutions $(\theta^\tau, \gamma^\tau)$, on a time interval $[0,T_\tau]$, for any $\tau > 0$. The primary step in taking the limit as $\tau \to 0$ is establishing a time interval for existence, which is independent of τ; note the similarity to Lemma 8.5.6. Just as Lemma 8.5.6 relied upon an estimate for solutions, which was indepedent of ε, the main ingredient when sending τ to zero must be an estimate for solutions, which is independent of τ.

In [16], we achieve such an estimate after first making a change of variables. We introduce a quantity, which we call the *modified tangential velocity*,

$$\delta = \frac{\gamma}{2s_\alpha} + \mathbf{W} \cdot \hat{\mathbf{t}} - V.$$

There are two useful ways to think about δ: first, it can be thought of as simply being a multiple of γ plus a correction. Thus, we will rewrite the evolution equations to replace γ with δ instead. Second, there is a physical interpretation, as δ is equal to the difference between the Lagrangian tangential velocity of a particle on the interface (this is $\frac{\gamma}{2s_\alpha} + \mathbf{W} \cdot \hat{\mathbf{t}}$) and the artificial tangential velocity, which maintains the normalized arclength parameterization, V. The situation for 3D water waves is similar [17].

We have mentioned that we replace γ with δ when writing the evolution equations for the capillary water wave system, but it is in fact more helpful to write the evolution equation for δ_α. The following is the evolution equation one finds for δ_α:

$$\delta_{\alpha t} = \tau \frac{2\pi^2}{L^2} \theta_{\alpha\alpha\alpha} - c\theta_\alpha - \frac{\pi}{L} \partial_\alpha^2 \delta^2 - \frac{L_t}{L} \delta_\alpha - \frac{L_{tt}}{2\pi} + \theta_t \mathbf{W}_\alpha \cdot \hat{\mathbf{n}} + \frac{\pi}{L} \theta_\alpha \gamma \theta_t$$
$$+ \frac{2\pi}{L} \mathbf{W}_\alpha \cdot \hat{\mathbf{n}} + \frac{2\pi^2}{L^2} \theta_\alpha^2 \delta\gamma. \tag{8.6.1}$$

The terms on the right-hand side of (8.6.1) can be grouped in convenient ways and the regularity of specific terms can be studied, leading to the following version:

$$\delta_{\alpha t} = \tau \frac{2\pi^2}{L^2} \theta_{\alpha\alpha\alpha} - c\theta_\alpha - \frac{2\pi}{L} \delta\delta_{\alpha\alpha} + \psi. \tag{8.6.2}$$

The evolution equation for θ is, naturally, unchanged, as it still incorporates the fact that the free surface moves with normal velocity $U = \mathbf{W} \cdot \hat{\mathbf{n}}$ and with tangential velocity V, determined by the normalized arclength parameterization. However, the equation for θ must be rephrased to be in terms of δ rather than γ; the result is

$$\theta_t = \frac{2\pi}{L} H(\delta_\alpha) - \frac{2\pi}{L} \delta\theta_\alpha + \phi. \tag{8.6.3}$$

Of course, the collections of terms ψ and ϕ on the right-hand sides of (8.6.2) and (8.6.3) can be written explicitly in terms of the quantities θ, δ, L, γ, $\mathbf{W}$, and so on. Both the collections ϕ and ψ are comprised entirely of lower-order terms, and may be treated routinely in the energy estimates. It is important to discuss the quantity c, which appears on the right-hand side of (8.6.1) and

(8.6.2). This c is not a constant, but is a function of α and t :

$$c = c(\alpha, t) := -\nabla p \cdot \hat{\mathbf{n}}.$$

Theorem 8.5.7 gives well-posedness of the initial value problem for (8.6.2), (8.6.3) when $\tau > 0$, but of course, we are now interested in the case $\tau = 0$. The gravity water wave problem (i.e., the problem when $\tau = 0$) is well-posed if c satisfies a condition, which is sometimes known as the Generalized Taylor Condition,

$$\exists \bar{C} > 0 \text{ s.t. } \forall \alpha, \quad c(\alpha, t) > \bar{C} > 0. \tag{8.6.4}$$

If c is allowed to be negative, the initial value problem is known to be ill-posed [87]. The condition (8.6.4) was shown by Wu to hold for the gravity water wave problem (in both 2D and 3D) as long as the free surface is non–self-intersecting [13, 14]. In the case of capillary-gravity water waves (i.e., with $\tau > 0$), one can see that c is indeed positive for sufficiently small values of τ.

As mentioned earlier, after formulating the problem in this way, we are able to pass to the limit as τ vanishes. The primary tool to enable this passage to the limit is an energy estimate that is uniform in τ; this is the content of Section 4 of [16]. This enables one to show that the water wave without surface tension is the zero surface tension limit of water waves with surface tension. If the initial value problem without surface tension were not well-posed, then one could not expect to be able to take the zero surface tension limit. Indeed, in the case of Hele–Shaw flow without surface tension, a condition similar to (8.6.4) is required for well-posedness. In [84], it is shown that the zero surface tension limit can be taken when the relevant stability condition is satisfied. When instead the stability condition is violated, it can be seen (using a mixture of analysis and computing) that the zero surface tension limit does not yield zero surface tension solutions [88–91].

8.7 Extensions to 3D

As previously mentioned, one great benefit of the HLS formulation for problems with surface tension in 2D fluids is that the problems turn out to be semilinear; this makes the energy estimates simpler than they might otherwise have been. However, as seen in Section 8.6, the formulation is still useful for problems that only turn out to be quasilinear rather than semilinear. For irrotational interfacial flow problems in 3D fluids, it turns out that quasilinear is the best that can be hoped for; while the proofs are more involved than for the same problems in 2D fluids, the method from 2D does have a suitable generalization.

The main ingredient in the formulation for our 3D flows, then, is the replacement for the parameterization by arclength. The approach taken in

[17, 27], and [38] is to use an isothermal parameterization (this was suggested to the author by Jalal Shatah). In these papers, the geometry considered was not periodic, but instead the free surface $\mathbf{X}(\alpha,\beta,t) = (x(\alpha,\beta,t), y(\alpha,\beta,t), z(\alpha,\beta,t))$ was taken to be asymptotic to the $z = 0$ plane as α and β go to infinity. In this setting, the isothermal parameterization requires

$$\mathbf{X}_\alpha \cdot \mathbf{X}_\alpha = \mathbf{X}_\beta \cdot \mathbf{X}_\beta, \qquad \mathbf{X}_\alpha \cdot \mathbf{X}_\beta = 0. \qquad (8.7.1)$$

The quantities in (8.7.1) are the first fundamental coefficients of the surface $\mathbf{X}$ (i.e., $E = \mathbf{X}_\alpha \cdot \mathbf{X}_\alpha$, $F = \mathbf{X}_\alpha \cdot \mathbf{X}_\beta$, and $G = \mathbf{X}_\beta \cdot \mathbf{X}_\beta$). Just as the choice of a normalized arclength parameterization fixes the definition of the tangential velocity (8.3.6) in the case of 2D fluids, the choices (8.7.1) fix the definition of the two tangential velocities in this case.

We consider the evolution of the surface $\mathbf{X}$ to be

$$\mathbf{X}_t = U\hat{\mathbf{n}} + V_1\hat{\mathbf{t}}^1 + V_2\hat{\mathbf{t}}^2, \qquad (8.7.2)$$

with

$$\hat{\mathbf{t}}^1 = \frac{\mathbf{X}_\alpha}{\sqrt{E}}, \qquad \hat{\mathbf{t}}^2 = \frac{\mathbf{X}_\beta}{\sqrt{G}}, \qquad \hat{\mathbf{n}} = \hat{\mathbf{t}}^1 \times \hat{\mathbf{t}}^2.$$

To find the equations for V_1 and V_2, one takes the relationships from (8.7.1), differentiates with respect to time, and plugs in from (8.7.2), assuming that (8.7.1) holds. This yields the following elliptic system for V_1 and V_2:

$$\Delta\left(\frac{V_1}{\sqrt{E}}\right) = \left(\frac{U(L-N)}{E}\right)_\alpha + \left(\frac{2UM}{E}\right)_\beta, \qquad (8.7.3)$$

$$\Delta\left(\frac{V_2}{\sqrt{E}}\right) = \left(\frac{2UM}{E}\right)_\alpha - \left(\frac{U(L-N)}{E}\right)_\beta, \qquad (8.7.4)$$

where the quantities L, M, and N on the right-hand sides of (8.7.3) and (8.7.4) are the second fundamental coefficients of the free surface,

$$L = -\mathbf{X}_\alpha \cdot \hat{\mathbf{n}}_\alpha, \quad M = -\mathbf{X}_\alpha \cdot \hat{\mathbf{n}}_\beta = -\mathbf{X}_\beta \cdot \hat{\mathbf{n}}_\alpha, \quad N = -\mathbf{X}_\beta \cdot \hat{\mathbf{n}}_\beta. \qquad (8.7.5)$$

Thus, if the initial surface is parameterized according to (8.7.1), and if the tangential velocities are chosen according to (8.7.3), (8.7.4), then (8.7.1) will continue to be satisfied.

In the doubly periodic case, we make a small modification to (8.7.1): we introduce a function $\lambda(t)$, and require $E = \lambda(t)G$. This is similar to using the notion of the "normalized arclength parameterization" rather than simply an arclength parameterization in the case of 2D fluids; there, rather than choosing $s_\alpha = 1$, we chose $s_\alpha = L(t)/2\pi$, thus ensuring that the parameter α could always be taken from the same interval, $[0, 2\pi]$, at every time. For this generalized isothermal parameterization, equations similar to (8.7.3) and (8.7.4) can be found [44].

As before, the normal velocity U must be taken to be the normal component of the Birkhoff-Rott integral. The Birkhoff-Rott integral, $\mathbf{W}$, in the 3D case is given by

$$\mathbf{W} = \frac{1}{4\pi i}\text{PV}\int_{\mathbb{R}^2} \left(\mu_\alpha(\vec{a}')\mathbf{X}_\beta(\vec{a}') - \mu_\beta(\vec{a}')\mathbf{X}_\alpha(\vec{a}')\right) \times \frac{\mathbf{X}(\vec{a}) - \mathbf{X}(\vec{a}')}{|\mathbf{X}(\vec{a}) - \mathbf{X}(\vec{a}')|} \, d\vec{a}',$$
(8.7.6)

where $\vec{a} = (\alpha, \beta)$. One very significant difference between the 2D and 3D cases is that there is no analogue of the formula (8.2.4) available in the present setting. This was very convenient, as (8.2.4) allows the integral to be computed by considering only one fundamental period. In analytical work, this poses little difficulty, but numerically, computing (8.7.6) poses a challenge. In [44], we introduce a fast method for this based on Ewald summation.

As before, μ represents jump in velocity potential across the interface, and is also the source strength in the double-layer potential representation of the potentials. The evolution equation for μ is found, as is described in Section 8.2 earlier, by considering Bernoulli equations for the velocity potentials on either side of the interface, taking the limit approaching the interface from either side, and computing the jump. The Laplace-Young jump condition again comes into play while carrying out this process; the Laplace-Young jump condition in this case indicates that the jump in pressure across the interface is proportional to the mean curvature of the interface. In isothermal coordinates, the mean curvature is

$$\kappa = \frac{L+N}{2E}.$$
(8.7.7)

An additional difficulty in the case of 3D fluids is that there is, in a sense, a conflict between $\mathbf{X}$ and κ. In the case of 2D fluids, we made estimates for θ, and when we needed to talk about the surface z, we used the relationship $z_\alpha = \frac{L}{2\pi}e^{i\theta}$ [16, 26, 31, 84, 85]. Numerically, the situation is similar, in that θ is evolved in the HLS numerical method, and reconstructing z is straightforward [29, 30]. In the case of 3D fluids, we are able to make good energy estimates for κ, but the relationship between $\mathbf{X}$ and κ is much more complicated. As a result, in Section 5.2.1 of [27], significant care is taken with the relationship between $\mathbf{X}$ and κ when constructing the approximation scheme for the system; this approximation scheme is an iterative scheme, unlike the mollifiers used above, but this is in the same spirit. For the numerical method developed in [44], this was dealt with by evolving the surface, $\mathbf{X}$, rather than κ. There, a version of the SSD is developed directly for the evolution of $\mathbf{X}$. As mentioned earlier, this is not semilinear, so the use of the implicit-explicit scheme is more complicated than in the case of 2D fluids, but it nevertheless can be made to work.

Acknowledgments

The author is grateful to the National Science Foundation for support through grant DMS-1016267. The author is also grateful to the Isaac Newton Institute for hospitality in the summer of 2014 during the Theory of Water Waves program, and to the organizers of the Theory of Water Waves program and summer school.

References

[1] Caflisch, R.E., and Orellana, O.F. 1989. Singular solutions and ill-posedness for the evolution of vortex sheets. *SIAM J. Math. Anal.*, **20**(2), 293–307.

[2] Sulem, C., Sulem, P.-L., Bardos, C., and Frisch, U. 1981. Finite time analyticity for the two- and three-dimensional Kelvin-Helmholtz instability. *Comm. Math. Phys.*, **80**(4), 485–516.

[3] Duchon, J., and Robert, R. 1988. Global vortex sheet solutions of Euler equations in the plane. *J. Diff. Eq.*, **73**(2), 215–224.

[4] Ambrose, D.M., Bona, J.L., and Milgrom, T. 2014. Global solutions and ill-posedness for the Kaup system and related Boussinesq systems. Preprint.

[5] Milgrom, T., and Ambrose, D.M. 2013. Temporal boundary value problems in interfacial fluid dynamics. *Appl. Anal.*, **92**(5), 922–948.

[6] Beck, T., Sosoe, P., and Wong, P. 2014. Duchon-Robert solutions for the Rayleigh-Taylor and Muskat problems. *J. Diff. Eq.*, **256**(1), 206–222.

[7] Kamotski, V., and Lebeau, G. 2005. On 2D Rayleigh-Taylor instabilities. *Asymptot. Anal.*, **42**(1-2), 1–27.

[8] Lebeau, G. 2002. Régularité du problème de Kelvin-Helmholtz pour l'équation d'Euler 2d. *ESAIM Control Optim. Calc. Var.*, **8**, 801–825 (electronic). A tribute to J. L. Lions.

[9] Wu, S. 2006. Mathematical analysis of vortex sheets. *Comm. Pure Appl. Math.*, **59**(8), 1065–1206.

[10] Nalimov, V.I. 1974. The Cauchy-Poisson problem. *Dinamika Splošn. Sredy*, 104–210, 254.

[11] Yosihara, H. 1982. Gravity waves on the free surface of an incompressible perfect fluid of finite depth. *Publ. Res. Inst. Math. Sci.*, **18**(1), 49–96.

[12] Craig, W. 1985. An existence theory for water waves and the Boussinesq and Korteweg-de Vries scaling limits. *Comm. Partial Differential Equations*, **10**(8), 787–1003.

[13] Wu, S. 1997. Well-posedness in Sobolev spaces of the full water wave problem in 2-D. *Invent. Math.*, **130**(1), 39–72.

[14] Wu, S. 1999. Well-posedness in Sobolev spaces of the full water wave problem in 3-D. *J. Amer. Math. Soc.*, **12**(2), 445–495.

[15] Alazard, T., Burq, N., and Zuily, C. 2014. On the Cauchy problem for gravity water waves. *Invent. Math.*, **198**(1), 71–163.

[16] Ambrose, D.M., and Masmoudi, N. 2005. The zero surface tension limit of two-dimensional water waves. *Comm. Pure Appl. Math.*, **58**(10), 1287–1315.

[17] Ambrose, D.M., and Masmoudi, N. 2009. The zero surface tension limit of three-dimensional water waves. *Indiana Univ. Math. J.*, **58**(2), 479–521.

[18] Córdoba, A., Córdoba, D., and Gancedo, F. 2010. Interface evolution: water waves in 2-D. *Adv. Math.*, **223**(1), 120–173.

[19] Lannes, D. 2005. Well-posedness of the water-waves equations. *J. Amer. Math. Soc.*, **18**(3), 605–654 (electronic).

[20] Lannes, D. 2013. *The water waves problem*. Mathematical Surveys and Monographs, vol. 188. American Mathematical Society, Providence, RI. Mathematical analysis and asymptotics.

[21] Beyer, K., and Günther, M. 1998. On the Cauchy problem for a capillary drop. I. Irrotational motion. *Math. Methods Appl. Sci.*, **21**(12), 1149–1183.

[22] Iguchi, T. 2001. Well-posedness of the initial value problem for capillary-gravity waves. *Funkcial. Ekvac.*, **44**(2), 219–241.

[23] Yosihara, H. 1983. Capillary-gravity waves for an incompressible ideal fluid. *J. Math. Kyoto Univ.*, **23**(4), 649–694.

[24] Ambrose, D.M. 2007a. Regularization of the Kelvin-Helmholtz instability by surface tension. *Philos. Trans. R. Soc. Lond. Ser. A Math. Phys. Eng. Sci.*, **365**(1858), 2253–2266.

[25] Iguchi, T., Tanaka, N., and Tani, A. 1997. On the two-phase free boundary problem for two-dimensional water waves. *Math. Ann.*, **309**(2), 199–223.

[26] Ambrose, D.M. 2003. Well-posedness of vortex sheets with surface tension. *SIAM J. Math. Anal.*, **35**(1), 211–244 (electronic).

[27] Ambrose, D.M., and Masmoudi, N. 2007. Well-posedness of 3D vortex sheets with surface tension. *Commun. Math. Sci.*, **5**(2), 391–430.

[28] Ming, M., and Zhang, Z. 2009. Well-posedness of the water-wave problem with surface tension. *J. Math. Pures Appl. (9)*, **92**(5), 429–455.

[29] Hou, T.Y., Lowengrub, J.S., and Shelley, M.J. 1994. Removing the stiffness from interfacial flows with surface tension. *J. Comput. Phys.*, **114**(2), 312–338.

[30] Hou, T.Y., Lowengrub, J.S., and Shelley, M.J. 1997. The long-time motion of vortex sheets with surface tension. *Phys. Fluids*, **9**(7), 1933–1954.

[31] Ambrose, D.M. 2004. Well-posedness of two-phase Hele-Shaw flow without surface tension. *European J. Appl. Math.*, **15**(5), 597–607.

[32] Christianson, H., Hur, V.M., and Staffilani, G. 2010. Strichartz estimates for the water-wave problem with surface tension. *Comm. Partial Differential Equations*, **35**(12), 2195–2252.

[33] Córdoba, A., Córdoba, D., and Gancedo, F. 2011. Interface evolution: the Hele-Shaw and Muskat problems. *Ann. of Math. (2)*, **173**(1), 477–542.

[34] Düll, W.-P. 2012. Validity of the Korteweg-de Vries approximation for the two-dimensional water wave problem in the arc length formulation. *Comm. Pure Appl. Math.*, **65**(3), 381–429.

[35] Guo, Y., Hallstrom, C., and Spirn, D. 2007. Dynamics near unstable, interfacial fluids. *Comm. Math. Phys.*, **270**(3), 635–689.

[36] Ye, J., and Tanveer, S. 2011. Global existence for a translating near-circular Hele-Shaw bubble with surface tension. *SIAM J. Math. Anal.*, **43**(1), 457–506.

[37] Ye, J., and Tanveer, S. 2012. Global solutions for a two-phase Hele-Shaw bubble for a near-circular initial shape. *Complex Var. Elliptic Equ.*, **57**(1), 23–61.

[38] Ambrose, D.M. 2007b. Well-posedness of two-phase Darcy flow in 3D. *Quart. Appl. Math.*, **65**(1), 189–203.

[39] Córdoba, A., Córdoba, D., and Gancedo, F. 2013. Porous media: the Muskat problem in three dimensions. *Anal. PDE*, **6**(2), 447–497.

[40] Wang, W., Zhang, P., and Zhang, Z. 2012. Well-posedness of hydrodynamics on the moving elastic surface. *Arch. Ration. Mech. Anal.*, **206**(3), 953–995.

[41] Hou, T.Y., and Zhang, P. 2002. Convergence of a boundary integral method for 3-D water waves. *Discrete Contin. Dyn. Syst. Ser. B*, **2**(1), 1–34.

[42] Nie, Q. 2001. The nonlinear evolution of vortex sheets with surface tension in axisymmetric flows. *J. Comput. Phys.*, **174**(1), 438–459.

[43] Ambrose, D.M., and Siegel, M. 2012. A non-stiff boundary integral method for 3D porous media flow with surface tension. *Math. Comput. Simulation*, **82**(6), 968–983.

[44] Ambrose, D.M., Siegel, M., and Tlupova, S. 2013b. A small-scale decomposition for 3D boundary integral computations with surface tension. *J. Comput. Phys.*, **247**, 168–191.

[45] Cheng, C.-H.A., Coutand, D., and Shkoller, S. 2008. On the motion of vortex sheets with surface tension in three-dimensional Euler equations with vorticity. *Comm. Pure Appl. Math.*, **61**(12), 1715–1752.

[46] Coutand, D., and Shkoller, S. 2007. Well-posedness of the free-surface incompressible Euler equations with or without surface tension. *J. Amer. Math. Soc.*, **20**(3), 829–930.

[47] Lindblad, H. 2005. Well-posedness for the motion of an incompressible liquid with free surface boundary. *Ann. of Math. (2)*, **162**(1), 109–194.

[48] Ogawa, M., and Tani, A. 2002. Free boundary problem for an incompressible ideal fluid with surface tension. *Math. Models Methods Appl. Sci.*, **12**(12), 1725–1740.

[49] Schweizer, B. 2005. On the three-dimensional Euler equations with a free boundary subject to surface tension. *Ann. Inst. H. Poincaré Anal. Non Linéaire*, **22**(6), 753–781.

[50] Shatah, J., and Zeng, C. 2011. Local well-posedness for fluid interface problems. *Arch. Ration. Mech. Anal.*, **199**(2), 653–705.

[51] Zhang, P., and Zhang, Z. 2008. On the free boundary problem of three-dimensional incompressible Euler equations. *Comm. Pure Appl. Math.*, **61**(7), 877–940.

[52] Zhang, P., and Zhang, Z.-F. 2007. On the local wellposedness of 3-D water wave problem with vorticity. *Sci. China Ser. A*, **50**(8), 1065–1077.

[53] Alazard, T., and Delort, J.-M. 2013a. Global solutions and asymptotic behavior for two dimensional gravity water waves. Preprint. arXiv:1305.4090.

[54] Alazard, T., and Delort, J.-M. 2013b. Sobolev estimates for two dimensional gravity water waves. Preprint. arXiv:1307.3836.

[55] Germain, P., Masmoudi, N., and Shatah, J. 2009. Global solutions for the gravity water waves equation in dimension 3. *C. R. Math. Acad. Sci. Paris*, **347**(15-16), 897–902.

[56] Germain, P., Masmoudi, N., and Shatah, J. 2012. Global solutions for the gravity water waves equation in dimension 3. *Ann. of Math. (2)*, **175**(2), 691–754.

[57] Germain, P., Masmoudi, N., and Shatah, J. 2014. Global Existence for Capillary Water Waves. *Comm. Pure Appl. Math.*, n/a–n/a.

[58] Hunter, J., Ifrim, M., and Tataru, D. 2014. Two dimensional water waves in holomorphic coordinates. Preprint. arXiv:1401.1252.

[59] Ifrim, M., and Tataru, D. 2014a. The lifespan of small data solutions in two dimensional capillary water waves. Preprint. arXiv:1406.5471.

[60] Ifrim, M., and Tataru, D. 2014b. Two dimensional water waves in holomorphic coordinates II: global solutions. Preprint. arXiv:1404.7583.

[61] Ionescu, A.D., and Pusateri, F. 2013. Global solutions for the gravity water waves system in 2d. Preprint. arXiv:1303.5357.

[62] Wu, S. 2009. Almost global wellposedness of the 2-D full water wave problem. *Invent. Math.*, **177**(1), 45–135.

[63] Wu, S. 2011. Global wellposedness of the 3-D full water wave problem. *Invent. Math.*, **184**(1), 125–220.

[64] Castro, A., Cordoba, D., Fefferman, C., Gancedo, F., and Gomez-Serrano, J. 2012a. Finite time singularities for water waves with surface tension. *J. Math. Phys.*, **53**(11), –.

[65] Castro, A., Córdoba, D., Fefferman, C.L., Gancedo, F., and Gómez-Serrano, J. 2012b. Splash singularity for water waves. *Proc. Natl. Acad. Sci. USA*, **109**(3), 733–738.

[66] Coutand, D., and Shkoller, S. 2014a. On the Finite-Time Splash and Splat Singularities for the 3-D Free-Surface Euler Equations. *Comm. Math. Phys.*, **325**(1), 143–183.

[67] Coutand, D., and Shkoller, S. 2014b. On the impossibility of finite-time splash singularities for vortex sheets. Preprint. arXiv:1407.1479.

[68] Fefferman, C., Ionescu, A.D., and Lie, V. 2013. On the absence of "splash" singularities in the case of two-fluid interfaces. Preprint. arXiv.1312.2917.

[69] Fefferman, C.L. 2014. No-splash theorems for fluid interfaces. *Proc. Natl. Acad. Sci. USA*, **111**(2), 573–574.

[70] Alazard, T., Burq, N., and Zuily, C. 2010. Cauchy problem and Kato smoothing for water waves with surface tension. Pages 1–14 of: *Harmonic analysis and nonlinear partial differential equations*. RIMS Kôkyûroku Bessatsu, B18. *Res. Inst. Math. Sci.* (RIMS), Kyoto.

[71] Alazard, T., Burq, N., and Zuily, C. 2011. On the water-wave equations with surface tension. *Duke Math. J.*, **158**(3), 413–499.

[72] Christianson, H., Hur, V.M., and Staffilani, G. 2009. Local smoothing effects for the water-wave problem with surface tension. *C. R. Math. Acad. Sci. Paris*, **347**(3-4), 159–162.

[73] Kinsey, R.H., and Wu, S. 2014. A priori estimates for two-dimensional water waves with angled crests. Preprint. arXiv:1406.7573.

[74] Saffman, P.G. 1992. *Vortex dynamics*. Cambridge Monographs on Mechanics and Applied Mathematics. Cambridge University Press, New York.

[75] Majda, A.J., and Bertozzi, A.L. 2002. *Vorticity and incompressible flow*. Cambridge Texts in Applied Mathematics, vol. 27. Cambridge: Cambridge University Press.

[76] Caflisch, R.E., and Li, X.-F. 1992. Lagrangian theory for 3D vortex sheets with axial or helical symmetry. *Transport Theory Statist. Phys.*, **21**(4-6), 559–578.

[77] Muskhelishvili, N.I. 1992. *Singular integral equations*. Dover Publications, Inc., New York. Boundary problems of function theory and their application to mathematical physics, Translated from the second (1946) Russian edition and with a preface by J. R. M. Radok, Corrected reprint of the 1953 English translation.

[78] Ablowitz, M.J., and Fokas, A.S. 1997. *Complex variables: introduction and applications*. Cambridge Texts in Applied Mathematics. Cambridge University Press, Cambridge.

[79] Helson, H. 1983. *Harmonic analysis*. Reading, MA: Addison-Wesley Publishing Company Advanced Book Program.

David M. Ambrose

[80] Beale, J.T., Hou, T.Y., and Lowengrub, J.S. 1993. Growth rates for the linearized motion of fluid interfaces away from equilibrium. *Comm. Pure Appl. Math.*, **46**(9), 1269–1301.

[81] Ascher, U.M., Ruuth, S.J., and Wetton, B.T.R. 1995. Implicit-explicit methods for time-dependent partial differential equations. *SIAM J. Numer. Anal.*, **32**(3), 797–823.

[82] Ambrose, D.M. 2009. Singularity formation in a model for the vortex sheet with surface tension. *Math. Comput. Simulation*, **80**(1), 102–111.

[83] Ambrose, D.M., Kondrla, M., and Valle, M. 2013a. Computing time-periodic solutions of a model for the vortex sheet with surface tension. *Quart. Appl. Math.* To appear.

[84] Ambrose, D.M. 2014. The zero surface tension limit of two-dimensional interfacial Darcy flow. *J. Math. Fluid Mech.*, **16**(1), 105–143.

[85] Ambrose, D.M., and Siegel, M. 2014. Well-posedness of two-dimensional hydroelastic waves. Preprint.

[86] Baker, G.R., Meiron, D.I., and Orszag, S.A. 1982. Generalized vortex methods for free-surface flow problems. *J. Fluid Mech.*, **123**, 477–501.

[87] Ebin, D.G. 1987. The equations of motion of a perfect fluid with free boundary are not well posed. *Comm. Partial Differential Equations*, **12**(10), 1175–1201.

[88] Siegel, M., and Tanveer, S. 1996. Singular perturbation of smoothly evolving Hele-Shaw solutions. *Phys. Rev. Lett.*, **76**(Jan), 419–422.

[89] Siegel, M., Tanveer, S., and Dai, W.-S. 1996. Singular effects of surface tension in evolving Hele-Shaw flows. *J. Fluid Mech.*, **323**, 201–236.

[90] Ceniceros, H.D., and Hou, T.Y. 2000. The singular perturbation of surface tension in Hele-Shaw flows. *J. Fluid Mech.*, **409**, 251–272.

[91] Ceniceros, H.D., and Hou, T.Y. 2001. Numerical study of interfacial problems with small surface tension. Pages 63–92 of: *First International Congress of Chinese Mathematicians (Beijing, 1998)*. AMS/IP Stud. Adv. Math., vol. 20. Amer. Math. Soc., Providence, RI.

9

Wellposedness and Singularities of the Water Wave Equations

Sijue Wu

Abstract

A class of water wave problems concerns the dynamics of the free interface separating an inviscid, incompressible and irrotational fluid, under the influence of gravity, from a zero-density region. In this note, we present some recent methods and ideas developed concerning the local and global wellposedness of these problems, the focus is on the structural aspect of the equations.

9.1 Introduction

A class of water wave problems concerns the motion of the interface separating an inviscid, incompressible, irrotational fluid, under the influence of gravity, from a region of zero density (i.e., air) in n-dimensional space. It is assumed that the fluid region is below the air region. Assume that the density of the fluid is 1, the gravitational field is $-\mathbf{k}$, where $\mathbf{k}$ is the unit vector pointing in the upward vertical direction, and at time $t \geq 0$, the free interface is $\Sigma(t)$, and the fluid occupies region $\Omega(t)$. When surface tension is zero, the motion of the fluid is described by

$$\begin{cases} \mathbf{v}_t + (\mathbf{v} \cdot \nabla)\mathbf{v} = -\mathbf{k} - \nabla P & \text{on } \Omega(t),\, t \geq 0, \\ \operatorname{div} \mathbf{v} = 0, \quad \operatorname{curl} \mathbf{v} = 0, & \text{on } \Omega(t),\, t \geq 0, \\ P = 0, & \text{on } \Sigma(t) \\ (1, \mathbf{v}) \text{ is tangent to the free surface } (t, \Sigma(t)), \end{cases} \quad (9.1.1)$$

where $\mathbf{v}$ is the fluid velocity, P is the fluid pressure. There is an important condition for these problems:

$$-\frac{\partial P}{\partial \mathbf{n}} \geq 0 \quad (9.1.2)$$

pointwise on the interface, where $\mathbf{n}$ is the outward unit normal to the fluid interface $\Sigma(t)$ [1]; it is well known that when surface tension is neglected and the Taylor sign condition (9.1.2) fails, the water wave motion can be subject to the Taylor instability [1–3].

The study on water waves dates back centuries. Early mathematical works include Stokes [4], Levi-Civita [5], and Taylor [1]. Nalimov [6], Yosihara [7], and Craig [8] proved local in time existence and uniqueness of solutions for the 2D water wave equation (9.1.1) for small initial data. In [9, 10], we showed that for dimensions $n \geq 2$, the strong Taylor sign condition

$$-\frac{\partial P}{\partial \mathbf{n}} \geq c_0 > 0 \tag{9.1.3}$$

always holds for the infinite depth water wave problem (9.1.1), as long as the interface is non–self-intersecting and smooth; and the initial value problem of equation (9.1.1) is locally well-posed in Sobolev spaces H^s, $s \geq 4$ for arbitrary given data. Since then, local wellposedness for water waves with additional effects such as the surface tension, bottom and non-zero vorticity, under the assumption (9.1.3)[a] were obtained (c.f. [11–19]). Alazard, Burq & Zuily [20, 21] proved local wellposedness of (9.1.1) in low regularity Sobolev spaces where the interfaces are only in $C^{3/2}$. Recently, we proved the almost global and global wellposedness of (9.1.1) for small, smooth, and localized initial data for dimensions $n \geq 2$ [22, 23]; Germain, Masmudi & Shatah obtained global existence for 3D water waves for a different class of small, smooth and localized data [24]. Our 2D almost global existence result has now been extended to global by Ionescu & Pusateri [25] and independently Alazard & Delort [26], see [27, 28] for an alternative proof. Finally, we mention our most recent work on two-dimensional water waves with angled crests [29–31], in which we showed that for water waves with angled crests, only the degenerate Taylor sign condition (9.1.2) holds, with degeneracy at the singularities on the interface. We proved an a priori estimate [30], a blow-up criteria and the local existence [31] for the 2D water wave equation (9.1.1) in this framework.

The advances of water waves theory rely crucially on the understanding of the structure of the water wave equations. Indeed, (9.1.1) is a nonlinear equation defined on moving domains, it is difficult to study it directly. A classical approach is to reduce from (9.1.1) to an equation on the interface, and study solutions of the interface equation. Then use the incompressibility and irrotationality of the velocity field to recover the velocity in the fluid domain by solving a boundary value problem for the Laplace equation. However the fluid interface equation is itself a fully nonlinear and nonlocal equation, its structure is not easy to understand. It is by achieving better understandings of the structure of this equation that has enabled us to apply analytical tools to deduce informations on the nature of the fluid motion.

---

[a] When there is surface tension, or bottom, or vorticity, (9.1.3) does not always hold, it needs to be assumed.

In this note we describe the approach in [9, 10, 22, 23]. Our focus is on the structural aspect of the work. It is clear that the 2D case is structurally simpler than 3D. Our strategy has been to first understand the two-dimensional case, taking advantage of complex analysis tools, in particular the Riemann mapping theorem, then use Clifford analysis to extend the 2D results to 3D. We note that although Riemann mapping is not available in 3D, it is by using it that has enabled us to understand the 2D case well enough to develop an approach that extends to all dimensions.

We consider solutions of the water wave equation (9.1.1) in the setting where

$$\mathbf{v}(\xi, t) \to 0, \qquad \text{as } |\xi| \to \infty$$

and the interface $\Sigma(t)$ tends to the horizontal plane at infinity.[b]

In section 9.2, we discuss the local wellposedness of (9.1.1), the focus is on deriving the quasilinear structure of equation (9.1.1), c.f. [9, 10, 30]. In section 9.3, we consider the global in time behavior of solutions of (9.1.1) in the regime of small waves, the focus is on understanding the nature of the nonlinearity of equation (9.1.1), c.f. [22, 23]. We give some preparatory materials in the appendices. In Appendix A, we give some basic analysis tools such as estimates for commutators and operators involved in the equations. These inequalities tell us how a certain term behaves in terms of estimates, whether it is of higher order, or lower order etc., and guide us in our derivations of the structure of the equation. In Appendix B we give some commutator identities that is used in our derivations.

We use the following notations and conventions: $[A, B] := AB - BA$ is the commutator of operators A and B; compositions are always in terms of the spatial variables and we write for $f = f(\cdot, t)$, $g = g(\cdot, t)$, $f(g(\cdot, t), t) := f \circ g(\cdot, t) := U_g f(\cdot, t)$.

9.2 Local Wellposedness of the Water Wave Equations

From basic partial differential equation (PDE) theory we expect that in general, the Cauchy problem for a hyperbolic type PDE is locally solvable in Sobolev spaces, and we can solve it, for a short time, by energy estimates and an iterative argument; while the Cauchy problem for an elliptic type PDE is ill-posed. Hence in order to understand whether the water wave equation

---

[b] The problem with velocity $\mathbf{v}(\xi, t) \to (c, 0)$ as $|\xi| \to \infty$ can be reduced to one with $\mathbf{v}$ tends to 0 at infinity by the following observation: if $(\mathbf{v}, P)$ with $\Sigma(t) : \xi = \xi(\cdot, t)$ is a solution of (9.1.1), then

$$\mathbf{V} = \mathbf{v}(\zeta + (c, 0)t, t) - c, \quad \mathbf{P} = P(\zeta + (c, 0)t, t), \quad \text{with} \quad \Sigma(t) - (c, 0)t : \zeta = \xi(\cdot, t) - (c, 0)t$$

is also a solution of (9.1.1).

(9.1.1) is uniquely solvable for a positive time period for arbitrary given Cauchy data, it is crucial to understand its quasilinear structure. In this section, we derive the quasilinear structure of the water wave equation (9.1.1), our focus is on the 2D case, since it is through a thorough understanding of this case that has enabled us to extend our work to 3D. We only give a brief description of how to extend the 2D derivations to 3D. We show that the strong Taylor inequality (9.1.3) always holds for $C^{1,\gamma}$ interfaces, while for singular interfaces only the weak Taylor inequality (9.1.2) holds; and this implies that the quasilinear structure of the water wave equation (9.1.1) is of hyperbolic type in the regime of $C^{1,\gamma}$ interfaces and respectively of degenerate hyperbolic type in the regime that includes singular free surfaces. The derivation given here is based on that in [9, 10]; due to the scope of this lecture note, we will only discuss the structural aspect of the work [9, 10], and leave out the proof for the local wellposedness of (9.1.1). The interested reader may consult [9, 10] for the proof.

9.2.1 The Equation of the Fluid Interface in Two-Space Dimensions

In two-space dimensions, we identify (x,y) with the complex number $x + iy$; $\operatorname{Re} z$, $\operatorname{Im} z$ are the real and imaginary parts of z; $\bar{z} = \operatorname{Re} z - i\operatorname{Im} z$ is the complex conjugate.

Let the interface $\Sigma(t) : z = z(\alpha,t)$, $\alpha \in \mathbb{R}$ be given by Lagrangian parameter α, so $z_t(\alpha,t) = \mathbf{v}(z(\alpha,t);t)$ is the velocity of the fluid particles on the interface, $z_{tt}(\alpha,t) = \mathbf{v}_t + (\mathbf{v} \cdot \nabla)\mathbf{v}(z(\alpha,t);t)$ is the acceleration; notice that $P = 0$ on $\Sigma(t)$ implies that ∇P is normal to $\Sigma(t)$, therefore $\nabla P = -i\mathfrak{a}z_\alpha$, where $\mathfrak{a} = -\frac{1}{|z_\alpha|}\frac{\partial P}{\partial \mathbf{n}}$; we have from the first and third equation of (9.1.1) that

$$z_{tt} + i = i\mathfrak{a}z_\alpha. \tag{9.2.1}$$

The second equation of (9.1.1): div $\mathbf{v} = $ curl $\mathbf{v} = 0$ implies that $\bar{\mathbf{v}}$ is holomorphic in the fluid domain $\Omega(t)$; hence $\bar{z}_t$ is the boundary value of a holomorphic function in $\Omega(t)$.

Let $\Omega \subset \mathbb{C}$ be a domain with boundary $\Sigma : z = z(\alpha)$, $\alpha \in I$, oriented clockwise. Let $\mathfrak{H}$ be the Hilbert transform associated to Ω:

$$\mathfrak{H}f(\alpha) = \frac{1}{\pi i}\,\mathrm{pv.}\int \frac{z_\beta(\beta)}{z(\alpha) - z(\beta)}f(\beta)\,d\beta. \tag{9.2.2}$$

We have the following characterization of the trace of a holomorphic function on Ω.

Proposition 9.2.1 [32] a. Let $g \in L^p$ for some $1 < p < \infty$. Then g is the boundary value of a holomorphic function G on Ω with $G(z) \to 0$ at infinity if and only if

$$(I - \mathfrak{H})g = 0. \tag{9.2.3}$$

b. Let $f \in L^p$ for some $1 < p < \infty$. Then $\mathbb{P}_H f := \frac{1}{2}(I + \mathfrak{H})f$ is the boundary value of a holomorphic function $\mathfrak{G}$ on Ω, with $\mathfrak{G}(z) \to 0$ as $|z| \to \infty$.

c. $\mathfrak{H}1 = 0$.

Observe Proposition 9.2.1 gives that $\mathfrak{H}^2 = I$ on L^p.

From Proposition 9.2.1 the second equation of (9.1.1) is equivalent to $\bar{z}_t = \mathfrak{H}\bar{z}_t$. Therefore the motion of the fluid interface $\Sigma(t) : z = z(\alpha, t)$ is given by [c]

$$\begin{cases} z_{tt} + i = i a z_\alpha \\ \bar{z}_t = \mathfrak{H}\bar{z}_t \end{cases} \tag{9.2.4}$$

(9.2.4) is a fully nonlinear equation. To understand whether the equation is well posed, a usual strategy is to quasilinearize the equation by differentiating. Notice that it can be hard to analyze the Hilbert transform $\mathfrak{H}$ in the second equation since it depends nonlinear nonlocally on the interface $z = z(\alpha, t)$, this motivates us to use the Riemann mapping (c.f.[9]).[d]

9.2.2 Riemann Mappings and the Quasilinear Structure of 2D Water Waves

Let $\Phi(\cdot, t) : \Omega(t) \to P_-$ be the Riemann mapping taking $\Omega(t)$ to the lower half plane P_-, satisfying $\lim_{z \to \infty} \Phi_z(z, t) = 1$. Let

$$h(\alpha; t) := \Phi(z(\alpha, t), t),$$

so $h : \mathbb{R} \to \mathbb{R}$ is a homeomorphism. Let h^{-1} be defined by

$$h(h^{-1}(\alpha', t), t) = \alpha', \quad \alpha' \in \mathbb{R};$$

and

$$Z(\alpha', t) := z \circ h^{-1}(\alpha', t), \quad Z_t(\alpha', t) := z_t \circ h^{-1}(\alpha', t), \quad Z_{tt}(\alpha', t) := z_{tt} \circ h^{-1}(\alpha', t)$$

be the reparametrization of the position, velocity, and acceleration of the interface in the Riemann mapping variable α'. Let

$$Z_{,\alpha'}(\alpha', t) := \partial_{\alpha'} Z(\alpha', t), \qquad Z_{t,\alpha'}(\alpha', t) := \partial_{\alpha'} Z_t(\alpha', t), \quad \text{etc.}$$

We note that $\Phi^{-1}(\alpha', t) = Z(\alpha', t)$. Notice that $\bar{v} \circ \Phi^{-1} : P_- \to \mathbb{C}$ is holomorphic in the lower half plane P_- with $\bar{v} \circ \Phi^{-1}(\alpha', t) = \bar{Z}_t(\alpha', t)$. Precomposing (9.2.1)

---

[c] Equation (9.1.1) and equation (9.2.4) are equivalent, see [9, 10].

[d] In [6, 7], a quasilinear equation for the interface was derived for small and smooth waves in terms of the Lagrangian coordinates. In [9], by using the Riemann mapping, a more concise quasilinear equation was derived for all non–self-intersecting waves. The approach described here is inspired by that in [9], it was used when we extended the work in [9] to 3D [10]. It is similar to that in [9], with the difference that in [9] the derivation is in terms of the real component and here it is in both components.

with h^{-1} and applying Proposition 9.2.1 to $\bar{\mathbf{v}} \circ \Phi^{-1}$ in P_-, we have the free surface equation in the Riemann mapping variable:

$$\begin{cases} Z_{tt} + i = i\mathcal{A}Z_{,\alpha'} \\ \bar{Z}_t = \mathbb{H}\bar{Z}_t \end{cases} \tag{9.2.5}$$

where $\mathcal{A} \circ h = \mathfrak{a}h_\alpha$ and $\mathbb{H}$ is the Hilbert transform associated with the lower half plane P_-:

$$\mathbb{H}f(\alpha') = \frac{1}{\pi i}\text{pv.} \int \frac{1}{\alpha' - \beta'} f(\beta')\,d\beta'.$$

[e] As has been shown in [9], the quasilinearization of (9.2.4) or (9.2.5) can be accomplished by just taking one derivative to t to equation (9.2.1).

Taking one derivative with respect to t to (9.2.1), we get

$$\bar{z}_{ttt} + i\mathfrak{a}\bar{z}_{t\alpha} = -i\mathfrak{a}_t\bar{z}_\alpha = \frac{\mathfrak{a}_t}{\mathfrak{a}}(\bar{z}_{tt} - i), \tag{9.2.6}$$

the free surface equation is now (9.2.6) with the constraint $\bar{z}_t = \mathfrak{H}\bar{z}_t$. Precomposing with h^{-1} on both sides of (9.2.6) we have in the Riemann mapping variable the free surface equation[f]

$$\begin{cases} \bar{Z}_{ttt} + i\mathcal{A}\bar{Z}_{t,\alpha'} = \frac{\mathfrak{a}_t}{\mathfrak{a}} \circ h^{-1}(\bar{Z}_{tt} - i) \\ \bar{Z}_t = \mathbb{H}\bar{Z}_t. \end{cases} \tag{9.2.7}$$

We note that from the chain rule that for any function f,

$$U_{h^{-1}}\partial_t U_h f(\alpha', t) = (\partial_t + \mathcal{B}\partial_{\alpha'})f(\alpha', t)$$

where $\mathcal{B} = h_t \circ h^{-1}$. Hence $\bar{Z}_{tt} = (\partial_t + \mathcal{B}\partial_{\alpha'})\bar{Z}_t$ and $\bar{Z}_{ttt} = (\partial_t + \mathcal{B}\partial_{\alpha'})^2\bar{Z}_t$. We will find $\mathcal{B}$, $\mathcal{A}$, and $\frac{\mathfrak{a}_t}{\mathfrak{a}} \circ h^{-1}$ in terms of $\bar{Z}_t$ and $\bar{Z}_{tt}$ and show that (9.2.7) is a quasilinear equation for $\mathcal{U} = \bar{Z}_t$ with the right hand side consisting of lower order terms.[g,h]

[e] The advantage of using the Riemann mapping is that in P_-, the boundary value of a holomorphic function is characterized by $g = \mathbb{H}g$, see Proposition 9.2.1. The kernel of the Hilbert transform $\mathbb{H}$ is purely imaginary and is independent of the interface, it is easier to use $\mathbb{H}$ to understand the relations among various quantities and hence the quasilinear structure of the free surface equation.

[f] (9.2.6) and (9.2.7) are equivalent to (9.2.1) and (9.2.5) provided the initial data for (9.2.6) and (9.2.7) stafisfy (9.2.1) or (9.2.5) at $t = 0$.

[g] From Proposition 9.2.1, $(I - \mathfrak{H})f$ measures how un-holomorphic a function f is. The reason (9.2.6) is quasilinear with the right hand side of lower order is that the two terms on the left hand side are "almost holomorphic" in the sense that $(I - \mathfrak{H})(\bar{z}_{ttt} + i\mathfrak{a}\partial_\alpha\bar{z}_t)$ is a commutator – since by $(I - \mathfrak{H})\bar{z}_t = 0$, $(I - \mathfrak{H})(\bar{z}_{ttt} + i\mathfrak{a}\bar{z}_{t\alpha}) = [\partial_t^2 + i\mathfrak{a}\partial_\alpha, \mathfrak{H}]\bar{z}_t$, while the conjugate of the right hand side points in the normal direction. We know a commutator is of lower order (c.f. Appendices A and B); and a holomorphic function with real part zero must be a constant; similarly an anti-holomorphic function with zero tangential part on the boundary must be a constant. If the left hand side of (9.2.6) were holomorphic, then it would have to be zero. Now the left hand side is almost holomorphic with $(I - \mathfrak{H})(\bar{z}_{ttt} + i\mathfrak{a}\partial_\alpha\bar{z}_t)$ a lower order term, then the right hand side must be a lower order term. We get this insight after our work in [9], not before.

[h] We sometimes abuse notation and say a function f is holomorphic if f is the boundary value of a holomorphic function in $\Omega(t)$.

To this end, we need some basic estimates for commutators, we leave these and some other preparatory materials in the Appendices. The reader may want to consult the Appendices before continuing.

The quantity $\mathcal{A}$ and the Taylor sign condition

The basic idea of deriving the formulas for $\mathcal{A}$, $\mathcal{B}$ and $\frac{a_t}{a} \circ h^{-1}$ is to use the holomorphicity or almost holomorphicity of our quantities, and the fact that $\text{Re}(I - \mathbb{H})f = f$ for real valued functions f. We will often write $(I - \mathbb{H})(fg)$ as a commutator: $(I - \mathbb{H})(fg) = [f, \mathbb{H}]g$ when g satisfies $(I - \mathbb{H})g = 0$, since commutators are favorable in terms of estimates, see Appendix A.

Let $D_\alpha := \frac{1}{z_\alpha}\partial_\alpha$, and $D_{\alpha'} := \frac{1}{Z_{,\alpha'}}\partial_{\alpha'}$. Notice that for any holomorphic function G in $\Omega(t)$ with boundary value $g(\alpha, t) = G(z(\alpha, t), t)$,

$$D_\alpha g(\alpha, t) = G_z(z(\alpha, t), t)$$

and for any function f,

$$(D_\alpha f) \circ h^{-1} = D_{\alpha'}(f \circ h^{-1}).$$

We note that $\mathcal{A}$ is related to the important quantity $-\frac{\partial P}{\partial \mathbf{n}}$ by $\mathcal{A} \circ h = \mathfrak{a}h_\alpha$ and $\mathfrak{a} = -\frac{1}{|z_\alpha|}\frac{\partial P}{\partial \mathbf{n}}$, therefore

$$-\frac{\partial P}{\partial \mathbf{n}}\Big|_{z=z(\alpha,t)} = (\mathcal{A}|Z_{,\alpha}|) \circ h. \tag{9.2.8}$$

In this subsection we derive a formula for the quantity $\mathcal{A}$. This formula was first derived in [9] to show that the strong Taylor sign condition (9.1.3) always holds for smooth non–self-intersecting interfaces. It has also played a key role in our recent work on 2D water waves with angled crests [30, 31].

Taking complex conjugate of the first equation in (9.2.5) then multiply by $Z_{,\alpha'}$ yields

$$Z_{,\alpha'}(\bar{Z}_{tt} - i) = -i\mathcal{A}|Z_{,\alpha'}|^2 := -iA_1. \tag{9.2.9}$$

The left hand side of (9.2.9) is almost holomorphic since $Z_{,\alpha'}$ is the boundary value of the holomorphic function $(\Phi^{-1})_{z'}$ and $\bar{Z}_{tt}$ is the time derivative of the holomorphic function $\bar{z}_t$. We explore the almost holomorphicity of $\bar{z}_{tt}$ by expanding. Let $F = \bar{\mathbf{v}}$, we know F is holomorphic in $\Omega(t)$, and $\bar{z}_t = F(z(\alpha, t), t)$, so

$$\bar{z}_{tt} = F_t(z(\alpha, t), t) + F_z(z(\alpha, t), t)z_t(\alpha, t), \qquad \bar{z}_{t\alpha} = F_z(z(\alpha, t), t)z_\alpha(\alpha, t) \tag{9.2.10}$$

therefore

$$\bar{z}_{tt} = F_t \circ z + \frac{\bar{z}_{t\alpha}}{z_\alpha}z_t. \tag{9.2.11}$$

Precomposing with h^{-1}, subtracting $-i$, then multiplying by $Z_{,\alpha'}$, we have

$$Z_{,\alpha'}(\bar{Z}_{tt} - i) = Z_{,\alpha'}F_t \circ Z + Z_t\bar{Z}_{t,\alpha'} - iZ_{,\alpha'} = -iA_1.$$

Apply $(I - \mathbb{H})$ to both sides of the equation. Notice that $F_t \circ Z$ is the boundary value of the holomorphic function $F_t \circ \Phi^{-1}$, so $(I - \mathbb{H})(Z_{,\alpha'} F_t \circ Z) = 0$, $(I - \mathbb{H})Z_{,\alpha'} = 1$,[i] therefore

$$-i(I - \mathbb{H})A_1 = (I - \mathbb{H})(Z_t \bar{Z}_{t,\alpha'}) - i.$$

Taking imaginary parts on both sides and using the fact $(I - \mathbb{H})\bar{Z}_{t,\alpha'} = 0$ [j] to rewrite $(I - \mathbb{H})(Z_t \bar{Z}_{t,\alpha'})$ as $[Z_t, \mathbb{H}]\bar{Z}_{t,\alpha'}$ yields

$$A_1 = 1 - \text{Im}[Z_t, \mathbb{H}]\bar{Z}_{t,\alpha'}. \tag{9.2.12}$$

Compute

$$
\begin{aligned}
- \text{Im}[Z_t, \mathbb{H}]\bar{Z}_{t,\alpha'} &= -\frac{1}{\pi} \text{Re} \int \frac{(Z_t(\alpha',t) - Z_t(\beta',t))\partial_{\beta'}(\bar{Z}_t(\alpha',t) - \bar{Z}_t(\beta',t))}{\alpha' - \beta'} d\beta \\
&= -\frac{1}{2\pi} \int \frac{\partial_{\beta'}|Z_t(\alpha',t) - Z_t(\beta',t)|^2}{\alpha' - \beta'} d\beta \\
&= \frac{1}{2\pi} \int \frac{|Z_t(\alpha',t) - Z_t(\beta',t)|^2}{(\alpha' - \beta')^2} d\beta'
\end{aligned}
\tag{9.2.13}
$$

where in the last step we used integration by parts. We conclude

Proposition 9.2.2 (c.f. [9], Lemma 3.1) We have
1.

$$A_1 = 1 - \text{Im}[Z_t, \mathbb{H}]\bar{Z}_{t,\alpha'} = 1 + \frac{1}{2\pi} \int \frac{|Z_t(\alpha',t) - Z_t(\beta',t)|^2}{(\alpha' - \beta')^2} d\beta' \geq 1. \tag{9.2.14}$$

2.

$$-\frac{\partial P}{\partial \mathbf{n}} = \frac{A_1}{|Z_{,\alpha'}|}; \tag{9.2.15}$$

in particular if the interface $\Sigma(t) \in C^{1,\gamma}$ for some $\gamma > 0$, then the strong Taylor sign condition (9.1.3) holds.

From (9.2.9) we have

$$\mathcal{A} = \frac{A_1}{|Z_{,\alpha'}|^2} = \frac{|\bar{Z}_{tt} - i|^2}{A_1} \tag{9.2.16}$$

with A_1 given by (9.2.14).

The quantity $\mathcal{B} = h_t \circ h^{-1}$

The quantity $\mathcal{B} = h_t \circ h^{-1}$ can be calculated similarly. Recall $h(\alpha,t) = \Phi(z(\alpha,t),t)$, so

$$h_t = \Phi_t \circ z + (\Phi_z \circ z)z_t, \qquad h_\alpha = (\Phi_z \circ z)z_\alpha$$

---

[i] We know $(\Phi^{-1})_{z'} \to 1$ as $z' \to \infty$; we assume a priori that $(\Phi^{-1})_{z'} F_t \circ \Phi^{-1} \to 0$ as $z' \to \infty$. As proved in [9] equation (9.2.5) is well posed in this regime.
[j] Because $(I - \mathbb{H})\bar{Z}_t = 0$.

hence $h_t = \Phi_t \circ z + \frac{h_\alpha}{z_\alpha} z_t$. Precomposing with h^{-1} yields

$$h_t \circ h^{-1} = \Phi_t \circ Z + \frac{Z_t}{Z_{,\alpha'}}. \tag{9.2.17}$$

Now $(I - \mathbb{H})\Phi_t \circ Z = (I - \mathbb{H})(\frac{1}{Z_{,\alpha'}} - 1) = 0$ since $\Phi_t \circ Z$ and $\frac{1}{Z_{,\alpha'}} - 1$ are boundary values of the holomorphic functions $\Phi_t \circ \Phi^{-1}$ and $\frac{1}{(\Phi^{-1})_{z'}} - 1$, respectively.[k] Apply $(I - \mathbb{H})$ to both sides of (9.2.17) then take the real parts; rewriting $(I - \mathbb{H})(Z_t(\frac{1}{Z_{,\alpha'}} - 1))$ as $[Z_t, \mathbb{H}](\frac{1}{Z_{,\alpha'}} - 1)$, we get

$$B = h_t \circ h^{-1} = \mathrm{Re}\left([Z_t, \mathbb{H}](\frac{1}{Z_{,\alpha'}} - 1)\right) + 2\,\mathrm{Re}\,Z_t. \tag{9.2.18}$$

Here we used the fact that $\mathrm{Re}(I - \mathbb{H})Z_t = 2\,\mathrm{Re}\,Z_t$, since $(I + \mathbb{H})Z_t = \overline{(I - \mathbb{H})\bar{Z}_t} = 0$.

The quantity $\frac{a_t}{a} \circ h^{-1}$

We analyze $\frac{a_t}{a} \circ h^{-1}$ similarly. Start with (9.2.6) or the first equation of (9.2.7), and expand the left hand side using $\bar{z}_t = F(z(\alpha, t), t)$. Taking one more derivative to t to the first equation in (9.2.10), we get

$$\bar{z}_{ttt} = (F_{zz} \circ z)z_t^2 + 2(F_{tz} \circ z)z_t + (F_z \circ z)z_{tt} + F_{tt} \circ z.$$

From (9.2.11), the second equation of (9.2.10), and the holomorphicity of F_z, F_t, we have

$$F_z \circ z = D_\alpha \bar{z}_t, \quad F_{zz} \circ z = D_\alpha^2 \bar{z}_t, \quad \text{and } F_{tz} \circ z = D_\alpha(\bar{z}_{tt} - (D_\alpha \bar{z}_t)z_t)$$

therefore

$$\bar{z}_{ttt} = (D_\alpha^2 \bar{z}_t)z_t^2 + 2z_t D_\alpha(\bar{z}_{tt} - (D_\alpha \bar{z}_t)z_t) + (D_\alpha \bar{z}_t)z_{tt} + F_{tt} \circ z.$$

Precomposing with h^{-1} yields

$$\bar{Z}_{ttt} = (D_{\alpha'}^2 \bar{Z}_t)Z_t^2 + 2Z_t D_{\alpha'}(\bar{Z}_{tt} - (D_{\alpha'} \bar{Z}_t)Z_t) + (D_{\alpha'} \bar{Z}_t)Z_{tt} + F_{tt} \circ Z. \tag{9.2.19}$$

Now multiply the first equation of (9.2.7) by $Z_{,\alpha'}$, then substitute in (9.2.19). We have, by (9.2.9),

$$Z_{,\alpha'}\{(D_{\alpha'}^2 \bar{Z}_t)Z_t^2 + 2Z_t D_{\alpha'}(\bar{Z}_{tt} - (D_{\alpha'} \bar{Z}_t)Z_t) + (D_{\alpha'} \bar{Z}_t)Z_{tt} + \quad F_{tt} \circ Z + i A \bar{Z}_{t,\alpha'}\}$$

$$= -iA_1 \frac{a_t}{a} \circ h^{-1}. \tag{9.2.20}$$

Apply $(I - \mathbb{H})$ to both sides of (9.2.20). Using the fact that $(I - \mathbb{H})(\partial_{\alpha'} D_{\alpha'} \bar{Z}_t) = 0$, $(I - \mathbb{H})\partial_{\alpha'}(\bar{Z}_{tt} - (D_{\alpha'} \bar{Z}_t)Z_t) = 0$, $(I - \mathbb{H})\partial_{\alpha'} \bar{Z}_t = 0$ and $(I - \mathbb{H})F_{tt} \circ Z = 0$

---

[k] Again, we assume a priori that $\Phi_t \circ \Phi^{-1}(z', t) \to 0$ as $z' \to \infty$. It has been proved in [9] that such solutions exist.

because of the holomorphicity of $\partial_{z'}(F_z \circ \Phi^{-1})$, $\partial_{z'}(F_t \circ \Phi^{-1})$, $\partial_{z'}(F \circ \Phi^{-1})$, and $F_{tt} \circ \Phi^{-1}$, and the identity $Z_{tt} + i = i\mathcal{A}Z_{,\alpha'}$ (9.2.5), we rewrite each term on the left as commutator and get

$$[Z_t^2, \mathbb{H}]\partial_{\alpha'}D_{\alpha'}\bar{Z}_t + 2[Z_t, \mathbb{H}]\partial_{\alpha'}(\bar{Z}_{tt} - (D_{\alpha'}\bar{Z}_t)Z_t) \quad + 2[Z_{tt}, \mathbb{H}]\partial_{\alpha'}\bar{Z}_t$$
$$= (I - \mathbb{H})(-iA_1 \frac{a_t}{a} \circ h^{-1}). \tag{9.2.21}$$

For the sake of estimates we need to further rewrite the first term and the second part of the second term. Using integration by parts on each term, after cancelations we have

$$[Z_t^2, \mathbb{H}]\partial_{\alpha'}D_{\alpha'}\bar{Z}_t - 2[Z_t, \mathbb{H}]\partial_{\alpha'}((D_{\alpha'}\bar{Z}_t)Z_t)$$
$$= -\frac{1}{\pi i}\int \frac{(Z_t(\alpha',t) - Z_t(\beta',t))^2}{(\alpha' - \beta')^2}D_{\beta'}\bar{Z}_t(\beta',t)\,d\beta'$$
$$:= -[Z_t, Z_t; D_{\alpha'}\bar{Z}_t].$$

Therefore we can write (9.2.21) as

$$2[Z_t, \mathbb{H}]\bar{Z}_{tt,\alpha'} + 2[Z_{tt}, \mathbb{H}]\partial_{\alpha'}\bar{Z}_t - [Z_t, Z_t; D_{\alpha'}\bar{Z}_t] = (I - \mathbb{H})(-iA_1 \frac{a_t}{a} \circ h^{-1}). \tag{9.2.22}$$

Taking imaginary parts on both sides and dividing by $-A_1$ yields

$$\frac{a_t}{a} \circ h^{-1} = \frac{-\operatorname{Im}(2[Z_t, \mathbb{H}]\bar{Z}_{tt,\alpha'} + 2[Z_{tt}, \mathbb{H}]\partial_{\alpha'}\bar{Z}_t - [Z_t, Z_t; D_{\alpha'}\bar{Z}_t])}{A_1}. \tag{9.2.23}$$

The quasilinear equation

Sum up (9.2.7), (9.2.14), (9.2.16), (9.2.18), and (9.2.23), we have the equation for the free interface:

$$\begin{cases} (\partial_t + \mathcal{B}\partial_{\alpha'})^2\bar{Z}_t + i\mathcal{A}\partial_{\alpha'}\bar{Z}_t = g \\ \bar{Z}_t = \mathbb{H}\bar{Z}_t \end{cases} \tag{9.2.24}$$

where

$$\begin{cases} \mathcal{A} = \frac{|\bar{Z}_{tt} - i|^2}{A_1}, \quad Z_{tt} = (\partial_t + \mathcal{B}\partial_{\alpha'})Z_t \\ A_1 = 1 - \operatorname{Im}[Z_t, \mathbb{H}]\bar{Z}_{t,\alpha'} = 1 + \frac{1}{2\pi}\int \frac{|Z_t(\alpha',t) - Z_t(\beta',t)|^2}{(\alpha' - \beta')^2}\,d\beta' \\ \mathcal{B} = h_t \circ h^{-1} = \operatorname{Re}\left([Z_t, \mathbb{H}](\frac{1}{Z_{,\alpha'}} - 1)\right) + 2\operatorname{Re}Z_t \\ \frac{1}{Z_{,\alpha'}} = i\frac{\bar{Z}_{tt} - i}{A_1} \\ g = (\bar{Z}_{tt} - i)\frac{-\operatorname{Im}(2[Z_t, \mathbb{H}]\bar{Z}_{tt,\alpha'} + 2[Z_{tt}, \mathbb{H}]\partial_{\alpha'}\bar{Z}_t - [Z_t, Z_t; D_{\alpha'}\bar{Z}_t])}{A_1}. \end{cases} \tag{9.2.25}$$

From the inequalities in Appendix A, we know in the regime where the interface is $C^{1,\gamma}$, $\gamma > 0$, (9.2.24)–(9.2.25) is a quasilinear equation of the conjugate velocity $\bar{Z}_t$, with the right hand side of (9.2.24) consisting of lower order terms. (9.2.24) is of hyperbolic type since

$$\mathcal{A} = \frac{A_1}{|Z_{,\alpha'}|^2} \geq c_0 > 0$$

for $C^{1,\gamma}$ interfaces, and $i\partial_{\alpha'}\bar{Z}_t = |\partial_{\alpha'}|\bar{Z}_t$.[1] Local well-posedness of (9.2.24)–(9.2.25) for $(\bar{Z}_t, \bar{Z}_{tt}) \in C([0,T], H^{s+1/2} \times H^s)$, $s \geq 4$ has been proved by the energy method and an iterative argument, we refer the reader to [9] for details.

We make the following remarks concerning some recent works.

Remark 1. The difference between the quasilinear equation (9.2.24)–(9.2.25) and the quasilinear equation (4.6)–(4.7) of [9] is that (9.2.24)–(9.2.25) is in terms of both components of Z_t and (4.6)–(4.7) of [9] is in terms of the real component of Z_t. (4.6)–(4.7) of [9] has been written in a way so that it is easy to prove its equivalence with the interface equation (9.2.4) or equation (1.7)–(1.8) of [9]; see §6 of [9].

2. (9.2.24)–(9.2.25) is an equation for the conjugate velocity $\bar{Z}_t$ and conjugate acceleration $\bar{Z}_{tt}$, the interface doesn't appear explicitly, so a solution of (9.2.24)–(9.2.25) can exist even when $Z = Z(\cdot, t)$ becomes self-intersecting.[m] Checking through the derivation above we see that we arrived at (9.2.24)–(9.2.25) from (9.1.1) using only the following properties of the domain: 1. there is a conformal mapping taking the fluid region $\Omega(t)$ to P_-; 2. $P = 0$ on $\Sigma(t)$. We note that $z \to z^{1/2}$ is a conformal map that takes the region $\mathbb{C} \setminus \{z = x + i0, x > 0\}$ to the upper half plane; a domain with its boundary self-intersecting at the positive real axis can therefore be mapped conformally onto the lower half plane P_-. Taking such a domain as the initial fluid domain, assuming $P = 0$ on $\Sigma(t)$ even when $\Sigma(t)$ self-intersects,[n] one can still solve equation (9.2.24)–(9.2.25) for a short time. Indeed this is the main idea in the work of [33]. Using this idea and the time reversibility of the water wave equation, by choosing an appropriate initial velocity field that pulls the initial domain apart, [33] proved the existence of "splash" and "splat" singularities starting from a smooth non–self-intersecting fluid region.

3. The above derivation applies to fluid domains with arbitrary non-self-intersecting boundaries. We have from (9.2.15) and (9.2.14) that the Taylor sign condition (9.1.2) always holds, as long as the interface is non–self-intersecting. Assume that $\Sigma(t)$ is non–self-intersecting with angled crests, assume the interior angle at a crest is ν. We know the Riemann mapping Φ^{-1} (we move the singular point to the origin) behaves like

$$\Phi^{-1}(z') \approx (z')^r, \qquad \text{with } \nu = r\pi$$

---

[1] This is because $i\partial_{\alpha'}\bar{Z}_t = i\partial_{\alpha'}\mathbb{H}\bar{Z}_t = |\partial_{\alpha'}|\bar{Z}_t$.

[m] $Z = Z(\cdot, t)$ is defined by $z(\cdot, t) = z(\cdot, 0) + \int_0^t z_s(\cdot, s)\, ds$, where $z_t = Z_t \circ h$, $Z = z \circ h^{-1}$; and $h_t = \mathcal{B}(h, t)$.

[n] We note that when $\Sigma(t)$ self-intersects, the condition $P = 0$ on $\Sigma(t)$ is unphysical.

near the crest, so $Z_{,\alpha'} \approx (\alpha')^{r-1}$ near the crest. From (9.2.9) and the fact $A_1 \geq 1$, we can conclude that the interior angle at the crest must be $\leq \pi$ if the acceleration $|Z_{tt}| \neq \infty$; we can also conclude that $-\frac{\partial P}{\partial \mathbf{n}} = 0$ at the singularities where the interior angles are $< \pi$, therefore in the regime that includes singular free surfaces, the quasilinear equation (9.2.24)–(9.2.25) is degenerate hyperbolic, c.f. [30, 31]. In [30, 31] we proved an a priori estimate and the local existence for 2D water waves in the regime including interfaces with angled crests, showing that the water wave equation (9.1.1) admit such solutions.

4. The quasilinear equation for water waves in the periodic setting can be derived similarly, see [30].

5. The Riemann mapping variable is used in recent work [27, 28].

9.2.3 The Quasilinear Equation in Lagrangian Coordinates

In order to extend our work for 2D water waves to 3D, we need a derivation that does not rely on the Riemann mapping. Upon checking the derivation in section 9.2.2, we see that all we have done is to apply $\mathrm{Re}(I - \mathbb{H})$ to calculate the parameters in the equations. We certainly can do the same calculations with $(I - \mathfrak{H})$, Riemann mapping is not needed.

We now use this idea to analyze $\mathfrak{a}_t |z_\alpha|$ and show that indeed (9.2.6) is quasilinear with the right hand side of lower order.

First by (9.2.1) and Proposition 9.2.2 we have

$$\mathfrak{a}|z_\alpha| = |z_{tt} + i|, \qquad \text{and} \qquad -i\frac{\bar{z}_\alpha}{|z_\alpha|} = \frac{\bar{z}_{tt} - i}{|z_{tt} + i|}. \tag{9.2.26}$$

We apply $(I - \mathfrak{H})$ to (9.2.6). Using $\bar{z}_t = \mathfrak{H}\bar{z}_t$ and Proposition B.1 in Appendix B, we have

$$(I - \mathfrak{H})(-i\mathfrak{a}_t\bar{z}_\alpha) = (I - \mathfrak{H})(\bar{z}_{ttt} + i\mathfrak{a}\bar{z}_{t\alpha})$$

$$= [\partial_t^2 + i\mathfrak{a}\partial_\alpha, \mathfrak{H}]\bar{z}_t$$

$$= 2[z_{tt}, \mathfrak{H}]\frac{\bar{z}_{t\alpha}}{z_\alpha} + 2[z_t, \mathfrak{H}]\frac{\bar{z}_{tt\alpha}}{z_\alpha} - \frac{1}{\pi i}\int \left(\frac{z_t(\alpha,t) - z_t(\beta,t)}{z(\alpha,t) - z(\beta,t)}\right)^2 \bar{z}_{t\beta}\, d\beta. \tag{9.2.27}$$

Multiply both sides of (9.2.27) by $i\frac{z_\alpha}{|z_\alpha|}$ and take the real parts. Since $\mathfrak{a}$ and $\mathfrak{a}_t$ are real valued, we have

$$(I + \mathfrak{K}^*)(\mathfrak{a}_t|z_\alpha|)$$

$$= \mathrm{Re}\left(\frac{iz_\alpha}{|z_\alpha|}\{2[z_{tt}, \mathfrak{H}]\frac{\bar{z}_{t\alpha}}{z_\alpha} + 2[z_t, \mathfrak{H}]\frac{\bar{z}_{tt\alpha}}{z_\alpha} - \frac{1}{\pi i}\int \left(\frac{z_t(\alpha,t) - z_t(\beta,t)}{z(\alpha,t) - z(\beta,t)}\right)^2 \bar{z}_{t\beta}\, d\beta\}\right) \tag{9.2.28}$$

where

$$\mathfrak{K}^* f(\alpha,t) = p.v. \int \mathrm{Re}\{\frac{-1}{\pi i}\frac{z_\alpha}{|z_\alpha|}\frac{|z_\beta(\beta,t)|}{(z(\alpha,t) - z(\beta,t))}\} f(\beta,t)\, d\beta$$

is the adjoint of the double layer potential operator $\mathfrak{K}$ in $L^2(\Sigma(t),dS)$. We know $I + \mathfrak{K}^*$ is invertible on $L^2(\Sigma(t),dS)$ (cf. [34, 35]). The second equation of (9.2.26), (9.2.28) and the estimates in Appendix A show that $\mathfrak{a}_t|z_\alpha|$ has the same regularity as that of $\bar{z}_{tt}$ and $\bar{z}_t$.

We rewrite (9.2.6) with the constraint $\bar{z}_t = \mathfrak{H}\bar{z}_t$ as

$$\begin{cases} \bar{z}_{ttt} + i\mathfrak{a}\bar{z}_{t\alpha} = \frac{\bar{z}_{tt}-i}{|z_{tt}+i|}\mathfrak{a}_t|z_\alpha| \\ \bar{z}_t = \mathfrak{H}\bar{z}_t, \end{cases} \qquad (9.2.29)$$

where $\mathfrak{a}|z_\alpha|$ is given by (9.2.26) and $\mathfrak{a}_t|z_\alpha|$ is given by (9.2.28). Since $\bar{z}_t$ is holomorphic, $i\frac{1}{|z_\alpha|}\partial_\alpha\bar{z}_t = \nabla_{\mathbf{n}}\bar{z}_t$. By Green's identity, the Dirichlet–Neumann operator $\nabla_{\mathbf{n}}$ is a positive operator. From Proposition 9.2.2 we know $\mathfrak{a}|z_\alpha| = -\frac{\partial P}{\partial \mathbf{n}} \geq c_0 > 0$ in the regime of $C^{1,\gamma}$ interfaces. Therefore (9.2.29) is a quasilinear system of hyperbolic type, with the right hand side of the first equation in (9.2.29) consisting of terms of lower order derivatives of $\bar{z}_t$. The local in time wellposedness of (9.2.29)–(9.2.28)–(9.2.26) in Sobolev spaces (with $(z_t, z_{tt}) \in C([0,T], H^{s+1/2} \times H^s)$, $s \geq 4$) can then be proved by energy estimates and a fixed point iteration argument.[o]

This derivation has been extended to 3D using Clifford analysis, cf. [10], we will give a brief discussion on how to do this in section 9.2.4.

Before ending this subsection, we mention that the quasilinear system (9.2.29)–(9.2.28)–(9.2.26) is coordinate invariant.

For fixed t, let $k = k(\alpha, t) : \mathbb{R} \to \mathbb{R}$ be a diffeomorphism with $k_\alpha > 0$. Let k^{-1} be such that $k \circ k^{-1}(\alpha, t) = \alpha$. Define

$$\zeta := z \circ k^{-1}, \qquad b := k_t \circ k^{-1} \quad \text{and} \quad A \circ k := \mathfrak{a}k_\alpha. \qquad (9.2.30)$$

Let

$$D_t := U_k^{-1}\partial_t U_k := \partial_t + b\partial_\alpha \qquad (9.2.31)$$

be the material derivative. By a simple application of the chain rule, we have

$$U_k^{-1}(\partial_t^2 + i\mathfrak{a}\partial_\alpha)U_k = D_t^2 + iA\partial_\alpha,$$

and equation (9.2.29) becomes

$$\begin{cases} (D_t^2 + iA\partial_\alpha)\overline{D_t\zeta} = (\mathfrak{a}_t|z_\alpha|) \circ k^{-1} \dfrac{\overline{D_t^2\zeta} - i}{|D_t^2\zeta + i|} \\ \overline{D_t\zeta} = \mathcal{H}\overline{D_t\zeta} \end{cases} \qquad (9.2.32)$$

[o] A proof of the local wellposedness of the 3D counterpart of (9.2.29)–(9.2.28)–(9.2.26) is carried out in [10].

with

$$(I+\mathcal{K}^*)((\mathfrak{a}_t|z_\alpha|)\circ k^{-1}) = Re\left(\frac{i\zeta_\alpha}{|\zeta_\alpha|}\{2[D_t^2\zeta,\mathcal{H}]\frac{\partial_\alpha\overline{D_t\zeta}}{\zeta_\alpha} + 2[D_t\zeta,\mathcal{H}]\frac{\partial_\alpha\overline{D_t^2\zeta}}{\zeta_\alpha}\right.$$
$$\left.-\frac{1}{\pi i}\int(\frac{D_t\zeta(\alpha,t)-D_t\zeta(\beta,t)}{\zeta(\alpha,t)-\zeta(\beta,t)})^2\partial_\beta\overline{D_t\zeta}(\beta,t)\,d\beta\}\right)$$
$$(9.2.33)$$

and

$$\mathcal{H}f(\alpha,t) = U_k^{-1}\mathfrak{H}U_kf(\alpha,t) = \frac{1}{\pi i}p.v.\int\frac{f(\beta,t)\zeta_\beta(\beta,t)}{\zeta(\alpha,t)-\zeta(\beta,t)}\,d\beta, \qquad (9.2.34)$$

$$\mathcal{K}^*f(\alpha,t) = p.v.\int Re\{\frac{-1}{\pi i}\frac{\zeta_\alpha}{|\zeta_\alpha|}\frac{|\zeta_\beta(\beta,t)|}{(\zeta(\alpha,t)-\zeta(\beta,t))}\}f(\beta,t)\,d\beta. \qquad (9.2.35)$$

Notice the remarkable similarities between equations (9.2.29)–(9.2.28) and (9.2.32)–(9.2.33). In particular, the structures of the terms in (9.2.29)–(9.2.28) do not change under the change of variables. This makes it convenient for us to work in another coordinate system and to choose a different coordinate system when there is advantage to do so. In fact, this has been used in our study of the global in time behavior of water waves [22, 23].

9.2.4 The Quasilinear Equation for 3D Water Waves

We derive the quasilinear equation for 3D water waves by carrying out the same procedure as for 2D. We first need to write down the 3D counterpart of the interface equation (9.2.4). While equation (9.2.1) is readily available in 3D, to write down the second equation, we need a suitable counterpart in 3D of the equation for the trace on the interface of the velocity field **v** that satisfies div **v** = 0 and curl **v** = 0. This leads us to Clifford analysis.

Let's recall the basics of Clifford algebra, or in other words, the algebra of quaternions $\mathcal{C}(V_2)$ (c.f. [36]). Let $\{1,e_1,e_2,e_3\}$ be the basis of $\mathcal{C}(V_2)$, satisfying

$$e_i^2 = -1, \quad \text{for } i=1,2,3, \qquad e_ie_j = -e_je_i, \quad i\neq j, \qquad e_3 = e_1e_2. \quad (9.2.36)$$

Let $\mathcal{D} = \partial_x e_1 + \partial_y e_2 + \partial_z e_3$. By definition, a Clifford-valued function $F: \Omega \subset \mathbb{R}^3 \to \mathcal{C}(V_2)$ is Clifford analytic in domain Ω iff $\mathcal{D}F = 0$ in Ω. Therefore, $F = \sum_{i=1}^3 f_ie_i$ is Clifford analytic in Ω if and only if div$F = 0$ and curl$F = 0$ in Ω. Furthermore a function F is the trace of a Clifford analytic function in Ω if and only if $F = \mathfrak{H}_\Sigma F$, where

$$\mathfrak{H}_\Sigma g(\alpha,\beta) = p.v.\iint K(\eta(\alpha',\beta')-\eta(\alpha,\beta))(\eta'_{\alpha'}\times\eta'_{\beta'})g(\alpha',\beta')\,d\alpha'd\beta'$$
$$(9.2.37)$$

is the 3D version of the Hilbert transform on $\Sigma = \partial\Omega : \eta = \eta(\alpha, \beta)$, $(\alpha, \beta) \in \mathbb{R}^2$, with normal $\eta_\alpha \times \eta_\beta$ pointing out of Ω, and

$$\Gamma(\eta) = -\frac{1}{\omega_3 |\eta|}, \qquad K(\eta) = -2D\Gamma(\eta) = -\frac{2}{\omega_3} \frac{\eta}{|\eta|^3},$$

ω_3 is the surface area of the unit sphere in $\mathbb{R}^3$.[p] As in the 2D case, if $\xi = \xi(\alpha, \beta, t)$, $(\alpha, \beta) \in \mathbb{R}^2$ is the free interface $\Sigma(t)$ in Lagrangian coordinates (α, β) at time t, with $N = \xi_\alpha \times \xi_\beta$ pointing out of the fluid domain, we can rewrite the 3D water wave system (9.1.1) ($n = 3$) as

$$\begin{cases} \xi_{tt} + e_3 = \mathfrak{a}N \\ \xi_t = \mathfrak{H}_{\Sigma(t)}\xi_t \end{cases} \qquad (9.2.38)$$

where $\mathfrak{a} = -\frac{1}{|N|}\frac{\partial P}{\partial \mathbf{n}}$.

Differentiating the first equation with respect to t yields

$$\begin{cases} \xi_{ttt} - \mathfrak{a}N_t = \mathfrak{a}_t N \\ \xi_t = \mathfrak{H}_{\Sigma(t)}\xi_t. \end{cases} \qquad (9.2.39)$$

This is the 3D counterpart of the 2D quasilinear equation (9.2.6) with constraint $\bar{z}_t = \mathfrak{H}\bar{z}_t$. It has been proved in [10] that $N_t = -|N|\nabla_\mathbf{n}\xi_t$,[q] and similar to the calculation (9.2.27), (9.2.28) for 2D, an expression for $\mathfrak{a}_t|N|$ in terms of $(I - \mathfrak{H}_{\Sigma(t)})(\xi_{ttt} + \mathfrak{a}|N|\nabla_\mathbf{n}\xi_t) = [\partial_t^2 + \mathfrak{a}|N|\nabla_\mathbf{n}, \mathfrak{H}_{\Sigma(t)}]\xi_t$ has been derived.

The Taylor sign condition (9.1.3) also holds for C^2 interfaces of the water wave problem (9.1.1) in 3D. This was proved in [10] by an application of the Green's identity. Here we give a heuristic argument via the maximum principle:

Applying div to both sides of the Euler equation and using the assumption that $\text{curl}\mathbf{v} = 0$ yields

$$\Delta P = -|\nabla\mathbf{v}|^2 \le 0 \qquad \text{in } \Omega(t).$$

Therefore, from $P = 0$ on the interface $\Sigma(t)$ and the maximum principle, we have $-\frac{\partial P}{\partial \mathbf{n}} \ge 0$.

(9.2.39) is then a quasilinear equation of hyperbolic type with the right hand side consisting of lower order terms. The local in time wellposedness of (9.2.38) is proved in [10] by applying energy estimates and an iterative argument to (9.2.39).

---

[p] (9.2.37) is similar to the 2D Hilbert transform (9.2.2) in the sense that in 2D, the fundamental solution for Laplace equation is $\Gamma(z) = \frac{1}{2\pi}\ln|z|$, $K(z) = -2(\partial_x - i\partial_y)(\frac{1}{2\pi}\ln|z|) = -\frac{1}{\pi z}$, the outward normal is iz_α, and $K(z(\alpha) - z(\beta))(iz_\beta) = \frac{1}{\pi i}\frac{iz_\beta}{z(\alpha) - z(\beta)}$ is the kernel of the Hilbert transform (9.2.2).

[q] In 2D, we know iz_α is a normal vector to $\Sigma(t)$ pointing out of the fluid domain, and $\partial_t(\overline{iz_\alpha}) = -i\partial_\alpha\bar{z}_t = -|z_\alpha|\nabla_\mathbf{n}\bar{z}_t$.

Remark An analogous derivation of the quasilinear structure can now be performed directly on (9.1.1). Let $D_t = \partial_t + \mathbf{v} \cdot \nabla$ be the material derivative, we can rewrite the first equation in (9.1.1) as

$$D_t \mathbf{v} + \mathbf{k} = -\nabla P.$$

Apply D_t to both sides to find

$$D_t^2 \mathbf{v} + [D_t, \nabla] P = -\nabla D_t P. \tag{9.2.40}$$

From the third equation: $P = 0$ on $\Sigma(t)$, we have $D_t P = 0$ on $\Sigma(t)$, so $\nabla D_t P$ points in the normal direction on the interface. We note that

$$[D_t, \partial_{x_i}] P = -\partial_{x_i} \mathbf{v} \cdot \nabla P = -(\nabla P \cdot \nabla) \mathbf{v}_i = \mathfrak{a} |N| \nabla_\mathbf{n} \mathbf{v}_i$$

on the fluid interface $\Sigma(t)$,[r] so it corresponds to the term $-\mathfrak{a} N_t = \mathfrak{a} |N| \nabla_\mathbf{n} \zeta_t$ in equation (9.2.39), therefore (9.2.40) is the counterpart of our quasilinear equation (9.2.39) in the entire fluid region. In [12, 16], the authors used (9.2.40) to study a more general case where the vorticity curl$\mathbf{v}$ need not be zero.[s]

We now turn to the question of long time behavior of solutions for the water wave equation (9.1.1) for small initial data.

9.3 Global and almost Global Wellposedness of the Water Wave Equations

To understand the global time behavior of the water wave motion, we need to understand the dispersion, the nature of the nonlinearity of the water waves and their interaction.

In [22, 23] we studied the water wave equation (9.1.1) in two- and three-space dimensions for small data, we found a nonlinear transformation for the unknowns and a nonlinear change of the coordinates, so that the transformed quantities in the new coordinate system satisfy equations containing no quadratic nonlinear terms.[t] Using these canonical equations, we showed that for small, smooth and localized data of size ϵ, the solution of the 2D water wave equation (9.1.1) remain small and smooth for time $0 < t < e^{c/\epsilon}$ and for similar data, the solution of the 3D water wave equation (9.1.1) remain small and smooth for all time.

Let's give a brief explanation of the dispersion of the water waves, the structural advantage of our canonical equations and how we found the transformations.

---

[r] In the second equality we used curl$\mathbf{v} = 0$.

[s] In [12, 16], the strong Taylor condition (9.1.3) is assumed to hold.

[t] When understood appropriately.

Let $u = \bar{z}_t$ (or $u = \xi_t$). Linearizing the quasilinear system (9.2.29)–(9.2.28)–(9.2.26) (or (9.2.39)) at the zero solution gives

$$\partial_t^2 u + |D|u = F(u_t, |D|u), \qquad (\alpha, t) \in \mathbb{R}^{n-1} \times \mathbb{R} \qquad (9.3.1)$$

where $|D| = \sqrt{-\Delta}$, Δ is the Laplacian in $\mathbb{R}^{n-1}$ for n-dimensional water waves, F consists of the nonlinear terms. We know the dispersion relation of the linear water wave equation

$$\partial_t^2 u + |D|u = 0 \qquad (9.3.2)$$

is

$$\omega^2 = |k| \qquad (9.3.3)$$

for plane wave solution $u = e^{i(k \cdot \alpha + \omega t)}$, so waves of wave number k travel with phase velocity $\frac{\omega}{|k|}\hat{k}$, where $\hat{k} = \frac{k}{|k|}$, equation (9.3.1) is dispersive. For a large class $\mathfrak{B}$ of smooth initial data, the solution of the linear equation (9.3.2) exists for all time and remains smooth, and its L^∞ norm decays with rate $1/t^{\frac{n-1}{2}}$. The question is for small data in $\mathfrak{B}$, for how long does the solution of the nonlinear equation (9.3.1) remain smooth. We know nonlinear interactions can cause blow-up of the solutions at finite time. So to answer this question, we need to know for how long does the linear part of the equation (9.3.1) remain dominant. The weaker the nonlinear interaction, the longer the solution remains smooth. For small data, quadratic interactions are in general stronger than cubic and higher order interactions.

To understand these assertions in qualitative terms, let's consider the following model equation with a $(p+1)$th-order nonlinearity:

$$\partial_t^2 u + |D|u = (\partial_t u)^{p+1}, \qquad (\alpha, t) \in \mathbb{R}^{n-1} \times \mathbb{R}. \qquad (9.3.4)$$

Suppose we can prove decay estimates for the solution: for $i \leq s - 10$,

$$|\partial^i \partial_t u(t)|_{L^\infty} \lesssim (1+t)^{-\frac{n-1}{2}} E_s(t)^{1/2},$$

where ∂ is some kind of derivatives,

$$E_s(t) = \sum_{|j| \leq s} \int |\partial^j \partial_t u(\alpha, t)|^2 + |\partial^j |D|^{1/2} u(\alpha, t)|^2 \, d\alpha.$$

Then we can derive energy estimates for large enough s:

$$\frac{d}{dt} E_s(t) \lesssim (1+t)^{-\frac{(n-1)p}{2}} E_s(t)^{p/2+1},$$

therefore

$$E_s(0)^{-p/2} - E_s(T)^{-p/2} \lesssim \int_0^T (1+t)^{-\frac{(n-1)p}{2}} \, dt. \qquad (9.3.5)$$

Heuristically, we expect to prove existence of solutions of (9.3.4) for as long as the energy $E_s(t)$ remains finite; by (9.3.5), this can be achieved if

$$\int_0^T (1+t)^{-\frac{(n-1)p}{2}}\, dt \lesssim E_s(0)^{-p/2}.$$

Now if $p = 1$, i.e., if the nonlinear term in (9.3.4) is quadratic, then for both $n = 2$ and $n = 3$, the integral $\int_0^\infty (1+t)^{-\frac{(n-1)p}{2}}\, dt = \infty$, and we would not be able to conclude solutions exist for all time for small initial data from this analysis. In fact for $p = 1$ and $n = 2$, the expected existence time is of order $O(\epsilon^{-2})$ for data of size ϵ. If $p \geq 2$, i.e., if there is no quadratic nonlinearity, then we can expect to prove longer time existence for solutions of (9.3.4) for small initial data. In fact, we can expect for $n = 2$, $p = 2$ an existence time period of $[0, e^{c/\epsilon^2}]$; and for $n = 3$, $p = 2$ an existence time period $[0, \infty)$ for data of size ϵ, when ϵ is sufficiently small.

Now for the water wave equation (9.3.1), the nonlinearity $F(u_t, |D|u)$ contains quadratic terms, so appears too strong to conclude a global existence result. The question is whether there is another unknown $\mathfrak{v}$ that satisfies an equation of the type

$$\partial_t^2 \mathfrak{v} + |D|\mathfrak{v} = F_1(\mathfrak{v}_t, |D|\mathfrak{v}, u_t, |D|u) \tag{9.3.6}$$

with F_1 containing no quadratic nonlinearities and $\|\mathfrak{v}\| \approx \|u\|$ in various norms $\|\cdot\|$ involved in the analysis. The idea of finding such an unknown $\mathfrak{v}$ is the so-called method of normal forms, originally introduced by Poincaré to solve ordinary differential equations. Certainly in most cases, one should not expect such a new unknown $\mathfrak{v}$ exist. For quadratic Klein-Gordon equation however, Simon and Shatah [37, 38] succeeded in finding a bilinear normal form transformation of the type

$$\mathfrak{v} = u + B(u, u) \tag{9.3.7}$$

with $B(u, u)$ bilinear, canceling out the quadratic nonlinear terms in the Klein-Gordon equation, and satisfying the norm equivalence $\|u + B(u,u)\| \approx \|u\|$ for small $\|u\|$.

For water wave equations (9.2.29)–(9.2.28)–(9.2.26) (or [9.2.39]), however, a transformation of the type (9.3.7) doesn't quite work since it has a small divisor; working with the velocity potential and the Bernoulli equation[u], a bilinear transform of the type (9.3.7) has a loss of derivatives (for detailed calculations and discussions, see Appendix C of [23].). What we did in [22, 23] was to further introduce a change of the coordinates. Indeed this makes sense since when one applies a method from ODE to PDE, it is reasonable to also take into consideration the spatial variables.

---

[u] Or equivalently the Zakharov-Craig-Sulem equation for the interface.

We give in the next two subsections the transforms for the 2D and 3D water waves. The transforms are fully nonlinear. We found the transforms by first considering the 2D quasilinear equation (9.2.29)–(9.2.28)–(9.2.26), starting with the ansatz (9.3.7), taking into considerations of the coordinate invariance of (9.2.29), looking for a coordinate invariant transformation.[v] This entails much further efforts in understanding the bilinear transformation. What we finally arrived at is a fully nonlinear transform of the unknown function, coupled with a coordinate change. The process of finding the transforms is non-algorithmic. The transforms for the 3D water waves (9.2.39) is obtained by naturally extending the 2D version via Clifford analysis, c.f. [23].

9.3.1 The Transformations for the 2D Water Waves

We give here the transformations we constructed in [22] for the 2D water waves. Let $\Sigma(t) : z = z(\alpha, t)$, $\alpha \in \mathbb{R}$ be the interface in Lagrangian coordinate α.

Proposition 9.3.1 (Proposition 2.3 of [22])   Let $z = z(\alpha, t)$ be a solution of the 2D water wave system (9.2.4). Let $\Pi := (I - \mathfrak{H})(z - \bar{z})$; let $k = k(\cdot, t) : \mathbb{R} \to \mathbb{R}$ be an arbitrary diffeomorphism. Let $\zeta := z \circ k^{-1}$, $D_t\zeta := z_t \circ k^{-1}$ etc. be as in (9.2.30)–(9.2.31). Then

$$
(D_t^2 - iA\partial_\alpha)(\Pi \circ k^{-1})
$$
$$
= \frac{4}{\pi} \int \frac{(D_t\zeta(\alpha,t) - D_t\zeta(\beta,t))(\text{Im}\,\zeta(\alpha,t) - \text{Im}\,\zeta(\beta,t))}{|\zeta(\alpha,t) - \zeta(\beta,t)|^2} \partial_\beta D_t\zeta(\beta,t)d\beta
$$
$$
+ \frac{2}{\pi} \int \left(\frac{D_t\zeta(\alpha,t) - D_t\zeta(\beta,t)}{\zeta(\alpha,t) - \zeta(\beta,t)} \right)^2 \partial_\beta \text{Im}\,\zeta(\beta,t)d\beta.
$$
$$(9.3.8)$$

Notice that the right hand side of equation (9.3.8) is cubically small if the velocity $D_t\zeta$, the height function $\text{Im}\,\zeta$ of the interface and their derivatives are small, but the left hand side of (9.3.8) still contains quadratic nonlinearities. Naturally we ask if there is a coordinate change k, so that $b = k_t \circ k^{-1}$ and $A - 1 = (ak_\alpha) \circ k^{-1} - 1$ are quadratic. We need not look far, equation (9.2.18) suggests that we choose

$$
k(\alpha, t) = 2\,\text{Re}\,z(\alpha, t) - h(\alpha, t), \qquad \alpha \in \mathbb{R}. \qquad (9.3.9)
$$

---

[v] The bilinear transform of the type (9.3.7) is not coordinate invariant.

Proposition 9.3.2 (Proposition 2.4 of [22])   Let k be as given by (9.3.9), $b = k_t \circ k^{-1}$ and $A = (ak_a) \circ k^{-1}$. Let $\mathcal{H}$ be defined by (9.2.34). We have

$$
(I - \mathcal{H})b = -[D_t\zeta, \mathcal{H}]\frac{\bar{\zeta}_a - 1}{\zeta_a}
$$

$$
(I - \mathcal{H})(A - 1) = i[D_t\zeta, \mathcal{H}]\frac{\partial_a \overline{D_t\zeta}}{\zeta_a} + i[D_t^2\zeta, \mathcal{H}]\frac{\bar{\zeta}_a - 1}{\zeta_a}.
$$

(9.3.10)

Propositions 9.3.1 and 9.3.2 show that the quantity $\theta := \Pi \circ k^{-1} = (I - \mathcal{H})(\zeta - \bar{\zeta})$ with the coordinate change k given by (9.3.9) satisfies an equation of the type

$$
\begin{cases} (\partial_t^2 - i\partial_a)\theta = \mathcal{G} \\ (I + \mathcal{H})\theta = 0 \end{cases}
$$

(9.3.11)

with $\mathcal{G}$ containing only nonlinear terms of cubic and higher order. We make the following remarks:

Remark   1. The transformation $I - \mathfrak{H}$ and the coordinate change k as given in (9.3.9) are fully nonlinear in terms of the unknown function z and its derivatives.[w]

2. The bilinear part of the quantity $\Pi := (I - \mathfrak{H})(z - \bar{z})$ has a bounded Fourier symbol. The coordinate change k takes care of the small divisor in the Fourier symbol of the bilinear normal form transformation. For a detailed explanation see Appendix C of [23].

3. For θ satisfying $(I + \mathcal{H})\theta = 0$, $(\partial_t^2 - i\partial_a)\theta = \partial_t^2\theta + i\partial_a\mathcal{H}\theta = (\partial_t^2 + |D|)\theta +$ quadratic $+ \cdots$. The quadratic nonlinearity comes from $\mathcal{H}$, which depends nonlinearly on the unknown ζ.

4. Let ϕ be the velocity potential, $\psi(\alpha, t) := \phi(z(\alpha, t), t)$ be the trace of ϕ on the free interface. It has been shown in [22] that the quantities $U_{k-1}(I - \mathfrak{H})\psi$, $U_{k-1}\partial_t\Pi$ also satisfy equations of the type (9.3.11), with their equations given in Proposition 2.3 of [22]. $\partial_a U_{k-1}(I - \mathfrak{H})\psi$, $U_{k-1}\partial_t\Pi$, $\frac{i}{2}\partial_a U_{k-1}\Pi$ and $\operatorname{Im}\partial_a U_{k-1}\Pi$ are near identity transforms of the velocity $D_t\zeta$, $2D_t\zeta$, the acceleration $D_t^2\zeta$ and $2\operatorname{Im}\partial_a\zeta$, see Propositions 2.5 and 2.6 of [22].

5. (9.3.8)–(9.3.9) is used in [39] to give a rigorous justification of the NLS from the 2D water wave equation (9.1.1).

6. The idea of changing coordinates is subsequently used in [40] to remove the quadratic nonlinear terms in the Burgers-Hilbert equation.

We now give the proofs of Propositions 9.3.1, 9.3.2.[x]

Proof. We first prove (9.3.8). Let $z = z(\cdot, t)$ be a solution of the water wave equation (9.2.4). Apply $(\partial_t^2 - i\mathfrak{a}\partial_a)$ to $\Pi := (I - \mathfrak{H})(z - \bar{z})$ and commute

---

[w] That is, $I - \mathfrak{H}$ and k are not finite sums of multi-linear operators of z and its derivatives.
[x] The proofs are taken from [22].

$\partial_t^2 - i a \partial_\alpha$ with $I - \mathfrak{H}$ to find

$$(\partial_t^2 - i a \partial_\alpha)\{(I - \mathfrak{H})(z - \bar{z})\} = (I - \mathfrak{H})\{(\partial_t^2 - i a \partial_\alpha)(z - \bar{z})\} - [\partial_t^2 - i a \partial_\alpha, \mathfrak{H}](z - \bar{z}).$$
(9.3.12)

Use (9.2.4) to find $(\partial_t^2 - i a \partial_\alpha)(z - \bar{z}) = -2\bar{z}_{tt}$ then use $\bar{z}_t = \mathfrak{H}\bar{z}_t$ to write $(I - \mathfrak{H})\bar{z}_{tt}$ as the commutator $[\partial_t, \mathfrak{H}]\bar{z}_t$ and applying Lemma B.1 yields

$$(I - \mathfrak{H})\{(\partial_t^2 - i a \partial_\alpha)(z - \bar{z})\} = (I - \mathfrak{H})(-2\bar{z}_{tt}) = -2[z_t, \mathfrak{H}]\frac{\bar{z}_{t\alpha}}{z_\alpha}.$$
(9.3.13)

Applying Lemma B.1 to the second term gives

$$[\partial_t^2 - i a \partial_\alpha, \mathfrak{H}](z - \bar{z}) = 2[z_t, \mathfrak{H}]\frac{z_{t\alpha} - \bar{z}_{t\alpha}}{z_\alpha} - \frac{1}{\pi i}\int \left(\frac{z_t(\alpha, t) - z_t(\beta, t)}{z(\alpha, t) - z(\beta, t)}\right)^2$$
$$\times \partial_\beta(z(\beta, t) - \bar{z}(\beta, t))\, d\beta.$$
(9.3.14)

Subtract (9.3.14) from (9.3.13). After cancelation this leaves, from (9.3.12),

$$(\partial_t^2 - i a \partial_\alpha)\{(I - \mathfrak{H})(z - \bar{z})\}$$
$$= -2[z_t, \mathfrak{H}]\frac{z_{t\alpha}}{z_\alpha} + \frac{1}{\pi i}\int \left(\frac{z_t(\alpha, t) - z_t(\beta, t)}{z(\alpha, t) - z(\beta, t)}\right)^2 (z - \bar{z})_\beta\, d\beta.$$
(9.3.15)

Because $\bar{z}_t$ and $\dfrac{z_{t\alpha}}{z_\alpha}$ are holomorphic, from part 1 of Lemma B.2

$$[z_t, \mathfrak{H}\frac{1}{\bar{z}_\alpha}]z_{t\alpha} = [z_t, \mathfrak{H}]\frac{z_{t\alpha}}{\bar{z}_\alpha} = 0.$$

In the first term of the right hand side of (9.3.15) insert $[z_t, \bar{\mathfrak{H}}\frac{1}{\bar{z}_\alpha}]z_{t\alpha}$ to make it cubic. We have

$$(\partial_t^2 - i a \partial_\alpha)\{(I - \mathfrak{H})(z - \bar{z})\}$$
$$= -2[z_t, \mathfrak{H}\frac{1}{z_\alpha} + \bar{\mathfrak{H}}\frac{1}{\bar{z}_\alpha}]z_{t\alpha} + \frac{1}{\pi i}\int \left(\frac{z_t(\alpha, t) - z_t(\beta, t)}{z(\alpha, t) - z(\beta, t)}\right)^2 (z_\beta - \bar{z}_\beta)\, d\beta.$$
(9.3.16)

Precomposing with k^{-1} and expanding the two terms on the right gives (9.3.8).
$\square$

Proof. We prove Proposition 9.3.2. We have from (9.3.9)

$$k - \bar{z} = z - h.$$

Recall $h(\alpha, t) = \Phi(z(\alpha, t), t)$ where $\Phi(\cdot, t) : \Omega(t) \to P_-$ is the Riemann mapping satisfying $\lim_{z \to \infty} \Phi_z(z, t) = 1$. We have

$$h_t = \Phi_t \circ z + (\Phi_z \circ z)z_t, \qquad h_\alpha = (\Phi_z \circ z)z_\alpha,$$

therefore

$$\bar{z}_t - k_t = \Phi_t \circ z + (\Phi_z \circ z - 1)z_t, \qquad \bar{z}_\alpha - k_\alpha = (\Phi_z \circ z - 1)z_\alpha.$$
(9.3.17)

Apply $(I - \mathfrak{H})$ to the first equality in (9.3.17). Because Φ_t, Φ_z are holomorphic in $\Omega(t)$ with $\lim_{z \to \infty} \Phi_z(z,t) = 1$, using Proposition 9.2.1 and rewriting $(I - \mathfrak{H})\{(\Phi_z \circ z - 1)z_t\}$ as $[z_t, \mathfrak{H}](\Phi_z \circ z - 1)$ yields

$$-(I - \mathfrak{H})k_t = (I - \mathfrak{H})(\bar{z}_t - k_t)$$

$$= (I - \mathfrak{H})\{(\Phi_z \circ z - 1)z_t\} = [z_t, \mathfrak{H}](\Phi_z \circ z - 1) = [z_t, \mathfrak{H}]\frac{\bar{z}_\alpha - k_\alpha}{z_\alpha}.$$
$$(9.3.18)$$

Precomposing with k^{-1} gives the first equality of (9.3.10).

Now multiply ia then apply $(I - \mathfrak{H})$ to the second equality in (9.3.17). Using (9.2.4) and the fact that $(I - \mathfrak{H})(\Phi_z \circ z - 1) = 0$, we also have

$$(I - \mathfrak{H})(ia\bar{z}_\alpha - iak_\alpha) = (I - \mathfrak{H})(ia z_\alpha(\Phi_z \circ z - 1)) = [z_{tt}, \mathfrak{H}](\Phi_z \circ z - 1)$$

$$= [z_{tt}, \mathfrak{H}]\frac{\bar{z}_\alpha - k_\alpha}{z_\alpha}.$$

Use (9.2.4) and Lemma B.1 to calculate

$$(I - \mathfrak{H})(ia\bar{z}_\alpha) = (I - \mathfrak{H})(-\bar{z}_{tt} + i) = i - [\partial_t, \mathfrak{H}]\bar{z}_t = i - [z_t, \mathfrak{H}]\frac{\bar{z}_{t\alpha}}{z_\alpha}$$

so

$$-(I - \mathfrak{H})(iak_\alpha) = -(I - \mathfrak{H})(ia\bar{z}_\alpha) + [z_{tt}, \mathfrak{H}]\frac{\bar{z}_\alpha - k_\alpha}{z_\alpha}$$
$$(9.3.19)$$

$$= -i + [z_t, \mathfrak{H}]\frac{\bar{z}_{t\alpha}}{z_\alpha} + [z_{tt}, \mathfrak{H}]\frac{\bar{z}_\alpha - k_\alpha}{z_\alpha}.$$

Precomposing with k^{-1} yields the second equality of (9.3.10). $\qquad\square$

To extend the 2D coordinate change (9.3.9) to 3D, we need an expression that does not rely on the Riemann mapping. Observe that for the diffeomorphism k given by (9.3.9), $k - \bar{z} = z - h$ and $z - h$ is holomorphic with $\text{Im}(z - h) = \text{Im}\, z$, so we can replace $z - h$ by $\frac{1}{2}(I + \mathfrak{H})(I + \mathfrak{K})^{-1}(z - \bar{z})$, where $\mathfrak{K} = \text{Re}\, \mathfrak{H}$ is the double layer potential operator, and

$$k = \bar{z} + \frac{1}{2}(I + \mathfrak{H})(I + \mathfrak{K})^{-1}(z - \bar{z}) \qquad (9.3.20)$$

modulo a real constant.[y] The expression (9.3.20) is directly extendable to 3D.

9.3.2 The Transformation for the 3D Water Waves

We extend the 2D transformations to 3D in the framework of the Clifford algebra $\mathcal{C}(V_2)$. Besides those in subsection 9.2.4, we need some additional notations.

---

[y] $z - h$ and $\frac{1}{2}(I + \mathfrak{H})(I + \mathfrak{K})^{-1}(z - \bar{z})$ are holomorphic with the same imaginary part, so the difference between them is a constant in $\mathbb{R}$.

An element $\sigma \in C(V_2)$ can be represented uniquely by $\sigma = \sigma_0 + \sum_{i=1}^{3} \sigma_i e_i$, with $\sigma_i \in \mathbb{R}$ for $0 \le i \le 3$. Define $\operatorname{Re}\sigma := \sigma_0$ and call it the real part of σ. We call σ a vector if $\sigma_0 = 0$. If not specified, we always assume in an expression $\sigma = \sigma_0 + \sum_{i=1}^{3} \sigma_i e_i$ that $\sigma_i \in \mathbb{R}$ for $0 \le i \le 3$. Define $\bar{\sigma} := e_3 \sigma e_3$, the conjugate of σ. We identify a point or a vector $\xi = (x_1, x_2, y) \in \mathbb{R}^3$ with its $C(V_2)$ counterpart $\xi = x_1 e_1 + x_2 e_2 + y e_3$. For vectors ξ, $\eta \in C(V_2)$, we know

$$\xi\eta = -\xi \cdot \eta + \xi \times \eta, \tag{9.3.21}$$

where $\xi \cdot \eta$ is the dot product, $\xi \times \eta$ the cross product. For vectors ξ, ζ, η, $\xi(\zeta \times \eta)$ is obtained by first finding the cross product $\zeta \times \eta$, then regard it as a Clifford vector and calculating its multiplication with ξ by the rule (9.2.36). We write $\nabla = (\partial_{x_1}, \partial_{x_2}, \partial_y)$. We abbreviate notations such as

$$\mathfrak{H}_\Sigma f(\alpha, \beta) = \iint K(\eta(\alpha', \beta') - \eta(\alpha, \beta))\,(\eta'_{\alpha'} \times \eta'_{\beta'}) f(\alpha', \beta')\, d\alpha'\, d\beta'$$

$$:= \iint K(\eta' - \eta)\,(\eta'_{\alpha'} \times \eta'_{\beta'}) f'\, d\alpha'\, d\beta' := \iint KN' f'\, d\alpha'\, d\beta'.$$

As in the 2D case, $\mathfrak{H}_\Sigma^2 = I$ in L^2, and $\mathfrak{H}_\Sigma 1 = 0$.

We give the transformation for the 3D water wave equation (9.2.38). Let the free interface $\Sigma(t)$ be given by $\xi = \xi(\alpha, \beta, t) = x_1(\alpha, \beta, t)e_1 + x_2(\alpha, \beta, t)e_2 + y(\alpha, \beta, t)e_3$ in Lagrangian coordinates (α, β) with $N = \xi_\alpha \times \xi_\beta$ pointing out of the fluid domain $\Omega(t)$. For fixed t, let $k = k(\cdot, t) = k_1 e_1 + k_2 e_2 : \mathbb{R}^2 \to \mathbb{R}^2$ be a diffeomorphism with Jacobian $J(k(t)) > 0$. Let k^{-1} be such that $k \circ k^{-1}(\alpha, \beta, t) = \alpha e_1 + \beta e_2$. Define

$$\zeta := \xi \circ k^{-1}, \quad b := k_t \circ k^{-1}, \quad A \circ k e_3 := \mathfrak{a}J(k)e_3 := \mathfrak{a}k_\alpha \times k_\beta. \tag{9.3.22}$$

Let $D_t := U_k^{-1} \partial_t U_k$ be the material derivative, $\mathcal{N} := \zeta_\alpha \times \zeta_\beta$. By the chain rule, we know

$$D_t = \partial_t + b \cdot (\partial_\alpha, \partial_\beta), \qquad U_k^{-1}(\mathfrak{a}N \times \nabla)U_k = A\mathcal{N} \times \nabla = A(\zeta_\beta \partial_\alpha - \zeta_\alpha \partial_\beta), \tag{9.3.23}$$

and $U_k^{-1} \mathfrak{H}_{\Sigma(t)} U_k := \mathcal{H}_{\Sigma(t)}$, with

$$\mathcal{H}_{\Sigma(t)} f(\alpha, \beta, t) = \iint K(\zeta(\alpha', \beta', t) - \zeta(\alpha, \beta, t))(\zeta'_{\alpha'} \times \zeta'_{\beta'}) f(\alpha', \beta', t)\, d\alpha'\, d\beta'. \tag{9.3.24}$$

We have

Proposition 9.3.3 (Proposition 1.3 of [23])   Let $\xi = \xi(\alpha, \beta, t)$ be a solution of the 3D water wave system (9.2.38). Let $\Pi = (I - \mathfrak{H}_{\Sigma(t)})(\xi - \bar{\xi})$, and for fixed $t, k(\cdot, t) : \mathbb{R}^2 \to \mathbb{R}^2$ be a diffeomorphism. We have

$$(D_t^2 - A\mathcal{N} \times \nabla)(\Pi \circ k^{-1})$$

$$= 2 \iint K(\zeta' - \zeta)(D_t\zeta - D_t'\zeta') \times (\zeta'_{\beta'}\partial_{\alpha'} - \zeta'_{\alpha'}\partial_{\beta'})\overline{D_t'\zeta'}\, d\alpha' d\beta'$$

$$- \iint K(\zeta' - \zeta)(D_t\zeta - D_t'\zeta') \times ((D_t'\zeta')_{\beta'}\partial_{\alpha'} - (D_t'\zeta')_{\alpha'}\partial_{\beta'})(\zeta' - \bar{\zeta}')\, d\alpha' d\beta'$$

$$- \iint D_t K(\zeta' - \zeta)(D_t\zeta - D_t'\zeta') \times (\zeta'_{\beta'}\partial_{\alpha'} - \zeta'_{\alpha'}\partial_{\beta'})(\zeta' - \bar{\zeta}')\, d\alpha' d\beta'.$$

$$(9.3.25)$$

Observe that the second and third terms in the right hand side of (9.3.25) are cubicly small provided the velocity $D_t\zeta$ and the steepness of the height function $\partial_\alpha(\zeta - \bar{\zeta})$, $\partial_\beta(\zeta - \bar{\zeta})$ are small, while the first term appears to be only quadratically small. Unlike the 2D case, multiplications of Clifford analytic functions are not necessarily analytic, so we cannot reduce the first term at the right hand side of equation (9.3.25) into a cubic form. However we note that the first term is almost analytic in the fluid domain $\Omega(t)$, while the left hand side of (9.3.25) is almost analytic in the air region. The orthogonality of the projections $\frac{1}{2}(I - \mathcal{H}_{\Sigma(t)})$ and $\frac{1}{2}(I + \mathcal{H}_{\Sigma(t)})$ allows us to reduce the first term into a cubic in energy estimates, see [23].

Now the left hand side of (9.3.25) still contains quadratic terms. As in the 2D case, we resolve this difficulty by choosing an appropriate coordinate change k. Let

$$k = k(\alpha, \beta, t) = \xi(\alpha, \beta, t) - (I + \mathfrak{H}_{\Sigma(t)})y(\alpha, \beta, t)e_3 + \mathfrak{K}_{\Sigma(t)}y(\alpha, \beta, t)e_3. \quad (9.3.26)$$

Here $\mathfrak{K}_{\Sigma(t)} = \operatorname{Re}\mathfrak{H}_{\Sigma(t)}$:

$$\mathfrak{K}_{\Sigma(t)}f(\alpha, \beta, t) = -\iint K(\xi(\alpha', \beta', t) - \xi(\alpha, \beta, t)) \cdot N'f(\alpha', \beta', t)\, d\alpha' d\beta'$$

$$(9.3.27)$$

is the double layer potential operator. It is clear that the e_3 component of k as defined in (9.3.26) is zero. In addition, the real part of k is also zero. This is because

$$\iint K(\xi' - \xi) \times (\xi'_{\alpha'} \times \xi'_{\beta'}) y' e_3 \, d\alpha' \, d\beta'$$

$$= \iint (\xi'_{\alpha'} \xi'_{\beta'} \cdot K - \xi'_{\beta'} \xi'_{\alpha'} \cdot K) y' e_3 \, d\alpha' \, d\beta'$$

$$= -2 \iint (\xi'_{\alpha'} \partial_{\beta'} \Gamma(\xi' - \xi) - \xi'_{\beta'} \partial_{\alpha'} \Gamma(\xi' - \xi)) y' e_3 \, d\alpha' \, d\beta'$$

$$= 2 \iint \Gamma(\xi' - \xi) (\xi'_{\alpha'} y_{\beta'} - \xi'_{\beta'} y_{\alpha'}) e_3 \, d\alpha' \, d\beta'$$

$$= 2 \iint \Gamma(\xi' - \xi) (N'_1 e_1 + N'_2 e_2) \, d\alpha' \, d\beta'.$$

So,

$$\mathfrak{H}_{\Sigma(t)} y e_3 = \mathfrak{K}_{\Sigma(t)} y e_3 + 2 \iint \Gamma(\xi' - \xi) (N'_1 e_1 + N'_2 e_2) \, d\alpha' \, d\beta'. \qquad (9.3.28)$$

This shows that the mapping k defined in (9.3.26) has only the e_1 and e_2 components $k = (k_1, k_2) = k_1 e_1 + k_2 e_2$. If $\Sigma(t)$ is a graph with small steepness, i.e., if y_α and y_β are small, then the Jacobian of $k = k(\cdot, t)$: $J(k) = J(k(t)) = \partial_\alpha k_1 \partial_\beta k_2 - \partial_\alpha k_2 \partial_\beta k_1 > 0$ and $k(\cdot, t) : \mathbb{R}^2 \to \mathbb{R}^2$ defines a valid coordinate change (c.f. [23]).

The following proposition shows that if k is as given in (9.3.26), then b and $A - 1$ are quadratic. Let

$$\mathcal{K}_{\Sigma(t)} := \operatorname{Re} \mathcal{H}_{\Sigma(t)} =: U_k^{-1} \mathfrak{K}_{\Sigma(t)} U_k, \quad P := \alpha e_1 + \beta e_2, \quad \text{and} \quad \zeta := P + \lambda. \qquad (9.3.29)$$

Proposition 9.3.4 (Proposition 1.4 of [23]) Let k be as given in (9.3.26). Let $b = k_t \circ k^{-1}$ and $A \circ k = \mathfrak{a} J(k)$. We have

$$b = \frac{1}{2} (\mathcal{H}_{\Sigma(t)} - \overline{\mathcal{H}_{\Sigma(t)}}) \overline{D_t \zeta} - \frac{1}{2} [D_t, \mathcal{H}_{\Sigma(t)} - \mathcal{K}_{\Sigma(t)}](\zeta - \bar{\zeta})$$

$$+ \frac{1}{2} \mathcal{K}_{\Sigma(t)} (D_t \zeta - \overline{D_t \zeta})$$

$$(A - 1) e_3 = \frac{1}{2} (-\mathcal{H}_{\Sigma(t)} + \overline{\mathcal{H}_{\Sigma(t)}}) \overline{D_t^2 \zeta} + \frac{1}{2} ([D_t, \mathcal{H}_{\Sigma(t)}] D_t \zeta - \overline{[D_t, \mathcal{H}_{\Sigma(t)}] D_t \zeta})$$

$$+ \frac{1}{2} [A \mathcal{N} \times \nabla, \mathcal{H}_{\Sigma(t)}](\zeta - \bar{\zeta}) - \frac{1}{2} A \zeta_\beta \times (\partial_\alpha \mathcal{K}_{\Sigma(t)} (\zeta - \bar{\zeta}))$$

$$+ \frac{1}{2} A \zeta_\alpha \times (\partial_\beta \mathcal{K}_{\Sigma(t)} (\zeta - \bar{\zeta})) + A \partial_\alpha \lambda \times \partial_\beta \lambda. \qquad (9.3.30)$$

Here $\overline{\mathcal{H}_{\Sigma(t)}} = e_3 \mathcal{H}_{\Sigma(t)} e_3$.

Let $\chi = \Pi \circ k^{-1}$ with k be given by (9.3.26). The left hand side of equation (9.3.25) is

$$(\partial_t^2 - e_2\partial_\alpha + e_1\partial_\beta)\chi - \partial_\beta\lambda\partial_\alpha\chi + \partial_\alpha\lambda\partial_\beta\chi + \text{cubic and higher order terms.}$$

The quadratic term $\partial_\beta\lambda\partial_\alpha\chi - \partial_\alpha\lambda\partial_\beta\chi$ is new in 3D. Observe that this is one of the null forms studied in [41]. It is also null for our equation and can be written as the factor $1/t$ times a quadratic expression involving some "invariant vector fields" for $\partial_t^2 - e_2\partial_\alpha + e_1\partial_\beta$, see [23]. Therefore this term is cubic in nature and equation (9.3.25) is of the type "linear + cubic and higher order perturbations".

We refer the reader to [23] for the proof of Propositions 9.3.3 and 9.3.4.

We remark that the 3D transformations is recently used in [42] to give a rigorous justification of the modulation approximation for the 3D water wave equation (9.1.1). Besides the 3D transforms, [42] uses the method of normal form to handle the quadratic term $\partial_\beta\lambda\partial_\alpha\chi - \partial_\alpha\lambda\partial_\beta\chi$ since this term is truly quadratic in the modulation regime.

9.3.3 Global in Time Behavior of Solutions for the 2D and 3D Water Waves

In [22, 23] equations (9.3.8) and (9.3.25) together with the coordinate changes (9.3.9) and (9.3.26) are used to prove the almost global wellposedness of the 2D and global wellposedness of the 3D water wave equations (9.1.1) for small, smooth and localized initial data. The basic idea is what we illustrated with the model equation (9.3.4); it is made rigorous by the method of invariant vector fields. This involves constructing invariant vector fields for the operator $\partial_t^2 - e_2\partial_\alpha + e_1\partial_\beta$ (the invariant vector fields for $\partial_t^2 - i\partial_\alpha$ for the 2D case is available due to the well studied Schrödinger operator $i\partial_t - \partial_x^2$), proving generalized Sobolev inequalities that give $L^2 \to L^\infty$ estimates with the decay rate $1/t^{1/2}$ for the 2D and $1/t$ for the 3D water waves, using equations (9.3.8)–(9.3.9) and (9.3.25)–(9.3.26) to show that properly constructed energies that involve invariant vector fields remain bounded for the time period $[0, e^{c/\epsilon}]$ for the 2D and for all time for the 3D water waves for data of size $O(\epsilon)$. The projection $\frac{1}{2}(I - \mathfrak{H})$ is used in various ways to project away "quadratic noises" in the course of deriving the energy estimates. We remark that it is more natural to treat $D_t^2 - iA\partial_\alpha$ and $D_t^2 - A\mathcal{N} \times \nabla$ as the main operators for the 2D and 3D water wave equations than treating them as the perturbations of the linear operators $\partial_t^2 - i\partial_\alpha$ and $\partial_t^2 - e_2\partial_\alpha + e_1\partial_\beta$. The almost global well-posedness for the 2D and global well-posedness for the 3D water wave equations follow from the local well-posedness results, the uniform boundedness of the energies and continuity arguments. For details of the proofs see [22, 23]. We state the results.

Let $|D| = \sqrt{-\Delta}$, $H^s(\mathbb{R}^{n-1}) = \{f \mid (I+|D|)^s f \in L^2(\mathbb{R}^{n-1})\}$, with $\|f\|_{H^s(\mathbb{R}^{n-1})} = \|(I+|D|)^s f\|_{L^2(\mathbb{R}^{n-1})}$.

2D water waves

Let $s \geq 12$, $\max\{[\frac{s}{2}] + 3, 11\} \leq l \leq s - 1$. Assume

$$z(\alpha, 0) = (\alpha, y(\alpha)), \quad z_t(\alpha, 0) = \mathfrak{u}(\alpha), \quad z_{tt}(\alpha, 0) = \mathfrak{w}(\alpha) \quad \alpha \in \mathbb{R},$$
$$v(z, 0) = \mathbf{g}(z), \quad z \in \Omega(0) \tag{9.3.31}$$

and the data in (9.3.31) satisfies the 2D water wave system (9.2.4). In particular $\bar{\mathbf{g}}$ is a holomorphic function in the initial fluid domain $\Omega(0)$ and $\mathbf{g}(z(\alpha, 0)) = \mathfrak{u}(\alpha)$. Let $\Gamma = \partial_\alpha, \alpha\partial_\alpha$. Assume that

$$\sum_{|j| \leq s-1} (\|\Gamma^j y_\alpha\|_{H^{1/2}(\mathbb{R})} + \|\Gamma^j \mathfrak{u}\|_{H^{3/2}(\mathbb{R})} + \|\Gamma^j \mathfrak{w}\|_{H^1(\mathbb{R})}) < \infty.$$

Let

$$\epsilon = \sum_{|j| \leq l} (\|\Gamma^j y\|_{H^1(\mathbb{R})} + \|\Gamma^j \mathfrak{u}\|_{H^1(\mathbb{R})}) + \sum_{j \leq l-2} \|(z\partial_z)^j \bar{\mathbf{g}}\|_{L^2(\Omega(0))}.$$

Theorem 9.3.5 (2D Theorem, c.f.[22]). *There exist ϵ_0 and $c > 0$, such that for $\epsilon < \epsilon_0$, the initial value problem (9.2.4)–(9.3.31) has a unique classical solution for the time period $[0, e^{c/\epsilon}]$. During this time, the interface is a graph, the solution is as regular as the initial data and remains small. Moreover, the L^∞ norm of the steepness $\partial_\alpha(z - \bar{z})$, the velocity z_t and acceleration z_{tt} decay at rate $1/t^{1/2}$.*

3D water waves

Let $s \geq 27$, $\max\{[\frac{s}{2}] + 1, 17\} \leq l \leq s - 10$. Assume that initially

$$\xi(\alpha, \beta, 0) = (\alpha, \beta, y^0(\alpha, \beta)), \quad \xi_t(\alpha, \beta, 0) = \mathfrak{u}^0(\alpha, \beta), \quad \xi_{tt}(\alpha, \beta, 0) = \mathfrak{w}^0(\alpha, \beta),$$
$$\tag{9.3.32}$$

and the data in (9.3.32) satisfies the 3D water wave system (9.2.38). Let $\Gamma = \partial_\alpha, \partial_\beta, \alpha\partial_\alpha + \beta\partial_\beta, \alpha\partial_\beta - \beta\partial_\alpha$. Assume that

$$\sum_{\substack{|j| \leq s-1 \\ \partial = \partial_\alpha, \partial_\beta}} \||\Gamma^j|D|^{1/2}y^0\|_{L^2(\mathbb{R}^2)} + \|\Gamma^j\partial y^0\|_{H^{1/2}(\mathbb{R}^2)} + \|\Gamma^j\mathfrak{u}^0\|_{H^{3/2}(\mathbb{R}^2)} + \|\Gamma^j\mathfrak{w}^0\|_{H^1(\mathbb{R}^2)} < \infty.$$

$$\tag{9.3.33}$$

Let

$$\epsilon = \sum_{\substack{|j| \leq l+3 \\ \partial = \partial_\alpha, \partial_\beta}} \||\Gamma^j|D|^{1/2}y^0\|_{L^2(\mathbb{R}^2)} + \|\Gamma^j\partial y^0\|_{L^2(\mathbb{R}^2)} + \|\Gamma^j\mathfrak{u}^0\|_{H^{1/2}(\mathbb{R}^2)} + \|\Gamma^j\mathfrak{w}^0\|_{L^2(\mathbb{R}^2)}.$$

$$\tag{9.3.34}$$

Theorem 9.3.6 (3D Theorem, c.f. [23]). *There exists $\epsilon_0 > 0$, such that for $0 < \epsilon \leq \epsilon_0$, the initial value problem (9.2.38)–(9.3.32) has a unique classical solution globally in time. For each time $0 \leq t < \infty$, the interface is a graph, the solution has* *the same regularity as the initial data and remains small. Moreover the L^∞ norm of the steepness and the acceleration on the interface, the derivative of the velocity on the interface decay at rate $\frac{1}{t}$.*

Remark 1. The existence time in Theorem 9.3.5 is extended to global in [25, 26, 28] by further understanding the nature of the cubic nonlinearities in equation (9.1.1).

2. In [22] a quick dispersive estimate was proved by the vector field method. An in-depth analysis on the dispersion of the linear water wave operator $\partial_t^2 + |D|$ is performed in [43]; a threshold is found so that when the amount of small frequency waves in the initial data is below the threshold, the solution of the linear water wave equation (9.3.2) decays with rate $1/\sqrt{t}$, while above the threshold, there is a growth factor in the linear solution. Consequences in the nonlinear setting remain to be understood.

Acknowledgment

The author thanks Jeffrey Rauch and Shuang Miao for carefully reading through the manuscript and for their remarks and suggestions. The research reported in this chapter has been supported in part by NSF grants DMS-1101434 and DMS-1361791.

Appendix A: Basic Analysis Preparations

In this section we present some inequalities and identities on $\mathbb{R}$ that are used to guide the derivation of the quasilinear structure of the 2D water waves. Corresponding inequalities and identities are available in all dimensions $\mathbb{R}^d$, we refer the reader to [10, 23] for those.

Let $H \in C^1(\mathbb{R}; \mathbb{R}^l)$, $A_i \in C^1(\mathbb{R})$, $i = 1, \ldots m$, $F \in C^\infty(\mathbb{R}^l)$. Define

$$C_1(H, A, f)(x) = \text{p.v.} \int F\left(\frac{H(x) - H(y)}{x - y}\right) \frac{\Pi_{i=1}^m (A_i(x) - A_i(y))}{(x - y)^{m+1}} f(y) \, dy. \quad (A.1)$$

Proposition A.1 There exist constants $c_1 = c_1(F, \|H'\|_{L^\infty})$, $c_2 = c_2(F, \|H'\|_{L^\infty})$, such that
1. For any $f \in L^2$, $A_i' \in L^\infty$, $1 \le i \le m$,

$$\|C_1(H, A, f)\|_{L^2} \le c_1 \|A_1'\|_{L^\infty} \ldots \|A_m'\|_{L^\infty} \|f\|_{L^2}. \quad (A.2)$$

2. For any $f \in L^\infty$, $A_i' \in L^\infty$, $2 \le i \le m$, $A_1' \in L^2$,

$$\|C_1(H, A, f)\|_{L^2} \le c_2 \|A_1'\|_{L^2} \|A_2'\|_{L^\infty} \ldots \|A_m'\|_{L^\infty} \|f\|_{L^\infty}. \quad (A.3)$$

(A.2) is a result of Coifman, McIntosh, and Meyer [44]. (A.3) is a consequence of the Tb Theorem, a proof is given in [22].

Let H, A_i, F satisfy the same assumptions as in (A.1). Define

$$C_2(H,A,f)(x) = \int F\left(\frac{H(x)-H(y)}{x-y}\right)\frac{\Pi_{i=1}^m (A_i(x)-A_i(y))}{(x-y)^m}\partial_y f(y)\,dy. \qquad (A.4)$$

We have the following inequalities.

Proposition A.2 There exist constants $c_3 = c_3(F,\|H'\|_{L^\infty})$, $c_4 = c_4$
$(F,\|H'\|_{L^\infty})$, such that
  1. For any $f \in L^2$, $A_i' \in L^\infty$, $1 \leq i \leq m$,

$$\|C_2(H,A,f)\|_{L^2} \leq c_3\|A_1'\|_{L^\infty}\ldots\|A_m'\|_{L^\infty}\|f\|_{L^2}. \qquad (A.5)$$

  2. For any $f \in L^\infty$, $A_i' \in L^\infty$, $2 \leq i \leq m$, $A_1' \in L^2$,

$$\|C_2(H,A,f)\|_{L^2} \leq c_4\|A_1'\|_{L^2}\|A_2'\|_{L^\infty}\ldots\|A_m'\|_{L^\infty}\|f\|_{L^\infty}. \qquad (A.6)$$

Using integration by parts, the operator $C_2(H,A,f)$ can be easily converted into a sum of operators of the form $C_1(H,A,f)$. (A.5) and (A.6) follow from (A.2) and (A.3).

The following identities are useful to compute the derivatives of the integral operators.

Let

$$\mathbf{K}f(x,t) = p.v. \int K(x,y;t)f(y,t)\,dy$$

where either K or $(x-y)K(x,y;t)$ is continuous and bounded, and K is smooth away from the diagonal $\Delta = \{(x,y)\,|\,x=y\}$. We have for $f \in C^1(R^{1+1})$ vanishing as $|x| \to \infty$,

$$f(x,t) = \int \partial_t K(x,y;t)f(y,t)\,dy$$

$$[\partial_x,\mathbf{K}]f(x,t) = \int (\partial_x + \partial_y)K(x,y;t)f(y,t)\,dy. \qquad (A.7)$$

The first identity in (A.7) is straightforward, the second is obtained by integration by parts.

Using (A.7), $\partial_{\alpha'}^s[f,\mathbb{H}]\partial_{\alpha'}g$ equals to the sum of $[\partial_{\alpha'}^s f,\mathbb{H}]\partial_{\alpha'}g$, (Expand $[\partial_{\alpha'}^s f,\mathbb{H}]\partial_{\alpha'}g = \partial_{\alpha'}^s f\mathbb{H}(\partial_{\alpha'}g) - \mathbb{H}(\partial_{\alpha'}^s f\partial_{\alpha'}g)$ and estimate term by term.) $[f,\mathbb{H}]\partial_{\alpha'}^{s+1}g$ and some intermediate terms. By Propositions A.1 and A.2 and the Sobolev embedding, $[f,\mathbb{H}]\partial_{\alpha'}g$ has the same regularity as f and g.

When the Riemann mapping is not used, typically we work in the regime where the interface is chord-arc, that is there are constants $\mu_1 > 0$ and $\mu_2 > 0$, such that

$$\mu_1|\alpha - \beta| \leq |z(\alpha,t) - z(\beta,t)| \leq \mu_2|\alpha - \beta|, \qquad \forall \alpha, \beta \in \mathbb{R}.$$

By (A.7), derivatives of $[f,\mathfrak{H}]\frac{\partial_\alpha g}{z_\alpha}$ are of types (A.1) and (A.4), where in (A.1) or (A.4) $F \in C_0^\infty(\mathbb{C})$ is chosen to be $F(z) = \frac{1}{z^m}$ for $\mu_1 \leq |z| \leq \mu_2$ and $H(\alpha) = z(\alpha,t)$. An application of Proposition A.1 and A.2 and Sobolev embedding shows that $[f,\mathfrak{H}]\frac{\partial_\alpha g}{z_\alpha}$ has the same regularity as f, g and z.

Appendix B: Commutator Identities

We need the following identities in our derivation of the quasilinear structure (9.2.27) and Propositions 9.3.1 and 9.3.2. Let $z = z(\cdot, t)$ define a non-selfintersecting curve, and $\mathfrak{H}$ be the Hilbert transform as defined in (9.2.2).

Lemma B.1 (Lemma 2.1 of [22]) *Assume that $z_t, z_\alpha - 1 \in C^1([0, T], H^1(\mathbb{R})), f \in C^1(\mathbb{R} \times (0, T))$ satisfies $f_\alpha(\alpha, t) \to 0$, as $|\alpha| \to \infty$. We have*

$$[\partial_t, \mathfrak{H}]f = [z_t, \mathfrak{H}]\frac{f_\alpha}{z_\alpha}$$

$$[\partial_t^2, \mathfrak{H}]f = [z_{tt}, \mathfrak{H}]\frac{f_\alpha}{z_\alpha} + 2[z_t, \mathfrak{H}]\frac{f_{t\alpha}}{z_\alpha} - \frac{1}{\pi i}\int \left(\frac{z_t(\alpha, t) - z_t(\beta, t)}{z(\alpha, t) - z(\beta, t)}\right)^2 f_\beta \, d\beta$$

$$[a\partial_\alpha, \mathfrak{H}]f = [az_\alpha, \mathfrak{H}]\frac{f_\alpha}{z_\alpha}, \quad \partial_\alpha \mathfrak{H}f = z_\alpha \mathfrak{H}\frac{f_\alpha}{z_\alpha}$$

$$[\partial_t^2 - ia\partial_\alpha, \mathfrak{H}]f = 2[z_t, \mathfrak{H}]\frac{f_{t\alpha}}{z_\alpha} - \frac{1}{\pi i}\int \left(\frac{z_t(\alpha, t) - z_t(\beta, t)}{z(\alpha, t) - z(\beta, t)}\right)^2 f_\beta \, d\beta$$

$$(I - \mathfrak{H})(-ia_t \bar{z}_\alpha) = 2[z_{tt}, \mathfrak{H}]\frac{\bar{z}_{t\alpha}}{z_\alpha} + 2[z_t, \mathfrak{H}]\frac{\bar{z}_{tt\alpha}}{z_\alpha} - \frac{1}{\pi i}\int \left(\frac{z_t(\alpha, t) - z_t(\beta, t)}{z(\alpha, t) - z(\beta, t)}\right)^2 \bar{z}_{t\beta} \, d\beta.$$

$$\tag{B.1}$$

The proof of Lemma B.1 is straightforward by integration by parts. We omit the proof.

Let $\Omega \subset \mathbb{C}$ be a domain with boundary $\Sigma : z = z(\alpha), \alpha \in I$ oriented clockwise. Let $\mathfrak{H}$ be defined by (9.2.2).

Lemma B.2 (Lemma 2.2 of [22]) *1. If $f = \mathfrak{H}f$, $g = \mathfrak{H}g$, then $[f, \mathfrak{H}]g = 0$.*
2. For any $f, g \in L^2$, we have $[f, \mathfrak{H}]\mathfrak{H}g = -[\mathfrak{H}f, \mathfrak{H}]g$.

The first statement follows from the fact that the product of holomorphic functions is holomorphic. Observe that the first statement also holds for $f = -\mathfrak{H}f$ and $g = -\mathfrak{H}g$, the second statement follows from the first by applying the first identity to $(I \pm \mathfrak{H})f$ and $(I \pm \mathfrak{H})g$.

We do not give the commutator identities for 3D, but refer the reader to [10, 23] for details.

References

[1] G. I. Taylor *The instability of liquid surfaces when accelerated in a direction perpendicular to their planes I.* Proc. Roy. Soc. London A 201 (1950), pp. 192–196

[2] G. Birkhoff *Helmholtz and Taylor instability* Proc. Symp. in Appl. Math. XIII, pp. 55–76.

[3] T. Beale, T. Hou & J. Lowengrub *Growth rates for the linearized motion of fluid interfaces away from equilibrium* Comm. Pure Appl. Math. 46 (1993), no. 9, pp. 1269–1301.

[4] G. G. Stokes *On the theory of oscillatory waves.* Trans. Cambridge Philos. Soc., 8 (1847), pp. 441–455.

[5] T. Levi-Civita. *Détermination rigoureuse des ondes permanentes d'ampleur finie.* Math. Ann., 93(1), 1925. pp. 264–314

[6] V. I. Nalimov *The Cauchy-Poisson problem* (in Russian), Dynamika Splosh. Sredy 18, 1974, pp. 104–210.

[7] H. Yosihara *Gravity waves on the free surface of an incompressible perfect fluid of finite depth,* RIMS Kyoto 18 (1982), pp. 49–96

[8] W. Craig *An existence theory for water waves and the Boussinesq and Korteweg-devries scaling limits* Comm. in P. D. E. 10(8) (1985), pp. 787–1003

[9] S. Wu *Well-posedness in Sobolev spaces of the full water wave problem in 2-D* Invent. Math. 130 (1997), pp. 39–72

[10] S. Wu *Well-posedness in Sobolev spaces of the full water wave problem in 3-D* J. Amer. Math. Soc. 12 no. 2 (1999), pp. 445–495.

[11] D. Ambrose, N. Masmoudi *The zero surface tension limit of two-dimensional water waves* Comm. Pure Appl. Math. 58 (2005), no. 10, pp. 1287–1315

[12] D. Christodoulou, H. Lindblad *On the motion of the free surface of a liquid* Comm. Pure Appl. Math. 53 (2000), no. 12, pp. 1536–1602

[13] D. Coutand, S. Shkoller *Wellposedness of the free-surface incompressible Euler equations with or without surface tension* J. AMS. 20 (2007), no. 3, pp. 829–930.

[14] T. Iguchi *Well-posedness of the initial value problem for capillary-gravity waves* Funkcial. Ekvac. 44 (2001) no. 2, pp. 219–241.

[15] D. Lannes *Well-posedness of the water-wave equations* J. Amer. Math. Soc. 18 (2005), pp. 605–654

[16] H. Lindblad *Well-posedness for the motion of an incompressible liquid with free surface boundary* Ann. of Math. 162 (2005), no. 1, pp. 109–194.

[17] M. Ogawa, A. Tani *Free boundary problem for an incompressible ideal fluid with surface tension* Math. Models Methods Appl. Sci. 12, (2002), no. 12, pp. 1725–1740.

[18] J. Shatah, C. Zeng *Geometry and a priori estimates for free boundary problems of the Euler's equation* Comm. Pure Appl. Math. V. 61. no. 5 (2008), pp. 698–744.

[19] P. Zhang, Z. Zhang *On the free boundary problem of 3-D incompressible Euler equations.* Comm. Pure. Appl. Math. V. 61. no. 7 (2008), pp. 877–940

[20] T. Alazard, N. Burq & C. Zuily *On the Cauchy problem for gravity water waves* Invent. Math. 198 (2014), pp.71–163

[21] T. Alazard, N. Burq & C. Zuily *Strichartz estimates and the Cauchy problem for the gravity water waves equations* Preprint 2014, arXiv:1404.4276

[22] S. Wu *Almost global wellposedness of the 2-D full water wave problem* Invent. Math, 177 (2009), no. 1, pp. 45–135.

[23] S. Wu *Global wellposedness of the 3-D full water wave problem* Invent. Math. 184 (2011), no. 1, pp.125–220.

[24] P. Germain, N. Masmoudi, & J. Shatah *Global solutions of the gravity water wave equation in dimension 3* Ann. of Math (2). 175 (2012), no. 2, pp. 691–754.

[25] A. Ionescu & F. Pusateri. *Global solutions for the gravity water waves system in 2d* Invent. Math. 199 (2015), no. 3, p. 653804

[26] T. Alazard & J-M. Delort *Global solutions and asymptotic behavior for two dimensional gravity water waves* Ann. Sci. Éc. Norm. Supér., to appear

[27] J. Hunter, M. Ifrim & D. Tataru *Two dimensional water waves in holomorphic coordinates* Preprint 2014, arXiv:1401.1252

[28] M. Ifrim & D. Tataru *Two dimensional water waves in holomorphic coordinates II: global solutions* Preprint 2014, arXiv:1404.7583

[29] S. Wu *On a class of self-similar 2d surface water waves* Preprint 2012 arXiv1206:2208

[30] R. Kinsey & S. Wu *A priori estimates for two-dimensional water waves with angled crests* Preprint 2014, arXiv1406:7573

[31] S. Wu *A blow-up criteria and the existence of 2d gravity water waves with angled crests* Preprint 2015 arXiv1502:05342

[32] J-L. Journé *Calderon-Zygmund operators, pseudo-differential operators and the Cauchy integral of calderon*, vol. 994, Lecture Notes in Math. Springer, 1983.

[33] A. Castro, D. Córdoba, C. Fefferman, F. Gancedo & J. Gómez-Serrano *Finite time singularities for the free boundary incompressible Euler equations* Ann. of Math. (2) 178 (2013), no. 3, pp. 1061–1134

[34] G. Folland *Introduction to partial differential equations* Princeton University press, 1976.

[35] C. Kenig *Elliptic boundary value problems on Lipschitz domains* Beijing Lectures in Harmonic Analysis, ed. by E. M. Stein, Princeton Univ. Press, 1986, p. 131–183.

[36] J. Gilbert & M. Murray *Clifford algebras and Dirac operators in harmonic analysis* Cambridge University Press, 1991

[37] J. Simon *A wave operator for a nonlinear Klein-Gordon equation*, Letters. Math. Phys., 7 (1983), no. 5, pp. 387–398.

[38] J. Shatah *Normal forms and quadratic nonlinear Klein-Gordon equations* Comm. Pure Appl. Math. 38 (1985), pp. 685–696.

[39] N. Totz & S. Wu *A Rigorous justification of the modulation approximation to the 2D full water wave problem* Comm. Math. Phys. 310 (2012), pp. 817–883

[40] J. Hunter & M. Ifrim *Enhanced lifespan of smooth solutions of a Burgers-Hilbert equation* SIAM J. Math. Anal, 44 (3) 2012, pp. 1279–2235.

[41] S. Klainerman *The null condition and global existence to nonlinear wave equations* Lectures in Appl. Math. 23 (1986), pp. 293–325.

[42] N. Totz *A Rigorous justification of the modulation approximation to the 3D full water wave problem* accepted by Comm. Math. Phys.

[43] J. Beichman *Nonstandard estimates for a class of 1D dispersive equations and applications to linearized water waves* Preprint (2014) arXiv:1409.8088

[44] R. Coifman, A. McIntosh and Y. Meyer *L'integrale de Cauchy definit un operateur borne sur L^2 pour les courbes lipschitziennes* Annals of Math. 116 (1982), pp. 361–387.

[45] S. Chen & Y. Zhou *Decay rate of solutions to hyperbolic system of first order* Acta. Math. Sinica, English series. 15 (1999), no. 4, pp. 471–484

[46] R. Coifman, G. David and Y. Meyer *La solution des conjectures de Calderón* Adv. in Math. 48 (1983), pp.144–148.

10

Conformal Mapping and Complex Topographies

André Nachbin

Abstract

Many interesting research problems consider long water waves interacting with highly disordered bottom topographies. The topography profile can be of large amplitude, not smooth and rapidly varying. Disordered topographies do not have a well defined structure and therefore can be modelled as a random coefficient in the wave equations. A summary of results that arise through random modelling is provided. The main goal of this article is to present the Schwarz-Christoffel conformal mapping as a tool for dealing with these problems. Both from the theoretical point of view as well as regarding computational aspects that make use of the Schwarz-Christoffel Toolbox, developed by T. Driscoll.

10.1 Introduction and Mathematical Motivation

About 25 years ago we became interested in studying the effects of small scale features of a topography on long waves propagating on the water surface for large distances, such as a tsunami. Our goal was to study solitary waves over rapidly varying disordered topographies. At the time the theory for linear acoustic waves over rapidly varying (one dimensional-1D) layered media was maturing and becoming quite sophisticated [1, 2]. The theory developed by George Papanicolaou and collaborators considered linear hyperbolic systems with disordered rapidly varying coefficients as a model to understand linear pulse-shaped waves travelling in a medium with a random propagation speed. This setup is very useful for understanding the effect of uncertainty on a travelling wave, in direct and inverse problems related to the Earth's subsurface. Mathematically it called for new technology showing that random modelling produced some universal results of interest. These included effective wave propagation properties that were independent of a specific realisation. To study linear travelling waves in the presence of random multiple scattering, Papanicolaou and collaborators developed an asymptotic theory for stochastic ordinary differential equations (SODEs) regarding the transmitted and reflected signals. Limit theorems for randomly forced oscillators characterised asymptotically the expected value for the transmitted and reflected signals after the wave had propagated for large distances. Three scales are involved in this analysis: the medium's *microscale* ε^2, the pulse's characteristic width ε (the *mesoscale*) and the large propagation distance $O(1)$ (the *macroscale*). The theory's effective dynamics is obtained in the limit when $\varepsilon \to 0$ and is characterised by a *diffusion process* [3]. In our computations the scale ordering is, respectively, set as ε, $O(1)$ and $1/\varepsilon$. Similar theory and behaviour are observed for water waves where the effective dynamics is captured for long enough time intervals, which typically are of the order of tens of pulse-widths [4–6]. A remarkable feature, for example, is that the transmitted wave is under the effect of an *apparent diffusion* due to multiple scattering. In other words the scattering of energy promoted by the medium's heterogeneities can be viewed, to leading order, as a low-pass Gaussian filter [3, 7]. The transmitted mode is given by the convolution of the pulse's initial profile with a Gaussian (diffusion-like) kernel, where the diffusion-like coefficient is an integral of the random medium's auto-correlation function [8]. Energy is conserved and along the smooth wavefront the diffusion is apparent because energy is being converted from the pulse to fluctuations scattered in the forward and backward directions. We further discuss this problem in a following subsection. A detailed theory for waves in randomly layered media can be found in the book by Fouque, Garnier, Papanicolaou & Sølna [3].

Having the acoustic wave technology in mind our starting point was the study of linear surface water waves, in the long wave regime were the Shallow Water equations are known to be similar to the hyperbolic acoustic equations, but in principle not valid for non-smooth topography profiles such as the piecewise-constant one which is analogous to the layered Earth crust such as displayed in figure 10.1. Having all this in perspective, conformal mapping appeared as a technique very convenient for the asymptotic simplification of the potential theory [9] water wave formulation leading to a shallow water model valid in our regime of interest. We will briefly report on this.

Consider the acoustic layered medium as displayed in a schematic fashion on the left of figure 10.1. The equations studied are [1, 3]

$$(1/K)\, p_t + u_z = 0,$$
$$\rho\, u_t + p_z = 0,$$

$(10.1.1)$

where $p(z,t)$ is the pressure at depth z, within the layered medium, $u(z,t)$ is the velocity at that point at time t, $1/K(z/\varepsilon^2)$ is the medium's compressibility and $\rho(z/\varepsilon^2)$ is the layer's density at that depth. The small parameter $\varepsilon \ll 1$ indicates that the coefficients are rapidly varying. This system is hyperbolic and the

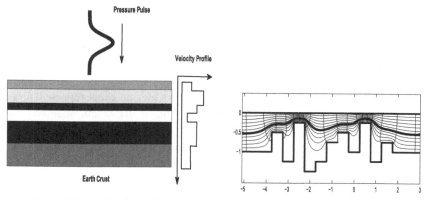

Figure 10.1. Left: a layered medium for the acoustic wave model in a randomly varying Earth crust. Right: a water channel with a piecewise constant bathymetry, in analogy with the layered medium. In both cases the propagation speed is piecewise constant. The heavy line depicts a terrain-following level curve of the (orthogonal) curvilinear coordinate system.

rapidly varying (piecewise-constant) propagation speed is $c(z/\varepsilon^2) = \pm\sqrt{K/\rho}$. The variable coefficients of this system of partial differential equations (PDE) can be modelled as realisations of a stochastic process. A simple example is when the random fluctuations are uniformly distributed about a given reference value. This is the layered medium. In order to import to water waves the technology developed for acoustic waves a good starting point are linear pulse-shaped waves in a shallow water system with a piecewise constant random depth leading to a piecewise-constant random speed. Once this first goal was achieved [10] the next step in our plan was to add dispersion in the linear regime [4, 6]. Next we added nonlinearity in the non-dispersive regime [5] to finally have both a weakly dispersive, weakly nonlinear model for solitary waves in a randomly varying channel [11, 12].

The water wave analog of the layered acoustic medium is depicted on the right of figure 10.1. The typical (long wave) asymptotic expansions are not valid in such a scenario [13]. A change to orthogonal curvilinear coordinates, through the conformal mapping, avoids this difficulty. If we follow the lines of the classical depth-averaged long wave model, we should then take the terrain-following "horizontal" velocity component (from potential theory) and average it over the depth. These velocity components are tangent to the terrain-following level curves displayed in figure 10.1. In [14] we found that using the velocity at a particular depth, as proposed by Nwogu [15] in 1993, gives best results in the variable bottom case also. As an example a "depth level" is highlighted in figure 10.1 (right). Recall that Nwogu proposed the optimal level $Z_o = 0.469$ arising from a mean-squared minimisation of phase errors over a certain wavenumber band. It is worth pointing out that for a flat bottom, the Boussinesq equations obtained at the depth level $Z_o = \sqrt{1/3}$ are the same as for the depth-averaged case [14, 15]. Having the depth parameter Z_o in mind, Kraenkel et al. [16] looked for an effective Korteweg-deVries equation in the case of weakly dispersive, weakly nonlinear waves over rapidly varying periodic topographies. Their asymptotic analysis yielded the "best" depth value $Z_o = (2/3 - \sqrt{1/5})^{1/2} \approx 0.469$, which provides an exact representation for the optimal value found by Nwogu using linear dispersion analysis and Padé approximation.

The curvilinear coordinate system was implemented into the nonlinear potential theory equations. The weakly nonlinear ($\alpha \ll 1$), weakly dispersive ($\beta \ll 1$) truncation of (potential theory) differential operators lead to the Boussinesq system [14, 17]

$$M(\xi)\,\eta_t + [(1 + \alpha\eta/M(\xi))u]_\xi + \beta/2\big[(Z_o^2 - 1/3)\big]_\xi = 0,$$
$$u_t + \eta_\xi + \alpha\left(u^2/(2M^2(\xi))\right)_\xi + \beta/2\big[(Z_o^2 - 1)\big]u_{\xi\xi t} = 0. \tag{10.1.2}$$

Consider the linear ($\alpha = 0$), nondispersive ($\beta = 0$) case in cartesian coordinates, as for example found in [18]. The linear shallow water system is given by

$$\eta_t + (h(x)\,u)_x = 0,$$
$$u_t + \eta_x = 0, \tag{10.1.3}$$

where $h(x)$ denotes the depth variations. This system is not valid for rapidly varying and non-smooth depth variations. These regimes can be accounted for when the conformal mapping takes place, namely in the $\xi\zeta$ coordinate system. When $\alpha = \beta = 0$ equations (10.1.2) simplify to the linear hyperbolic system

$$M(\xi/\varepsilon^2)\,\eta_t + u_\xi = 0,$$
$$u_t + \eta_\xi = 0, \tag{10.1.4}$$

valid, for example, for rapidly varying piecewise constant depth variations. This is analogous to the acoustic system (10.1.1), where the density is taken to be identically equal to 1 and the coefficient M plays the role of the compressibility. It will be shown that this coefficient is associated to the topography through the Jacobian of the conformal mapping evaluated along the free surface. Therefore, it is sometimes called the metric coefficient $M(\xi)$, it is smooth (C^∞) and is our main object of study in this article.

In the next section we present the potential theory formulation in curvilinear coordinates. We will then present two different formulations for the metric coefficient $M(\xi)$. But first we summarise some remarkable results for water waves over rapidly varying complex topographies.

10.1.1 Effective Water Wave Behaviour over Complex Bottom Topographies

In this subsection we further discuss the effective behaviour of waves in random media. Rapidly varying intricate topographies can be initially set into two categories: periodic and disordered. By *disordered* one obviously means a topography without a well defined structure as opposed to a periodic one. In many problems it is of interest to think of a disordered propagation medium as a random medium. This is a modelling option that enables one to deal with *uncertainty*, such as for acoustic waves in the Earth's subsurface, or to obtain *universal results* that are independent of a specific realisation of the medium. In the apparent diffusion case reported here the theoretical results depend only on certain statistical properties of the medium's fluctuations such as the average speed and the correlation function of related variable coefficients.

Laboratory experiments [19] have been performed to analyse, in particular, the transmission properties of long waves over randomly distributed steps. The main goal was to provide experimental evidence of *Anderson localisation* for water waves in a 1D random medium. In this context Anderson localisation establishes that, for a long enough random medium, linear waves will be completely reflected back by the medium. Consider a randomly varying topography over the positive real axis and a pulse-shaped wave propagating with constant unit speed from the left to the right incoming along the negative axis. Each right propagating Fourier mode of the pulse-shaped wave has an associated localisation length, which is a typical penetration depth into the random medium before its exponential decay has essentially reached zero. Eventually all of the wave energy is reflected back to the homogeneous negative semi-axis. The localisation length is related to the Lyapunov exponent of a random oscillator [3]. The Anderson localisation results for the laboratory experiments were presented by Belzons et al. [19] and a related theory by Devillard [20]. The theory was specialised to piecewise constant topographies and made use of transfer matrices between each constant depth region. The theory presented in Nachbin and Papanicolaou [21] and Nachbin [22] also established Anderson localisation but for more general topographies. This was our first work where conformal mapping played a role. Details can also be found in [23].

Effective medium theory is valid for short propagation distances on the scale of the wavelength. The effective medium (homogenised) model corresponds to replacing the random speed by the *effective medium's speed* [3, 8], which is the reciprocal of the averaged slowness, namely the harmonic mean of the speed. The reciprocal of the speed is called the *slowness*.

In Figure 10.2 we display results for the linear hyperbolic wave system. These results are qualitatively valid in other wave problems as well, such

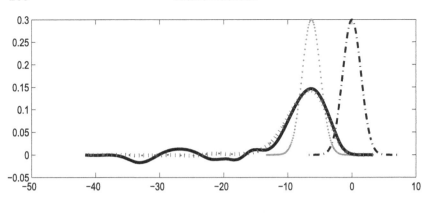

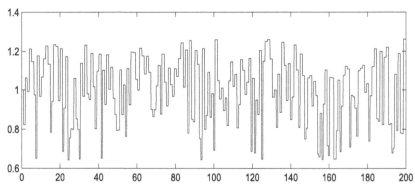

Figure 10.2. Bottom: A realisation for a piecewise-constant random propagation speed centred about the background unit-speed. There are 200 layers. Top: The horizontal axis is in terms of *delay-time*. Zero delay-time means the wave arrived on time regarding the underlying homogeneous system. The dash-dotted profile depicts the travelling wave solution of the underlying homogeneous hyperbolic system. The solution of the homogenised system is given by the dotted pulse profile. The solid line is for the solution of the hyperbolic system, with the small scale random fluctuations resolved by the grid. The profile with vertical dashes and dots is the result produced by a central limit theorem [3, 8].

as dispersive or weakly nonlinear. In the bottom part of the figure a piecewise constant randomly varying propagation speed profile is displayed. The background speed is normalised to 1. In other words we have a random propagation speed $c = c_o + \delta \cdot$ (mean-zero fluctuations), where the reference (background) speed c_o is taken to be equal to 1. The fluctuation level is controlled through the parameter $\delta < 1$. The speed fluctuation at a given layer arises from a (mean-zero) uniformly distributed random-depth variable. Details are found in [8]. Here we want to only discuss qualitative features. At the bottom of figure 10.2 the long propagation medium has 200 layers at a ratio of 10 layers per pulse-width. The horizontal axis in the top part of figure 10.2 is given in terms of delay time τ. The delay-time is defined as time minus travel-time [6, 7] as given by

$$\tau = t - \int_0^x \frac{1}{c(s)} ds.$$

When the medium is homogeneous with $C(S) = c_o \equiv 1$ there is no delay and, therefore, the pulse appears centered at zero. It is our reference for the arrival time of a signal at a given location, here specified by where the layers terminate. When the medium is heterogenous and the averaged speed (through the harmonic mean [3]) is less than one, as it happened by chance in the example presented, the travel time is larger than the reference arrival time and therefore a (negative valued) delay will be observed for the pulse. In this case it will be centred at a negative delay-time as in figure 10.2 because it arrived behind/after the reference pulse. For the realisation sampled with a random number generator the effective speed happened to be less than one. The solution of the homogenised system is given by the dotted pulse profile. It is still a travelling wave but it is delayed with respect to our reference solution as discussed.

Now if we solve the hyperbolic system, resolving the small scale random fluctuations, we observe a weak scattering of the wave energy that accumulates over larger distances. In other words, in the case of variable coefficients a scattering mechanism takes place and part of the energy of the wavefront is converted to fluctuations behind it, which propagate in both directions. As a consequence of this energy conversion the wave front appears to be attenuated as displayed by the solid-line profile. Therefore, the effective medium/homogenisation approximation is not well suited for large propagation distances. The system has constant (averaged) coefficients and does not promote scattering related phenomena. This diffusion-like attenuation is well approximated by a higher order theory, which arises from the application of a limit theorem for SODEs [3]. The result from this higher order theory is expressed through a Gaussian filter acting on the initial wave profile, hence the effect of an *apparent diffusion*. This more accurate asymptotic result is depicted by the vertical dashes and dots, which are in very good agreement with the (numerical) solid line.

In the next section we show how the conformal change of coordinates affects the nonlinear potential theory system and how we express the relevant metric coefficient $M(\xi)$.

10.2 Formulation and Scaling

Let variables with physical dimensions be denoted with a tilde. We introduce the length scales σ (a typical pulse width or wavelength), h_0 (a typical depth), a (a typical wave amplitude), l_b (the horizontal length scale for bottom irregularities) and L (the total length of the rough region or the total propagation distance). The acceleration due to gravity is denoted by g and the reference shallow water speed is $c_0 = \sqrt{gh_0}$. Dimensionless variables are then defined in a standard fashion [9, 24] by having

$$\tilde{x} = \sigma x, \ \ \tilde{y} = h_0 y, \ \ \tilde{t} = \left(\frac{\sigma}{c_0} \right) t, \ \ \tilde{\eta} = a\eta, \ \ \tilde{\phi} = \left(\frac{g\sigma a}{c_0} \right) \phi, \ \ \tilde{h} = h_0 H \left(\frac{\tilde{x}}{l_b} \right).$$

The velocity potential $\phi(x,y,t)$ and wave elevation $\eta(x,t)$ satisfy the dimensionless equations [9, 24]:

$$\beta \, \phi_{xx} + \phi_{yy} = 0 \ \ \text{for} \ -H(x/\gamma) < y < a\eta(x,t),$$

with the nonlinear free surface conditions

$$\eta_t + \alpha\phi_x\eta_x - \frac{1}{\beta}\phi_y = 0,$$

$$\eta + \phi_t + \frac{\alpha}{2}\left(\phi_x^2 + \frac{1}{\beta}\phi_y^2 \right) = 0,$$

at $y = \alpha\eta(x,t)$. The Neumann condition at the impermeable bottom is

$$\phi_y + \frac{\beta}{\gamma} H'(x/\gamma)\phi_x = 0.$$

The bottom boundary is at $y = -H(x/\gamma)$ where

$$H(x/\gamma) = \begin{cases} 1 + n(x/\gamma) & \text{when } 0 < x < L \\ 1 & \text{when } x \leq 0 \ \text{ or } \ x \geq L. \end{cases}$$

The bottom profile is described by $-n(x/\gamma)$. The topography is rapidly varying when $\gamma \ll 1$. The undisturbed depth is given by $y = -1$ and the topography can be of large amplitude provided that $|n| < 1$. We do not need to assume that the fluctuations n are small, nor continuous, nor slowly varying. The following dimensionless parameters arise:

$\alpha = a/h_0$ (nonlinearity parameter),

$\beta = h_0^2/\sigma^2$ (dispersion parameter),

$\gamma = l_b/\sigma$ (bottom irregularities compared to the wavescale).

We now introduce the conformal mapping and the associated change of coordinates.

10.2.1 Changing to Curvilinear Coordinates

A mapping from an uniform strip onto the fluid domain at rest is constructed. Let the former canonical domain be defined in the complex w-plane and the rough undisturbed channel (physical domain) be defined in the complex z-plane. By undisturbed channel we mean that no waves/disturbances are present along the free surface. Properties of the $z = f(w)$ mapping will be outlined below. It will be shown that working with a symmetric domain is very convenient for weakly nonlinear waves. We solve the (harmonic) conformal mapping problem in a symmetric configuration such as displayed at the top of figure 10.3, where we superimpose the symmetric domain in the z-plane with the curvilinear coordinates' level curves, originating from the w-plane. Hence, instead of referring to the w-plane we can interpret our formulation as working with orthogonal curvilinear coordinates in the physical domain. The dashed polygonal line at the top of figure 10.3 is a schematic representation of the reflected topography.

Following Hamilton [13] we reflect the topography about the undisturbed free surface (c.f. figure 10.3). We denote this domain by Ω_z where $z = x + i\sqrt{\beta}y$ and consider it as the image of the strip Ω_w where $w = \xi + i\tilde{\zeta}$ with $|\tilde{\zeta}| \leq \sqrt{\beta}$. Then $z = x(\xi,\tilde{\zeta}) + i\sqrt{\beta}y(\xi,\tilde{\zeta}) = x(\xi,\tilde{\zeta}) + i\tilde{y}(\xi,\tilde{\zeta})$, with x and $\tilde{y}$ a pair of harmonic functions on Ω_w. Working with x and $\tilde{y}$ is convenient for computing harmonic functions associated with the mapping because the parameter β drops from the Laplacian.

It is useful to point out that the velocity potential $\phi(\xi,\tilde{\zeta},t)$ when represented in the curvilinear coordinate system is such that $\phi_\xi = \phi_x\, x_\xi(\xi,\tilde{\zeta}) + \phi_{\tilde{y}}\, \tilde{y}_\xi(\xi,\tilde{\zeta})$ and $\phi_{\tilde{\zeta}} = \phi_x\, x_{\tilde{\zeta}}(\xi,\tilde{\zeta}) + \phi_{\tilde{y}}\, \tilde{y}_{\tilde{\zeta}}(\xi,\tilde{\zeta})$. In particular for symmetric flow regions, at the undisturbed free surface $\phi_\xi(\xi,0) = M(\xi)\phi_x$ and $\phi_{\tilde{\zeta}}(\xi,0) = M(\xi)\phi_{\tilde{y}}$, where $M(\xi) \equiv \tilde{y}_{\tilde{\zeta}}(\xi,0)$. Note that we have used the Cauchy–Riemann equations. Also we have that

$$\phi_x = \frac{1}{|J|}\left[\tilde{y}_{\tilde{\zeta}}\phi_\xi - \tilde{y}_\xi\phi_{\tilde{\zeta}}\right] \qquad (10.2.1)$$

and

$$\phi_{\tilde{y}} = \frac{1}{|J|}\left[-x_{\tilde{\zeta}}\phi_\xi + x_\xi\phi_{\tilde{\zeta}}\right], \qquad (10.2.2)$$

where $|J| = x_\xi\tilde{y}_{\tilde{\zeta}} - \tilde{y}_\xi x_{\tilde{\zeta}} = \tilde{y}_{\tilde{\zeta}}^2 + \tilde{y}_\xi^2$, and $\phi_x^2 + \phi_{\tilde{y}}^2 = (\phi_\xi^2 + \phi_{\tilde{\zeta}}^2)/|J|$.

The water wave equations in the orthogonal curvilinear coordinates $(\xi,\tilde{\zeta})$ are:

$$\phi_{\xi\xi} + \phi_{\tilde{\zeta}\tilde{\zeta}} = 0, \qquad -\sqrt{\beta} < \tilde{\zeta} < \alpha\sqrt{\beta}N(\xi,t), \qquad (10.2.3)$$

with free surface conditions

$$|J|N_t + \alpha\phi_\xi N_\xi - \frac{1}{\sqrt{\beta}}\phi_{\tilde{\zeta}} = 0, \qquad (10.2.4)$$

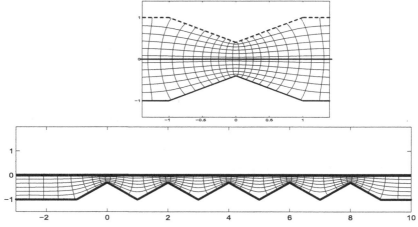

Figure 10.3. Top: The symmetric domain in the complex z-plane, where $z = x(\xi, \tilde{\zeta}) + i\tilde{y}(\xi, \tilde{\zeta})$. The lower half ($x \in [-1.5, 1.5]$, $y \in [-1, 0]$) is the physical channel with $y = \tilde{\zeta} = 0$ indicating the undisturbed free surface. Superimposed in this complex z-plane domain are the (curvilinear) coordinate level curves from the w-plane system $\xi\tilde{\zeta}$. The polygonal line at the bottom of the figure is a schematic representation of the topography (where $\tilde{\zeta} = $ constant). Bottom: Even for topographies of large amplitudes the Jacobian quickly goes to a constant away from the mountain range. The curvilinear coordinate system (quickly and smoothly) reduces to the usual cartesian system. This figure was generated using Schwarz-Chistoffel Toolbox (SCT) [25].

and

$$\phi_t + \eta + \frac{\alpha}{2|J|}\left(\phi_\xi^2 + \phi_{\tilde{\zeta}}^2\right) = 0 \qquad (10.2.5)$$

at $\tilde{\zeta} = \alpha\sqrt{\beta}N(\xi, t)$. The bottom condition is

$$\phi_{\tilde{\zeta}} = 0, \quad \text{at } \tilde{\zeta} = -\sqrt{\beta}. \qquad (10.2.6)$$

In the free surface equations we should have in mind the relation $x + i\eta(x, t) = f(\xi + iN(\xi, t))$. Regarding the initial conditions note that by starting with a pulse over a region of uniform depth, the initial data are not affected by the conformal mapping (c.f. bottom figure 10.3 at $x \leq -2$ and $x \geq 10$). If the flow domain has the same height in both the z and w-planes then we only need to replace x by ξ in the initial wave conditions. Or else it is a simple rescaling since the Jacobian matrix is a multiple of the identity.

Note that by using the reflected domain we have by construction $\tilde{y}(\xi, \tilde{\zeta})$ as an odd function in $\tilde{\zeta}$. A simple Taylor series in $\tilde{\zeta}$ about the undisturbed surface yields $\tilde{y}(\xi, \tilde{\zeta}) = \tilde{\zeta} \cdot \tilde{y}_{\tilde{\zeta}}(\xi, 0) + O(\tilde{\zeta}^3)$. For small amplitude waves we have that $\eta(x, t) \approx N(\xi, t) \cdot M(\xi)$, where $\tilde{y}_{\tilde{\zeta}}(\xi, 0) \equiv M(\xi)$ is the coefficient we will calculate in the next section. Also we have that $|J|(\xi, 0) \approx \tilde{y}_{\tilde{\zeta}}^2(\xi, 0)$. We clearly see from figure 10.3 that away from the rough region $|J| \approx constant$, where this adjustment is very fast. If the initial condition is over a region where the topography varies then we need to compute $N(\xi,0) = \mathbf{Im}(f^{-1}(x+i\eta(x,0)))$ in order to obtain the initial wave profile in the new coordinate system. But as will be shown there are cases where the above approximation $N(\xi,0) = \eta(x(\xi,0),0)/M(\xi)$ is valid and avoids computing $f^{-1}(z)$.

10.3 The Conformal Mapping of Complex Topographies

In this section we present the formulation of the metric coefficient $M(\xi)$ through two different frameworks. One by a PDE formulation, which provides a great deal of intuition, namely *qualitative information* regarding fast and slow topographies. The second formulation is through a numerical Schwarz-Christoffel mapping. This formulation is more adequate for extracting accurate *quantitative information* regarding the metric coefficient, but not on the intuitive side.

10.3.1 Using Green's Third Identity

The imaginary part of the conformal map is the harmonic function $\tilde{y}(\xi,\tilde{\zeta})$ that satisfies [13, 17]

$$\Delta \, \tilde{y}(\xi,\tilde{\zeta}) = 0 \quad \text{in } \Omega_w, \tag{10.3.1}$$

with Dirichlet boundary conditions

$$\tilde{y}(\xi,\pm\sqrt{\beta}) = \pm h(x(\xi)) \equiv \pm\sqrt{\beta}H\left(\frac{x(\xi,\pm\sqrt{\beta})}{\gamma}\right). \tag{10.3.2}$$

It is the harmonic conjugate of $x(\xi,\tilde{\zeta})$. In this symmetric setting the undisturbed free surface level is at $\tilde{\zeta} \equiv \tilde{\zeta}_0 = 0$. The Green's function for problem (10.3.1)-(10.3.2) is slightly different from Hamilton's [13] because we introduce dimensionless variables and keep depth effects through the parameter $\sqrt{\beta}$. The Green's function, vanishing along the boundaries $\tilde{\zeta} = \pm \sqrt{\beta}$, is given by

$$G(w;w_0) = \mathbf{Re} \, log \left((e^{\pi w/2\sqrt{\beta}} - e^{\pi w_0/2\sqrt{\beta}})/(e^{\pi w/2\sqrt{\beta}} + e^{\pi \overline{w_0}/2\sqrt{\beta}})\right), \tag{10.3.3}$$

where $\mathbf{Re}$ stands for the real part and the overbar denotes complex conjugation. Note that this Green's function is constructed easily using complex variables. We know that $\ln \mathbf{r}$ is the (free space) fundamental solution for the Laplacian in two dimensions, where $\mathbf{r}^2 = (\xi - \xi_0)^2 + (\tilde{\zeta} - \tilde{\zeta}_0)^2$. This is the real part of $log(w - w_0)$. By mapping a strip onto the unit circle we get a Green's function with the desired vanishing boundary value. The change of variables in the argument of expression (10.3.3) performs this task, namely at the boundaries

$\tilde{\zeta} = \pm\sqrt{\beta}$, $G \equiv 0$, because $\mathbf{r} \equiv 1$ by construction. Near a source point w_0, $G(w; w_0) \sim \mathbf{Re}\ log(w - w_0)$, meaning that it behaves like the free-space Green's function.

Recall Green's third identity. It arises from Green's identity using a harmonic function and the fundamental solution of Laplace equation's leading, through integration by parts, to

$$2\pi\ \tilde{y}(\xi_0, \tilde{\zeta}_0) = \oint_{\partial\Omega_w} \left(\tilde{y}(\xi, \tilde{\zeta}) \frac{dG}{dn}(w; w_0) - \frac{d\tilde{y}}{dn}(\xi, \tilde{\zeta})\ G(w; w_0) \right)\ d\xi,$$

where $\Delta G = 2\pi\ \delta$. The Green's function was constructed to vanish on the boundaries and therefore

$$2\pi\ \tilde{y}(\xi_0, \tilde{\zeta}_0) = \oint_{\partial\Omega_w} \tilde{y}(\xi, \tilde{\zeta}) \frac{dG}{dn}(w; w_0)\ d\xi.$$

By the Dirichlet condition this simplifies to the double-layer potential, boundary representation

$$2\pi\ \tilde{y}(\xi_0, \tilde{\zeta}_0) = \int_{-\infty}^{\infty} h(x(\xi))\ \left(G_{\tilde{\zeta}}^+ + G_{\tilde{\zeta}}^- \right)\ d\xi,$$

where

$$G_{\tilde{\zeta}}^+ = \frac{\partial G}{\partial\tilde{\zeta}}(\xi, +\sqrt{\beta}; \xi_0, \tilde{\zeta}_0)\quad \text{and}\quad G_{\tilde{\zeta}}^- = \frac{\partial G}{\partial\tilde{\zeta}}(\xi, -\sqrt{\beta}; \xi_0, \tilde{\zeta}_0).$$

Differentiating this identity with respect $\tilde{\zeta}_0$ and evaluating at $\tilde{\zeta}_0 = 0$ we get our quantity of interest:

$$\tilde{y}_{\tilde{\zeta}_0}(\xi_0, 0) = \frac{1}{2\pi} \int_{-\infty}^{\infty} h(x(\xi)) \left(G_{\tilde{\zeta}\tilde{\zeta}_0}^+ + G_{\tilde{\zeta}\tilde{\zeta}_0}^- \right) d\xi. \qquad (10.3.4)$$

The kernel along the undisturbed free surface is expressed as

$$G_{\tilde{\zeta}\tilde{\zeta}_0}(\xi, -\sqrt{\beta}; \xi_0, 0) + G_{\tilde{\zeta}\tilde{\zeta}_0}(\xi, \sqrt{\beta}; \xi_0, 0) = \frac{2\pi^2}{\beta}\ \frac{e^{\frac{\pi}{\sqrt{\beta}}(\xi+\xi_0)}}{(e^{\frac{\pi}{\sqrt{\beta}}\xi} + e^{\frac{\pi}{\sqrt{\beta}}\xi_0})^2} =$$

$$= \frac{\pi^2/\beta}{2\ \cosh^2 \frac{\pi}{2\sqrt{\beta}}(\xi - \xi_0)}, \qquad (10.3.5)$$

and we finally obtain

$$\tilde{y}_{\tilde{\zeta}_0}(\xi_0, 0) = \frac{\pi}{4\sqrt{\beta}} \int_{-\infty}^{\infty} \frac{H(x(\xi, -\sqrt{\beta})/\gamma)}{\cosh^2 \frac{\pi}{2\sqrt{\beta}}(\xi - \xi_0)}\ d\xi. \qquad (10.3.6)$$

This expression was obtained by Hamilton [13], but without the $\beta^{1/2}$-scaling. This intermediate $\beta^{1/2}$-scale plays an important role indicating the degree of topography smoothing. The rate of smoothing actually depends on $\beta^{1/2}/\gamma$ seen in the numerator of the convolution in (10.3.6). Numerical examples

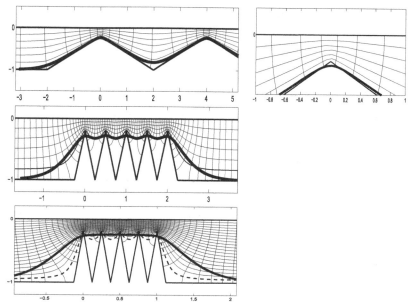

Figure 10.4. Large amplitude mountain range varying on three different length scales. Amplitude is always 80% of the total depth. Top-to-bottom: the triangular mountain has a base equal to 4, 0.5 and 0.25, respectively. The homogenisation through the metric coefficient $m(\xi)$ is displayed by the solid line running over the mountains. At the top right is a detail showing the smoothing of the topography's sharp corners.

of the topography smoothing are presented in figure 10.4. We call $\beta^{1/2}$ an intermediate scale because the water wave models work with integer powers in β as, for example, Whitham [9] in the derivation of Boussinesq's equations, the Korteweg-de Vries equation, among others. Note that

$$\int_{-\infty}^{\infty} \frac{\pi}{4\sqrt{\beta}} \operatorname{sech}^2\left[\frac{\pi}{2\sqrt{\beta}}(x-y)\right] dx = \frac{1}{2} \tanh\left[\frac{\pi}{2\sqrt{\beta}}x\right]_{-\infty}^{\infty} = 1. \quad (10.3.7)$$

Hence as $\sqrt{\beta} \downarrow 0$ the kernel in equation (10.3.6) goes to a delta function.

Using expressions (10.3.6) and (10.3.7) we obtain a very useful expression for the *metric term*:

$$M(\xi) \equiv \tilde{y}_{\zeta}(\xi,0) = 1 + m(\xi)$$

where

$$m(\xi;\sqrt{\beta},\gamma) \equiv \frac{\pi}{4\sqrt{\beta}} \int_{-\infty}^{\infty} \frac{n(x(\xi_0,-\sqrt{\beta})/\gamma)}{\cosh^2 \frac{\pi}{2\sqrt{\beta}}(\xi_0-\xi)} d\xi_0 = (K*(n \circ x))(\xi).$$

$$(10.3.8)$$

Lets discuss some useful properties of the metric term as given by Green's third identity. The kernel in (10.3.8) has a length scale associated with $\sqrt{\beta}$.

The interplay between the scales $\sqrt{\beta}$ and γ plays an important role in what will be discussed. When the (pulse-shaped) sech²-kernel is narrow with respect to the topography's length-scale then $m(\xi)$ is basically is a mollified version of $-n(x)$.

This can be seen at the top of figure 10.4 where we have a (submerged) periodic mountain range, of triangular mountains with a base of length equal to 4. This problem is purely geometrical and no wavelengths have been set. So we can think of β being associated only with the depth as if λ has been normalised to 1, which is the case in our computations with waves in random media. At the top-right we have a detail showing how $m(\xi)$ shaves off the sharp corner at the summit of $-n(x)$. Both the m and $-n$ profiles are very similar.

At the middle picture of figure 10.4 we have a rapidly varying mountain range where the mountain base is now equal to 0.5. We clearly see the start of a homogenisation-like effect in $m(\xi)$. The sharp triangular mountain range is now being effectively felt along the undisturbed free surface as a smooth bump with (small amplitude) smooth oscillations superimposed. Finally at the bottom picture we see the complete homogenisation effect when the mountain base is equal to 0.25. A smooth step is seen through $m(\xi)$. Therefore, expression (10.3.8) has built-in both aspects of a mollifier as well as of an averaging operator. Note that at the bottom figure 10.4 the topography

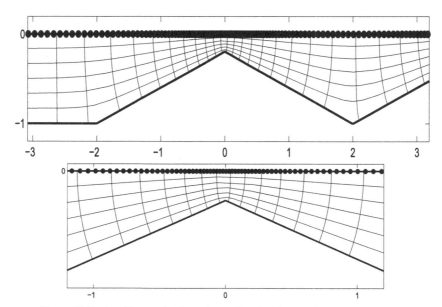

Figure 10.5. A uniform grid along the undisturbed free-surface in the canonical w-plane is mapped onto the physical domain. This provides a visualisation of the $x(\xi)$ dependence. The bottom graph is a detail near the mountain's summit.

varied on an even faster scale. We graphed 20 level curves of constant $\tilde{\zeta}$. The lowest level curve has been highlighted with a dashed line. For depth averaged Boussinesq-type models the region below the dashed line contributes with only 5% to the average. As mentioned in [17] this is good because in the region near the bottom the model has limitations, as for example due to vorticity generation. On the other hand Rey et al. [26] performed laboratory experiments for the propagation of surface gravity waves over a rectangular submerged bar and showed that potential theory is a good approximation if the vorticity generated near the sharp edges of the rectangular bar stays near the bottom and does not interact with the free surface.

We have been able to interpret many useful properties with expression (10.3.8) even though our work is unfinished. Note that it is very hard to extract numbers from expression (10.3.8). Why? Actually these two points are related. The answer is that at this stage we have no idea whatsoever of the function $x(\xi)$. For example, this dependence along the undisturbed free surface is very complicated. We display such nontrivial dependence in figure 10.5. A uniform grid along the undisturbed free surface in the w-plane is mapped onto the z-plane. As can be seen in this figure it is highly nonuniform. The bottom picture is a close up near the mountain's summit and we clearly see the non-trivial node compression there.

10.3.2 Using the Schwarz-Christoffel Toolbox

The Schwarz-Christoffel formula for the strip map $z = f(w)$ is

$$\frac{dz}{dw} = c \prod_{j=1}^{n} \left[\sinh \frac{\pi}{2} (w - a_j) \right]^{\alpha_j - 1}. \qquad (10.3.9)$$

This expression represents the derivative of the map $z = f(w)$ [25] from the canonical w-plane, where we have a uniform strip, to the physical z-plane where we have a corrugated strip, representing the flow domain, as depicted in figure 10.6, or the reflected flow domain when this case is considered, as at the top of figure 10.3. The pre-image of each vertex is denoted by a_j. The internal angle at each vertex is $\alpha_j \pi$ where the angle is measured counter-clockwise and $\alpha_j \in (0, 2]$. The value $\alpha_j = 0$ is used for a point at infinity as a convention of the Schwarz-Christoffel Toolbox (SCT). An example will be provided later. The constant c is a scaling factor related to the map. The SCT calculates c as well as the nontrivial nonlinear parameter problem, namely finding the values of the pre-images a_j. In most cases these are found with very good accuracy. In our water wave problem we need the Jacobian, which is $|J| = |dz/dw|^2$. Therefore, once the parameter problem is solved expression (10.3.9) is readily available for computing the Jacobian or our metric coefficient $M(\xi)$. Using

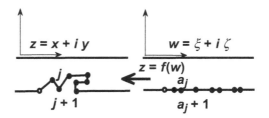

Figure 10.6. A schematic figure for the Schwarz-Christoffel conformal mapping of a uniform strip onto a corrugated strip. The origin (in both domains) is depicted by the white dot. The points a_j (black dots) are the pre-images of the vertices (j) in the physical domain.

the SCT we easily compute $M(\xi) = 1 + m(\xi) = Re(dz/dw)(\xi, \zeta_0)$, where ζ_0 is the undisturbed free surface level in the w-plane. This value will depend on how the SCT sets the mapping. In particular in the symmetric flow domain this level is at $\zeta_0 = 0.5$. In previous figures we depicted for comparison the topography profile given by $-n(x_k)$ and the metric fluctuations $m(\xi(x_k, 0))$, which was interpreted as how the topography is seen from the free surface, through the mapping. The points x_k are on a grid in the physical domain where $\xi(x_k) = \mathbf{Re}(f^{-1}(x_k + i*0))$.

The remaining part of this section contains a straightforward outline for the numerical computation of the Jacobian and the free surface coefficient $M(\xi)$. These are the variable coefficients in the non-local formulation (2.33)-(2.34) presented in [27] or in the corresponding Boussinesq-type equations that arise [14, 17, 28]. We also present the few steps needed for numerically calculating the physical wave elevation $\eta(x, t)$ from its conformally mapped counterpart $N(\xi, t)$ and vice-versa.

The calculations described are performed using the SCT, an open software available through Toby Driscoll's webpage:

http://www.math.udel.edu/~driscoll/software/.

Once the package is downloaded, an easy way to start is by running MATLAB in the directory containing all m-files. Details on the numerical schemes and issues regarding the Schwarz-Christoffel mapping are found in the book by Driscoll and Trefethen [25].

To become familiar with the conformal mapping and the associated level curves for the $\xi - \zeta$ coordinate system, the reader might want to first explore the GUI (graphics user interface) provided by the SCT, which is very easy to work with. Type the command >>scgui; and a new MATLAB window will open with buttons that allow the easy execution of commands. From now on, the symbol >> indicates MATLAB's prompt. The manual provided at Driscoll's webpage explains the main features of this GUI.

Very *few commands* are necessary to compute the desired variable coefficient on a computational grid specified by the SCT user. First recall that the Cauchy-Riemann equations yield the relation $|J| = x_\zeta^2(\xi, \zeta) + y_\zeta^2(\xi, \zeta) = |df/dw|^2$. As pointed out earlier in the symmetric channel mapping configuration, for weakly nonlinear waves $|J| \approx M^2(\xi) + O(\alpha^2)$, where $M(\xi) \equiv y_\zeta(\xi, \zeta_o) = \mathbf{Re}(df/dw(\xi, \zeta_o))$. These free surface variable coefficients are therefore constructed numerically by evaluating df/dw at mesh-points, which is achieved through the command **evaldiff**. To be more specific, take the symmetric channel configuration and let the free surface mesh-points locations in the canonical domain be stored in the array **w**, containing entries of the form $w_j = \xi_j + 0.5 * i$. The command

```
>> M = real(evaldiff(f,w));
```

returns mesh-point values of $M(\xi_j)$ in the vector **M**. We should now explain how to construct the mapping **f** using the SCT.

Let us consider the case of a submerged triangular mountain range as depicted in figure 10.3. Let the mountain height be $h = 0.7$ and half the base be given by $d_2 = 2$. The data needed for the Schwarz-Christoffel mapping are the location of the corners in the z-plane and the values of the internal angles at these corners. These are stored in the arrays **z** and **alpha** given below. The SCT convention is that **inf** indicates a polygon corner "at infinity" with angle $\alpha = 0$. Also there are 2 points that help define the upper boundary of the domain, namely the points at $z = 4 + i$ and $z = -4 + i$. There are no corners at these points so the angle for both is π, which leads to $\alpha = 1$. In summary, the commands and input data for constructing the mapping function $f(w)$ are:

```
>> h=0.7;
>> d2=2.;
>> ii=1*i;
>> a1=1-atan(h/d2)/pi;
>> z=[-inf, -d2, h*ii, d2, 2*d2+h*ii, 3*d2, 4*d2+h*ii, 5*d2,
      6*d2+h*ii,
7*d2, 8*d2+h*ii, 9*d2, inf, 4+ii, -4+ii];
>> alpha=[0., a1, 3-2*a1, 2*a1-1, 3-2*a1, 2*a1-1, 3-2*a1,
         2*a1-1,
3-2*a1, 2*a1-1,3-2*a1, a1, 0., 1., 1. ];
>> p=polygon(z,alpha);
>> f=stripmap(p)
```

The command **polygon** defines, within the SCT framework, a polygonal object **p** to be conformally mapped by **f** from a canonical strip through the command **stripmap**. A window will open asking the user to confirm the two points at

infinity, as indicated in red on the screen. Click on the left "infinity-node" first
and then on the "right-infinity" node. The SCT executes the mapping and prints
on the screen the values of the relevant parameters, including each vertex's
pre-image, referred to as a *prevertex*:

```
f =
  SC stripmap:
        vertex                  alpha               prevertex
  --------------------------------------------------------------
        -Inf + 0.00000i       0.00000                     -Inf
  -2.00000 + 0.00000i         0.89283      0.000000000000e+00
   0.00000 + 0.70000i         1.21433      3.624655549413e+00
   2.00000 + 0.00000i         0.78567      7.182608643277e+00
   4.00000 + 0.70000i         1.21433      1.074056122486e+01
   6.00000 + 0.00000i         0.78567      1.429851380644e+01
   8.00000 + 0.70000i         1.21433      1.785646638800e+01
  10.00000 + 0.00000i         0.78567      2.141441896957e+01
  12.00000 + 0.70000i         1.21433      2.497237155113e+01
  14.00000 + 0.00000i         0.78567      2.853032413270e+01
  16.00000 + 0.70000i         1.21433      3.208827722653e+01
  18.00000 + 0.00000i         0.89283      3.571293277590e+01
       Inf + 0.00000i         0.00000                      Inf
   4.00000 + 1.00000i         1.00000      1.074056123873e+01 + i
  -4.00000 + 1.00000i         1.00000     -1.938944214559e+00 + i
  c = 1 + 0i
  Apparent accuracy is 5.34e-09
```

The bottom figure 10.3 is created through the command >> plot(f,30,
10). The numbers indicate how many level curves in ξ and ζ appear in
the graph. For more complex topograhies the vectors **z** and **alpha** should
be generated automatically by another code as, for example, in a random
topography case.

We now discuss the computation of the metric coefficient $M(\xi)$ along a mesh
on the free surface. Consider the undisturbed free surface in the w-plane by
its grid-point values ($w_j = \xi_j + i\zeta_j$ stored in the vector **ww**). Computing the
Jacobian or $M(\xi)$ along this curve reduces to

```
>> ww=[-1:0.1:48]+i;
>> J=abs(evaldiff(f,ww))).^2;
```

or the metric coefficient $m(\xi)$

```
>> m=1-real(evaldiff(f,ww));
```

Note that in the symmetric case we should have

```
>> ww=[-1:0.1:48]+0.5*i;
```

To plot the comparison between $-n(x)$ and $m(\xi(x))$ perform the following
steps:

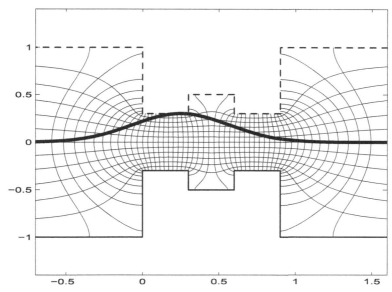

Figure 10.7. Symmetric channel with a submarine structure. The wave profile is indicated in the solid line near $y = 0$.

```
>> plot(f,40,10)
>> hold
>> zz=eval(f,ww);
>> plot(real(zz),m,'r')
```

Let the initial wave profile be $\eta(x,0) = \eta_o(x)$. In most reflection-transmission problems, such as those we studied in [4–6, 11, 12, 14, 29] the initial pulse-shaped wave is propagating from a region of constant unit depth towards the topography. In these cases no pre-processing of the initial profile is needed (i.e., $N_o \equiv \eta_o$) because away from the variable bottom very quickly the Jacobian approaches 1. But when this is not the case the initial wave profile $N_o \equiv N(\xi,0)$ has to be computed from η_o. Inversion of the mapping $x + i\eta(x,t) = f(\xi + iN(\xi,t))$ is denoted here as

$$\xi + iN_o(\xi) = f^{-1}(x + i\eta_o(x)), \qquad (10.3.10)$$

and is obtained through the command

```
>> N = imag(evalinv(f,Z));
```

where $\mathbf{Z}$ is a vector containing the free surface values $x_j + i\eta_o(x_j)$. At any time $t = t_n$ the wave profile $\eta(x_j, t_n)$ is recovered in the z-plane through the command

```
>> eta = imag(eval(f,W));
```

222

André Nachbin

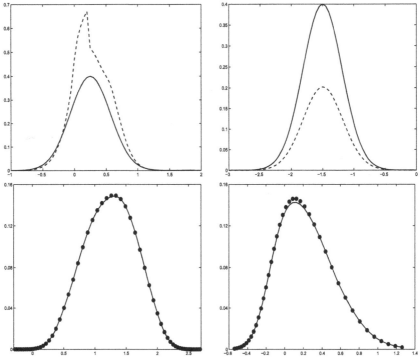

Figure 10.8. Symmetric channel as in figure 10.7. The dashed line indicates the wave profile $\eta(x)$ given by a Gaussian with an amplitude equal to a. The solid line represents $N(\xi(x)) = Im(f^{-1}(x + i\eta(x)))$. Top left: amplitude $a = 0.4$; near the singularities. Top right: $a = 0.4$; far from the singularities. Along the bottom row the wave profile approximation for N is tested. The solid line is given by $N(\xi(x)) = \mathbf{Im}(f^{-1}(x + i\eta(x)))$ while the dots are by $N(\xi_j, t) \approx \eta(x_j, t)/M(\xi_j)$. Bottom left: $a = 0.1$; the approximation is very accurate. Bottom right: $a = 0.2$; the approximation starts to display a small discrepancy.

where $\mathbf{W}$ is a vector containing the free surface values $\xi_j + iN(\xi_j, t_n)$ in the w-plane.

In the weakly nonlinear case, with the symmetric channel configuration, one can use the approximation

$$N(\xi_j, t) \approx \frac{\eta(x_j, t)}{M(\xi_j)} \qquad (10.3.11)$$

without having to access MATLAB several times. Evaluate $M(\xi_j)$ at the mesh-points ξ_j. In figure 10.8 we illustrate the quality of this approximation in the presence of a very abrupt bottom variation as shown in figure 10.7. For waves of moderate amplitude the approximation is quite effective. Some examples are displayed. At the top left of figure 10.8 the solid line represents our initial wave profile $\eta(x, 0)$, a Gaussian wave profile of amplitude equal to

0.4. The inverse map f^{-1} takes this profile onto the dashed line representing $N(\xi, 0)$. This curve is definitely not precise due to two aspects of our choice: the initial wave profile $\eta(x, 0)$ is too close to the "man-made" singularities in the upper part of the reflected domain. The corners along the dashed-line of the domain are not physical. Moreover the initial wave profile escapes the domain initially defined. The interested reader should explore these two facts with the SCT. When we centre our initial wave profile at $x_o = -1.5$ there is no problem whatsoever since it is away from the (reflected) singularity. See the graph at the top right. One clearly sees that $N(\xi, 0)$ is simply a rescaling of $\eta(x, 0)$ since the symmetric channel has total width equal to 2 in the physical domain and equal to 1 in the canonical domain. The Jacobian in this region is (approximately) a multiple of the identity. Note that for the first case discussed N was taller than η because at the region chosen the average width of the channel, in the physical domain, was smaller than 1.

Now we centre the initial condition at $x_o = 0.5$ but with a smaller Gaussian pulse. At the bottom left of figure 10.8 we do not display η but only N, obtained with the inverse map as before (in the solid line). Also we display N obtained with the approximation (10.3.11), which does not use $f^{-1}(z)$ and therefore does not need the SCT. The approximate values are displayed in dots and confirm its accuracy. We increase the initial wave amplitude from $a = 0.1$ to $a = 0.2$ and repeat the approximation at the bottom right graph. The approximation is still useful but not as accurate as for the previous case. In conclusion the user can construct a vector with values of $M(\xi_j)$ for the desired topography, save it in a regular text file and import this data into a code written in any other convenient programming language (that is not MATLAB) and use expression (10.3.11).

At this point the reader has all the information needed to explore the models and effective dynamics presented in [4–6, 10–12, 14, 27, 29, 30] for water waves interacting with complex topography.

Acknowledgment

The author is grateful to the organisers of the programme *Theory of Water Waves* held at the Isaac Newton Institute, Cambridge, U.K. In particular to Paul Milewski and to Tom Bridges, the organiser of the *Theory of Water Waves: Summer School*. The author is also grateful to Carlos Galeano-Rios and David Andrade for useful comments regarding the text.

This work was supported in part by: CNPq under (PQ-1B) #304381/2013-6 and FAPERJ *Cientistas do Nosso Estado* project #102.917/2011.

References

[1] R. Burridge, G. Papanicolaou and B. White, Statistics for pulse reflection from a randomly layered medium, SIAM J. Appl. Math., **47** (1987), 146–168.

[2] R. Burridge, G. Papanicolaou, P. Sheng and B. White, Probing a random medium with a pulse, SIAM J. Appl. Math., **49** (1989), 582–607.

[3] J. P. Fouque, J. Garnier, G. C.Papanicolaou and K. Sølna, *Wave Propagation and Time Reversal in Randomly Layered Media* (Springer Verlag, 2007).

[4] J. P. Fouque, J. Garnier and A. Nachbin, Time reversal for dispersive waves in random media, SIAM J. Appl. Math., **64** (2004), 1810–1838.

[5] J. P. Fouque, J. Garnier and A. Nachbin, Shock structure due to stochastic forcing and the time reversal of nonlinear waves, Physica D, **195** (2004), 324–346.

[6] J. C. Muñoz Grajales and A. Nachbin, Dispersive wave attenuation due to orographic forcing, SIAM J. Appl. Math., **64** (2004), 977–1001.

[7] R. Burridge and L. Berlyand, The accuracy of the O'Doherty-Anstey approximation for wave propagating in highly disordered stratified media, Wave Motion, **21** (1995), 357–373.

[8] A. Nachbin and K. Sølna, Apparent diffusion due to topographic microstructure in shallow waters, Phys. Fluids, **15** (2003), 66–77.

[9] G.B. Whitham, *Linear and Nonlinear Waves* (John Wiley, 1974).

[10] J. P. Fouque and A. Nachbin, Time reversed refocusing of surface water waves, Multiscale Modell. Simul., **1**, (2003) 609–629

[11] J.P. Fouque, J. Garnier, J.C. Muñoz Grajales and A. Nachbin, Time reversing solitary waves, *Phys. Rev. Lett.* **92** (2004), 094502.

[12] J. Garnier, J. C. Muñoz Grajales and A. Nachbin, Effective behavior of solitary waves over random topography, Multiscale Modell. Simul., **6** (2007), 995–1025.

[13] J. Hamilton, Differential equations for long-period gravity waves on a fluid of rapidly varying depth, J. Fluid Mech., **vol. 83** (1977), 289–310.

[14] J. C. Muñoz Grajales and A. Nachbin, Improved Boussinesq-type equations for highly-variable depths, IMA J. Appl. Math., **71** (2006), 600–633.

[15] O. Nwogu, Alternative form of Boussinesq equations for nearshore wave propagation, J. Waterway, Port, Coastal and Ocean Engineering, 119 (1993), 618–638.

[16] R.A. Kraenkel, J. Garnier and A. Nachbin, Optimal Boussinesq model for shallow-water waves interacting with a microstructure, Phys. Rev. E., **vol. 76** (2007), 046311–1.

[17] A. Nachbin, A terrain-following Boussinesq system, SIAM J. Appl. Math., **63** (2003), 905–922.

[18] R.S. Johnson, *A Modern Introduction to the Mathematical Theory of Water Waves* (Cambridge University Press, 1997).

[19] M. Belzons, E. Guazzelli and O. Parodi, Gravity waves on a rough bottom: experimental evidence of one-dimensional localization, J. Fluid Mech., **186** (1988), 539–558.

[20] P. Devillard, F. Dunlop and B. Souillard, Localization of gravity waves on a channel with a random bottom, J. Fluid Mech., **186** (1988), 521–538.

[21] A. Nachbin and G.C. Papanicolaou, Water waves in shallow channels of rapidly varying depth, J. Fluid Mech., **241** (1992), 311–332.

[22] A. Nachbin, The localization length of randomly scattered water waves, J. Fluid Mech., **296** (1995), 353–372.

[23] A. Nachbin, *Modelling of Water Waves in Shallow Channels* (Computational Mechanics Publications, Southampton, U.K., 1993).

[24] R.R. Rosales and G.C. Papanicolaou, Gravity waves in a channel with a rough bottom, Studies in Appl. Math., **68** (1983), 89–102.

[25] T.A. Driscoll, T.A. and L. Trefethen, *Schwarz-Christoffel Mapping* (Cambridge University Press, Cambridge, 2002 UK).

[26] V. Rey, M. Belzons and E. Guazzelli, Propagation of surface gravity waves over a rectangular submerged bar, J. Fluid Mech., **235**, (1992), 453–479.

[27] A.S. Fokas and A. Nachbin, Water waves over a variable bottom: a non-local formulation and conformal mappings J. Fluid Mech., **695** (2012), 288–309.

[28] W. Artiles and A. Nachbin, Nonlinear evolution of surface gravity waves over highly variable depth, Phys. Rev. Lett., **93**, (2004), 234501.

[29] J. C. Muñoz Grajales and A. Nachbin, Stiff microscale forcing and solitary wave refocusing, SIAM Multiscale Model. Simul., **3** (2005), 680–705.

[30] J. Garnier and A. Nachbin, Eddy viscosity for gravity waves propagating over turbulent surfaces, Phys. Fluids, **18** (2006), 055101.

11

Variational Water Wave Modelling: from Continuum to Experiment

Onno Bokhove & Anna Kalogirou

Abstract

Variational methods are investigated asymptotically and numerically to model water waves in tanks with wave generators. As a validation, our modelling results using (dis)continuous Galerkin finite element methods will be compared to a soliton splash event resulting after a sluice gate is removed during a finite time in a long water channel with a contraction at its end.

11.1 Introduction

A popular approach in the modelling of nonlinear water waves is to make the approximations that the three-dimensional fluid velocity $\mathbf{u}$ is irrotational and divergent free, such that $\mathbf{u} = \nabla\phi$ and $\nabla \cdot \mathbf{u} = \nabla^2\phi = 0$, and that the dynamics is inviscid, such that the dynamics is governed by variational and Hamiltonian dynamics [1, 2]. At least symbolically one can invert this Laplace equation for the interior potential ϕ and reduce the dynamics to the free surface, expressed in terms of the potential ϕ_s at the free surface and the position of this free surface. For non-overturning waves, this free surface dynamics can be expressed in terms of the water depth $h = h(x,y,t)$ and $\phi_s(x,y,t) = \phi(x,y,z = b + h,t)$ with horizontal coordinates x and y as well as time t. Here the fixed topography is denoted by $b = b(x,y)$. The free surface thus lies at the vertical level $z = b(x,y) + h(x,y,t)$, parametrised by x and y.

One then often considers the initial value problem governed by autonomous Hamiltonian dynamics for h and ϕ_s with initial conditions $h(x,y,0)$ and $\phi_s(x,y,0)$ without any forcing or dissipation. In practical situations, however, waves are generated continuously by wave makers or temporarily by opening a sluice gate, both involving time dependent internal or boundary conditions. This implies that the dynamics is non-autonomous, including explicit dependence of the equations on time. Sometimes, these non-autonomous aspects can be included in the variational principles governing the wave dynamics.

We will therefore start to formulate finite-dimensional variational dynamics in which the variational principle indeed depends explicitly on time. The forced-dissipative nonlinear pendulum with the harmonic oscillator as linearisation is a *first example* of such a non-autonomous variational principle. It is expressed in terms of the coordinate or angle $q = q(t)$ and its conjugate momentum $p = p(t)$. We will show that a fairly generic non-autonomous variational principle can be transformed into an autonomous one by lifting it to a larger phase space.

The variational principles for water waves can be derived from the incompressible Euler equations with a dynamic free surface, using both the velocity potential and the pressure as Lagrange multipliers to impose two constraints. In this approach [1, 3], the pressure variable eventually disappears from the formulation. Both driven wave makers and the removal of a sluice gate can be included in the variational principle for water waves. It adds explicit time-dependence to the principle. In Bokhove et al. [4] and Thornton et al. [5], this was investigated for water wave dynamics in a vertical Hele–Shaw cell with a moving interface between water and air, concerning damped motion with a wave pump and with damped wave sloshing, respectively. The Hele–Shaw cell is a narrow wave tank with a wave-maker, e.g., it has a width of 2 mm, length of 0.6 to 1 m, and depth of circa 0.1 m, between two closely spaced glass plates. The bulk damping due to the dominant friction against these glass plates leads to an exponential time dependent factor $e^{\gamma t}$ in the variational principle. Its inverse expresses the exponential decay due to friction with $\gamma > 0$ proportional to both the viscosity and the square inverse of the gap width between the glass plates (e.g., Bokhove et al. [4]).

Such explicit time dependence makes the numerical discretisation of the water wave problem more complicated and prone to numerical instabilities. In Gagarina et al. [6], this challenge was overcome, verified and validated for driven potential flow water wave dynamics. As a *second example* and to minimise overlap with that research, we illustrate the numerical modelling of variational water wave dynamics with a reduced water wave model that includes a simple, new model of a sluice gate. By modifying work of Pego & Quintero [7], we derive a model similar to Benney and Luke's [8] but remain entirely within the variational framework from the onset. The velocity potential at the bottom topography Φ and free surface deviation $\eta = h - H_0$ herein play a similar role as q and p for the nonlinear pendulum, with $z = H_0$ the water level at rest. The above two examples allow us to illustrate our time discretisation techniques for non-autonomous water wave dynamics. Finally, we validate the new model discretisation against a soliton splash phenomenon in a water wave channel with a removable sluice gate and a V-shaped channel contraction at its end.

While some techniques have been published before [4–6, 9], the derivation of the modified Benney–Luke equations, the systematic way to lift non-autonomous to autonomous variational principles in time, and the use of this lifting in finding extended time discretisations are novel. Also new is the soliton splash simulation.

11.2 Variational Mechanics Continuous in Time

11.2.1 General Formulation

We will consider non-autonomous Hamiltonian dynamics cast in the variational principle

$$0 = \delta \int_0^T L(\boldsymbol{Q}, \boldsymbol{P}, t)\, dt = \delta \int_0^T \boldsymbol{P} \cdot \frac{d\boldsymbol{Q}}{dt} - H(\boldsymbol{Q}, \boldsymbol{P}, t)\, dt, \qquad (11.2.1)$$

with time t in the time interval $t \in [0, T]$ for some $T > 0$, generalised coordinate (vector) $\boldsymbol{Q} = \boldsymbol{Q}(t)$ and conjugate momentum (vector) $\boldsymbol{P} = \boldsymbol{P}(t)$, and Hamiltonian $H = H(\boldsymbol{P}, \boldsymbol{Q}, t)$. Note that the Lagrangian density $L = L(\boldsymbol{Q}, \boldsymbol{P}, t)$ depends explicitly on time, in order to model forcing as well as (simplified) damping within the variational principle.

The variations in (11.2.1) are defined as

$$0 = \delta S[\boldsymbol{Q}, \boldsymbol{P}] \equiv \delta \int_0^T L(\boldsymbol{Q}, \boldsymbol{P}, t)\, dt \qquad (11.2.2)$$

$$= \lim_{\epsilon \to 0} \int_0^T \frac{L(\boldsymbol{Q} + \epsilon \delta \boldsymbol{Q}, \boldsymbol{P} + \epsilon \boldsymbol{P}, t) - L(\boldsymbol{Q}, \boldsymbol{P}, t)}{\epsilon}\, dt, \qquad (11.2.3)$$

Hence, the variation in (11.2.1) becomes

$$0 = \int_0^T \left(\frac{d\boldsymbol{Q}}{dt} - \frac{\partial H(\boldsymbol{Q}, \boldsymbol{P}, t)}{\partial \boldsymbol{P}} \right) \cdot \delta \boldsymbol{P} - \left(\frac{d\boldsymbol{P}}{dt} + \frac{\partial H(\boldsymbol{Q}, \boldsymbol{P}, t)}{\partial \boldsymbol{Q}} \right) \cdot \delta \boldsymbol{Q}\, dt + \boldsymbol{P} \cdot \delta \boldsymbol{Q} \Big|_0^T, \qquad (11.2.4)$$

in which end-point contributions cancel because of the conditions $\delta \boldsymbol{Q}(0) = \delta \boldsymbol{Q}(T) = 0$. The latter follow since $\boldsymbol{Q}(0) = \boldsymbol{Q}_0$ is a given value and by using time reversibility $\boldsymbol{Q}(T) = \boldsymbol{Q}_T$ can, likewise, be considered as given. Their variations are thus zero. Since variations $\delta \boldsymbol{Q}(t)$ and $\delta \boldsymbol{P}(t)$ are arbitrary and pointwise in time, Hamilton's equations follow from (11.2.4) as

$$\frac{d\boldsymbol{Q}}{dt} = \frac{\partial H}{\partial \boldsymbol{P}} \quad \text{and} \quad \frac{d\boldsymbol{P}}{dt} = -\frac{\partial H}{\partial \boldsymbol{Q}}. \qquad (11.2.5)$$

Finally, energy is generally not conserved when H also depends explicitly on time, since

$$\frac{dH}{dt} = \frac{\partial H(\boldsymbol{Q}, \boldsymbol{P}, t)}{\partial \boldsymbol{Q}} \cdot \frac{d\boldsymbol{Q}}{dt} + \frac{\partial H(\boldsymbol{Q}, \boldsymbol{P}, t)}{\partial \boldsymbol{P}} \cdot \frac{d\boldsymbol{P}}{dt} + \frac{\partial H}{\partial t} = \frac{\partial H}{\partial t} \neq 0. \qquad (11.2.6)$$

11.2.2 Kamiltonian Formulation

The non-autonomous variational principle (11.2.1) can be lifted to an autonomous variational principle (cf. Goldstein [10]) by considering time as an auxiliary variable with a new time coordinate. First, we rename time as τ in (11.2.1) to obtain

$$0 = \delta \int_0^T \boldsymbol{P} \cdot \frac{\mathrm{d}\boldsymbol{Q}}{\mathrm{d}\tau} - H(\boldsymbol{Q},\boldsymbol{P},\tau)\,\mathrm{d}\tau. \qquad (11.2.7)$$

Subsequently, we take $\tau = \tau(t)$ with a new time coordinate t, such that $\mathrm{d}\tau/\mathrm{d}t = 1$ and $\tau(0) = 0$ for convenience, and introduce a conjugate variable p for τ. A so–called "Kamiltonian" is then defined as $K(\boldsymbol{Q},\boldsymbol{P},\tau,p) \equiv H(\boldsymbol{Q},\boldsymbol{P},\tau) + p$, and the corresponding variational principle and variations are

$$0 = \delta \int_0^T \boldsymbol{P} \cdot \frac{\mathrm{d}\boldsymbol{Q}}{\mathrm{d}t} + p\frac{\mathrm{d}\tau}{\mathrm{d}t} - K(\boldsymbol{Q},\boldsymbol{P},\tau,p)\,\mathrm{d}t \qquad (11.2.8a)$$

$$= \int_0^T \left(\frac{\mathrm{d}\boldsymbol{Q}}{\mathrm{d}t} - \frac{\partial H(\boldsymbol{Q},\boldsymbol{P},t)}{\partial \boldsymbol{P}}\right) \cdot \delta\boldsymbol{P} - \left(\frac{\mathrm{d}\boldsymbol{P}}{\mathrm{d}t} + \frac{\partial H(\boldsymbol{Q},\boldsymbol{P},t)}{\partial \boldsymbol{Q}}\right) \cdot \delta\boldsymbol{Q}$$

$$+ \left(\frac{\mathrm{d}\tau}{\mathrm{d}t} - 1\right)\delta p - \left(\frac{\mathrm{d}p}{\mathrm{d}t} + \frac{\partial H}{\partial \tau}\right)\delta\tau\,\mathrm{d}t + \boldsymbol{P}\cdot\delta\boldsymbol{Q}\Big|_0^T + p\delta\tau\Big|_0^T, \qquad (11.2.8b)$$

where we used $\delta\boldsymbol{Q}(0) = \delta\boldsymbol{Q}(T) = 0$ as well as $\delta\tau(0) = \delta\tau(T) = 0$. Hence, the extended Hamilton's or "Kamilton's" equations become

$$\frac{\mathrm{d}\boldsymbol{Q}}{\mathrm{d}t} = \frac{\partial H}{\partial \boldsymbol{P}}, \quad \frac{\mathrm{d}\boldsymbol{P}}{\mathrm{d}t} = -\frac{\partial H}{\partial \boldsymbol{Q}}, \quad \frac{\mathrm{d}\tau}{\mathrm{d}t} = 1 \quad \text{and} \quad \frac{\mathrm{d}p}{\mathrm{d}t} = -\frac{\partial H}{\partial \tau}. \qquad (11.2.9)$$

We note that this last equation is decoupled from the rest of the equations. Moreover, the Kamiltonian K is by construction conserved in time: $\mathrm{d}K/\mathrm{d}t = 0$. The above reformulation into a "Kamiltonian" system indicates that a variational discretisation of an autonomous variational principle can thus, in principle, be extended systematically to one for a non-autonomous system.

11.2.3 Linearisation

Hamilton's equations (11.2.5) with $\partial H/\partial \boldsymbol{P} = \boldsymbol{P}$ for general (nonlinear) dynamics can be linearised and diagonalised in a series of uncoupled oscillators. This will be used later to interpret the numerical verification.

Autonomous case: Hamilton's equations (11.2.5) have fixed points $(\bar{\boldsymbol{Q}},\bar{\boldsymbol{P}})$ obeying $\partial H(\bar{\boldsymbol{Q}},\bar{\boldsymbol{P}})/\partial\bar{\boldsymbol{Q}} = 0$ and $\bar{\boldsymbol{P}} = 0$. When we substitute the decomposition $\boldsymbol{Q} = \bar{\boldsymbol{Q}} + \boldsymbol{Q}'$ and $\boldsymbol{P} = \bar{\boldsymbol{P}} + \boldsymbol{P}'$ into the variational principle (11.2.1) and keep terms quadratic into the (small-amplitude) perturbation variables $\boldsymbol{Q}'$ and $\boldsymbol{P}'$, we obtain a series of coupled linear oscillators. After an eigen-analysis, diagonalisation and under certain assumptions, a system of uncoupled harmonic oscillators remains. The one with the largest eigenfrequency $\omega = \omega_1$ with $q_1 = Q_1'$ and $p_1 = P_1'$ will therefore satisfy the following principle

$$0 = \delta \int_0^T p_1 \frac{dq_1}{dt} - H(q_1, p_1) \, dt \equiv \delta \int_0^T p_1 \frac{dq_1}{dt} - \left(\frac{1}{2}p_1^2 + \frac{1}{2}\omega^2 q_1^2\right) dt,$$
(11.2.10)

the variation of which yields the classic harmonic oscillator upon using the usual end point conditions. That is $dq_1/dt = p_1$ and $dp_1/dt = -\omega^2 q_1$.

Non-autonomous case: In contrast to the autonomous case, $\{\partial H(\bar{Q}, \bar{P})/\partial \bar{Q} = 0, \ \bar{P} = 0, \ p = 0\}$ is not a fixed point of the Kamiltonian system (11.2.9). However, the Kamiltonian form can be Taylor expanded around $\tau = 0$ and the fixed points of the autonomous case, focussing on one oscillator as before. Motivated by our examples, this heuristically yields the following "archetypical case"

$$0 = \delta \int_0^T p_1 \frac{dq_1}{dt} + p \frac{d\tau}{dt} - \left(\frac{1}{2}p_1^2 + \frac{1}{2}\omega^2 q_1^2 + aq_1\tau + p\right) dt \qquad (11.2.11a)$$

$$= \delta \int_0^T \tilde{p}_1 \frac{d\tilde{q}_1}{dt} + p \frac{d\tau}{dt} - \left(\frac{1}{2}\tilde{p}_1^2 + \frac{1}{2}\omega^2 \tilde{q}_1^2 + a\tilde{p}_1/\omega^2 + p - \frac{1}{2}a^2\tau^2/\omega^2\right) dt,$$
(11.2.11b)

for $a = \partial^2 V(q_1, \tau)/\partial q_1 \partial \tau |_{q_1 = \bar{Q}_1, \tau = 0}$ and $H = p_1^2/2 + V(q_1, \tau)$ and by using the transformation $\omega \tilde{q}_1 = \omega q_1 + a\tau/\omega$, $\tilde{p}_1 = p_1$ in the second line, cf. [9]. Consequently, the Kamiltonian system decouples with a new autonomous Hamiltonian

$$\frac{1}{2}\tilde{p}_1^2 + \frac{1}{2}\omega^2 \tilde{q}_1^2 + a\tilde{p}_1/\omega^2, \qquad (11.2.12)$$

as an invariant in time t. This regularises the Kamiltonian to one with at most quadratic terms, including a term linear in $\tilde{p}_1$.

11.3 Forced-Dissipative Nonlinear Oscillator

As *first, low-order example*, consider a forced and damped nonlinear oscillator

$$0 = \delta \int_0^T \left(p\frac{dq}{dt} - \left(\frac{1}{2}p^2 + V(q) - qF(t)\right)\right) e^{\gamma t} \, dt \qquad (11.3.1a)$$

$$= \int_0^T \left(\frac{dq}{dt} - p\right) e^{\gamma t} \delta p - \left(\frac{d}{dt}(pe^{\gamma t}) + V'(q)e^{\gamma t} - F(t)e^{\gamma t}\right) \delta q \, dt + p \, \delta q \, e^{\gamma t} \Big|_0^T,$$
(11.3.1b)

with generalised coordinate q and momentum p, potential $V = V(q)$, time dependent forcing function $F = F(t)$ and damping coefficient $\gamma \geq 0$ (for a damped oscillator it is given in Olver [11]). We assume that the leading order term in a Taylor expansion of $V(q)$ is quadratic in q. The dynamics become

$$\delta(pe^{\gamma t}): \quad \frac{dq}{dt} = p \tag{11.3.2a}$$

$$\delta q: \quad \frac{d}{dt}(pe^{\gamma t}) + V'(q)e^{\gamma t} = F(t)e^{\gamma t} \quad \Longleftrightarrow \quad \frac{dp}{dt} = -V'(q) + F(t) - \gamma p. \tag{11.3.2b}$$

The variational principle (11.3.1a) can be put in the form (11.2.1) by using

$$P = pe^{\gamma t/2} \quad \text{and} \quad Q = qe^{\gamma t/2}, \tag{11.3.3}$$

yielding

$$0 = \delta \int_0^T P\frac{dQ}{dt} - \tilde{H} + QF(t)e^{\gamma t/2}\,dt \tag{11.3.4a}$$

$$\equiv \int_0^T P\frac{dQ}{dt} - \left(\frac{1}{2}P^2 + \frac{1}{2}\gamma PQ + V(Qe^{-\gamma t/2})e^{\gamma t}\right) + QF(t)e^{\gamma t/2}\,dt, \tag{11.3.4b}$$

in which the leading order term $\frac{1}{2}aQ^2$ (for some $a > 0$) in a Taylor expansion of $V(Qe^{-\gamma t/2})e^{\gamma t}$ is now quadratic in Q, while higher order terms contain decaying exponentials in (nonzero) powers of $e^{-\gamma t/2}$. As a consequence, when in the unforced case $F(t) = 0$ or when $\lim_{t\to\infty} F(t)e^{\gamma t/2} \to 0$, the new Hamiltonian $\lim_{t\to\infty} \tilde{H} = \frac{1}{2}P^2 + \frac{1}{2}\gamma PQ + \frac{1}{2}aQ^2$ will become a quadratic invariant at long times. A Kamiltonian form (11.2.8) of the above variational principle, with conjugate $\tilde{p}$ for τ, is

$$0 = \delta \int_0^T P\frac{dQ}{dt} + \tilde{p}\frac{d\tau}{dt} - \tilde{H} + QF(\tau)e^{\gamma \tau/2} - \tilde{p}\,dt \tag{11.3.5a}$$

$$\equiv \delta \int_0^T P\frac{dQ}{dt} + \tilde{p}\frac{d\tau}{dt} - \left(\frac{1}{2}P^2 + \frac{1}{2}\gamma PQ + V(Qe^{-\gamma \tau/2})e^{\gamma \tau}\right)$$
$$+ QF(\tau)e^{\gamma \tau/2} - \tilde{p}\,dt. \tag{11.3.5b}$$

Another, second transformation is

$$P = pe^{\gamma t} \quad \text{and} \quad Q = q. \tag{11.3.6}$$

It yields the alternative Kamiltonian variational principle

$$0 = \delta \int_0^T P\frac{dQ}{dt} + \tilde{p}\frac{d\tau}{dt} - \tilde{H} + QF(\tau)e^{\gamma \tau} - \tilde{p}\,dt \tag{11.3.7a}$$

$$\equiv \delta \int_0^T P\frac{dQ}{dt} + \tilde{p}\frac{d\tau}{dt} - \left(\frac{1}{2}P^2 e^{-\gamma \tau} + V(Q)e^{\gamma \tau}\right) + QF(\tau)e^{\gamma \tau} - \tilde{p}\,dt. \tag{11.3.7b}$$

The above two transformations are important because they show us how to find time discretisations for the non-autonomous case based on ones for the autonomous variational principle.

A linear forced and damped harmonic oscillator emerges for $V(q) = \frac{1}{2}\omega^2 q^2$ with $F(t) = a\sin(\Omega t)$, forcing frequency Ω and amplitude a, as well as natural frequency ω. The system then becomes $dq/dt = p$ and $dp/dt = -\omega^2 q - \gamma p + a\sin(\Omega t)$. It has an analytical solution, comprised of the sum of homogeneous and quasi-steady state particular solutions,

$$q(t) = \big(A\cos(\beta t) + B\sin(\beta t)\big)e^{-\gamma t/2}$$

$$+ \frac{a}{(\omega^2 - \Omega^2)^2 + \Omega^2\gamma^2}\big((\omega^2 - \Omega^2)\sin(\Omega t) - \Omega\gamma\cos(\Omega t)\big), \quad (11.3.8a)$$

for $\beta = \sqrt{\omega^2 - \frac{1}{4}\gamma^2}$ real, initial conditions $q(0) = q_0$ and $p(0) = p_0$, and with

$$A = q_0 + \frac{a\Omega\gamma}{(\omega^2 - \Omega^2)^2 + \Omega^2\gamma^2} \quad \text{and}$$

$$B = \frac{1}{\beta}\left(p_0 + \frac{1}{2}\gamma A - \frac{a\Omega(\omega^2 - \Omega^2)}{(\omega^2 - \Omega^2)^2 + \Omega^2\gamma^2}\right). \quad (11.3.8b)$$

11.4 Variational Water Waves à la Benney–Luke

As a *second, high-order example*, we derive a reduced water wave model and subsequently discretise it in space using a Galerkin finite element expansion.

11.4.1 Principle for Finite, Shallow Depth

The deviation from a still water surface is denoted by $\eta = \eta(x,y,t)$ and the velocity potential as $\phi = \phi(x,y,z,t)$ with horizontal coordinates x and y, vertical coordinate z and time t. The still water rest level lies at $z = H_0$ and the flat bottom is at $z = 0$ with vertical velocity $\partial_z\phi = \partial\phi/\partial z = 0$. The free surface and bottom potentials are defined as $\phi_s(x,y,t) = \phi(x,y,H_0 + \eta(x,y,t),t)$ and $\Phi(x,y,t) = \phi(x,y,0,t)$, respectively. Horizontal gradients are $\nabla = (\partial_x, \partial_y)^T$ with transpose $(\cdot)^T$. The domain has horizontal extent Ω_h with vertical walls at $\partial\Omega_h$, where the normal gradient $\mathbf{n}\cdot\nabla\phi = 0$.

The potential flow water wave equations

$$\nabla^2\phi + \partial_{zz}\phi = 0 \quad \text{in} \quad \Omega_h, \quad (11.4.1a)$$

$$\partial_z\phi = 0 \quad \text{at} \quad z = 0, \quad (11.4.1b)$$

$$\partial_t\eta + \nabla\phi\cdot\nabla\eta - \partial_z\phi = 0 \quad \text{at} \quad z = H_0 + \eta, \quad (11.4.1c)$$

$$\partial_t\phi + \frac{1}{2}|\nabla\phi|^2 + \frac{1}{2}(\partial_z\phi)^2 + g(\eta - \eta_R) = 0 \quad \text{at} \quad z = H_0 + \eta, \quad (11.4.1d)$$

can be derived from Luke's variational principle

$$0 = \delta \int_0^T \iint_{\Omega_h} \int_0^{H_0+\eta} \partial_t \phi + \frac{1}{2} |\nabla \phi|^2 + \frac{1}{2} (\partial_z \phi)^2 + g(z - H_0 - \eta_R) \, dz \, dx \, dy \, dt,$$

(11.4.2)

with constant acceleration of gravity g, see Luke [1]. We added a (gravitational) potential $\eta_R = \eta_R(x,t)$ to model a sluice gate release problem. For flow at rest, $\phi = 0$ and $\eta = \eta_R(x)$. By adjusting $\eta_R(x,t)$ to become independent of x in a finite time, such a release can be modelled approximately in a relatively straightforward manner.

The following transformations (from Pego & Quintero [7]) are first applied

$$x = H_0 \hat{x}, \quad y = H_0 \hat{y}, \quad z = H_0 \hat{z}, \quad t = H_0 \hat{t}/\sqrt{gH_0},$$

$$\phi = \epsilon H_0 \sqrt{gH_0} \, \hat{\phi}, \quad \eta = \epsilon H_0 \hat{\eta}, \quad \eta_R = \epsilon H_0 \hat{\eta}_R, \tag{11.4.3}$$

with amplitude parameter $\epsilon = \alpha_0/H_0 \ll 1$ (for small wave amplitudes α_0, see also Fig. 11.1), which transform the variational principle after dropping the hats into

$$0 = \delta \int_0^T \iint_{\Omega_h} \int_0^{1+\epsilon\eta} \epsilon \partial_t \phi + \frac{\epsilon^2}{2} |\nabla \phi|^2 + \frac{\epsilon^2}{2} (\partial_z \phi)^2 + (z - 1 - \epsilon \eta_R) \, dz \, dx \, dy \, dt.$$

(11.4.4)

Adopting the second set of scalings (from Pego & Quintero [7])

$$x = \sqrt{\mu}^{-1} \tilde{x}, \quad y = \sqrt{\mu}^{-1} \tilde{y}, \quad z = \tilde{z}, \quad t = \sqrt{\mu}^{-1} \tilde{t}, \quad \phi = \sqrt{\mu}^{-1} \tilde{\phi}, \tag{11.4.5}$$

with small dispersion parameter $\mu = (H_0/\ell_0)^2 \ll 1$ (for long waves with wavelength ℓ_0, see Fig. 11.1), gives after dropping the tildes and a constant

$$0 = \delta \int_0^T \iint_{\Omega_h} \int_0^{1+\epsilon\eta} \epsilon \partial_t \phi + \frac{\epsilon^2}{2} |\nabla \phi|^2 + \frac{1}{2} \frac{\epsilon^2}{\mu} (\partial_z \phi)^2 \, dz$$

$$+ \epsilon^2 \left(\frac{1}{2} \eta^2 - \eta_R \eta \right) dx \, dy \, dt. \tag{11.4.6}$$

The system resulting from (11.4.6) is

$$\mu \nabla^2 \phi + \partial_{zz} \phi = 0 \quad \text{in} \quad \Omega_h, \tag{11.4.7a}$$

$$\partial_z \phi = 0 \quad \text{at} \quad z = 0, \tag{11.4.7b}$$

$$\partial_t \eta + \epsilon \nabla \phi \cdot \nabla \eta - \frac{1}{\mu} \partial_z \phi = 0 \quad \text{at} \quad z = 1 + \epsilon \eta, \tag{11.4.7c}$$

$$\partial_t \phi + \frac{\epsilon}{2} |\nabla \phi|^2 + \frac{1}{2} \frac{\epsilon}{\mu} (\partial_z \phi)^2 + \eta - \eta_R = 0 \quad \text{at} \quad z = 1 + \epsilon \eta. \tag{11.4.7d}$$

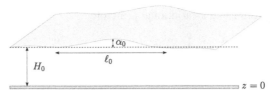

Figure 11.1. Typical wave length scales: wave amplitude α_0, average depth H_0 and wavelength ℓ_0.

We then expand ϕ in terms of the small parameter μ about the bottom potential,

$$\phi(x,y,z,t) = \Phi(x,y,t) + \mu\Phi_1(x,y,z,t) + \mu^2\Phi_2(x,y,z,t) + \cdots . \qquad (11.4.8)$$

Substituting this expansion into the Laplace equation (11.4.7a) (and using $\Delta = \nabla^2$), results in $\mu\Delta\Phi + (\mu^2\Delta\Phi_1 + \mu\partial_{zz}\Phi_1) + (\mu^3\Delta\Phi_2 + \mu^2\partial_{zz}\Phi_2) + \cdots = 0$. In addition, $\phi|_{z=0} = \Phi$ by definition and $\partial_z\phi|_{z=0} = (\mu\partial_z\Phi_1 + \mu^2\partial_z \Phi_2 + \cdots)|_{z=0} = 0$ from equation (11.4.7b), therefore Φ_1, Φ_2 and their vertical gradient vanish at $z = 0$. At leading order in μ, the equation $\partial_{zz}\Phi_1 = -\Delta\Phi$ with conditions $\partial_z\Phi_1 = 0$ and $\Phi_1 = 0$ at $z = 0$, yields $\Phi_1 = -\frac{1}{2}z^2\Delta\Phi$. Likewise, at the next order we have to solve $\partial_{zz}\Phi_2 = -\Delta\Phi_1 = \frac{1}{2}z^2\Delta^2\Phi$, giving $\Phi_2 = \frac{1}{24}z^4\Delta^2\Phi$. Therefore up to second order the expansion becomes

$$\phi = \Phi - \frac{\mu}{2}z^2\Delta\Phi + \frac{\mu^2}{24}z^4\Delta^2\Phi + \cdots . \qquad (11.4.9)$$

We use this expansion in the variational principle (11.4.6) and retain terms up to order $O(\epsilon^2\mu, \epsilon^3)$. After integration in z, the resulting leading-order principle becomes

$$0 = \delta\int_0^T \iint_{\Omega_h} \eta\partial_t\Phi - \frac{\mu}{2}\eta\partial_t\Delta\Phi + \frac{1}{2}(1+\epsilon\eta)|\nabla\Phi|^2 + \frac{\mu}{3}(\Delta\Phi)^2$$

$$+ \frac{1}{2}\eta^2 - \eta_R\eta\,\mathrm{d}x\,\mathrm{d}y\,\mathrm{d}t \qquad (11.4.10\mathrm{a})$$

$$= \delta\int_0^T \iint_{\Omega_h} \eta\partial_t\Phi + \frac{\mu}{2}\nabla\eta\cdot\partial_t\nabla\Phi + \frac{1}{2}(1+\epsilon\eta)|\nabla\Phi|^2 + \frac{\mu}{3}(\Delta\Phi)^2$$

$$+ \frac{1}{2}\eta^2 - \eta_R\eta\,\mathrm{d}x\,\mathrm{d}y\,\mathrm{d}t \qquad (11.4.10\mathrm{b})$$

$$= \int_0^T \iint_{\Omega_h} \left(\partial_t\Phi - \frac{\mu}{2}\partial_t\Delta\Phi + \frac{\epsilon}{2}|\nabla\Phi|^2 + \eta - \eta_R\right)\delta\eta$$

$$- \left(\partial_t\eta - \frac{\mu}{2}\partial_t\Delta\eta + \nabla\cdot\left((1+\epsilon\eta)\nabla\Phi\right) - \frac{2}{3}\mu\Delta^2\Phi\right)\delta\Phi\,\mathrm{d}x\,\mathrm{d}y\,\mathrm{d}t,$$

$$\qquad (11.4.10\mathrm{c})$$

in which we used the (somewhat ad hoc) boundary conditions $\mathbf{n} \cdot \nabla \Phi = 0$ and $\mathbf{n} \cdot \nabla \Delta \Phi = 0$ at $\partial \Omega_h$ with outward horizontal normal $\mathbf{n}$, in the integration by parts, as well as suitable end-point conditions at $t = 0$ and $t = T$. The variation of (11.4.10b) with respect to η and Φ in (11.4.10c) yields Hamilton's field equations

$$\delta \eta: \quad \partial_t \Phi - \frac{\mu}{2} \partial_t \Delta \Phi + \frac{\epsilon}{2} |\nabla \Phi|^2 + \eta - \eta_R = 0, \tag{11.4.11a}$$

$$\delta \Phi: \quad \partial_t \eta - \frac{\mu}{2} \partial_t \Delta \eta + \nabla \cdot \left((1 + \epsilon \eta) \nabla \Phi \right) - \frac{2}{3} \mu \Delta^2 \Phi = 0. \tag{11.4.11b}$$

An alternative formulation lowering the highest order derivatives is

$$0 = \delta \int_0^T \iint_{\Omega_h} \eta \partial_t \Phi + \frac{\mu}{2} \nabla \eta \cdot \partial_t \nabla \Phi + \frac{1}{2} (1 + \epsilon \eta) |\nabla \Phi|^2 + \frac{1}{2} \eta^2 - \eta_R \eta$$

$$+ \mu \left(\nabla q \cdot \nabla \Phi - \frac{3}{4} q^2 \right) dx\, dy\, dt \tag{11.4.12a}$$

$$= \int_0^T \iint_{\Omega_h} \left(\partial_t \Phi - \frac{\mu}{2} \partial_t \Delta \Phi + \frac{\epsilon}{2} |\nabla \Phi|^2 + \eta - \eta_R \right) \delta \eta$$

$$- \left(\partial_t \eta - \frac{\mu}{2} \partial_t \Delta \eta + \nabla \cdot \left((1 + \epsilon \eta) \nabla \Phi \right) + \mu \Delta q \right) \delta \Phi$$

$$- \mu \left(\Delta \Phi + \frac{3}{2} q \right) \delta q\, dx\, dy\, dt, \tag{11.4.12b}$$

which formulation is advantageous in a C^0-finite element formulation, used later. The resulting equations are

$$\delta \eta: \quad \partial_t \Phi - \frac{\mu}{2} \partial_t \Delta \Phi + \frac{\epsilon}{2} |\nabla \Phi|^2 + \eta - \eta_R = 0, \tag{11.4.13a}$$

$$\delta \Phi: \quad \partial_t \eta - \frac{\mu}{2} \partial_t \Delta \eta + \nabla \cdot \left((1 + \epsilon \eta) \nabla \Phi \right) + \mu \Delta q = 0, \tag{11.4.13b}$$

$$\delta q: \quad q = -\frac{2}{3} \Delta \Phi. \tag{11.4.13c}$$

Upon elimination of the auxiliary variable q, we again find (11.4.11).

11.4.2 Spatial Galerkin–Ritz Finite Element Approach

The two variabes η and Φ are expanded in terms of compact (finite element) C^0-basis functions $\varphi_l(x, y)$, yielding approximations

$$\eta_h(x, y, t) = \eta_l(t) \varphi_l(x, y) \quad \text{and} \quad \Phi_h(x, y, t) = \phi_k(t) \varphi_k(x, y), \tag{11.4.14}$$

in which the Einstein summation convention is used with indices k and l running over the finite degrees of freedom. The domain Ω_h is tessellated with quadrilaterals. Direct substitution of (11.4.14) into the variational principle (11.4.10b) leads to extra jump terms, arising from products of terms with second-order spatial derivatives because the basis functions are only continuous. Instead, we substitute (11.4.14), as well as

$$q_h(x,y,t) = q_l(t)\varphi_l(x,y), \tag{11.4.15}$$

into the variational principle (11.4.12a), to obtain

$$0 = \delta \int_0^T \left(M_{kl} + \frac{\mu}{2} S_{kl} \right) \eta_l \dot{\phi}_k + \frac{1}{2} S_{kl} \phi_k \phi_l + \frac{\epsilon}{2} \eta_m \phi_k \phi_l \iint_{\Omega_h} \varphi_m \nabla \varphi_k \cdot \nabla \varphi_l \, dx \, dy$$

$$+ \frac{1}{2} M_{kl} \eta_k \eta_l - R_l \eta_l + \mu \left(S_{kl} q_k \phi_l - \frac{3}{4} M_{kl} q_k q_l \right) dt. \tag{11.4.16}$$

This is a so-called Galerkin–Ritz approach, with matrices M_{kl}, S_{kl} and vector R_l given by

$$M_{kl} = \iint_{\Omega_h} \varphi_k \varphi_l \, dx \, dy, \quad S_{kl} = \iint_{\Omega_h} \nabla \varphi_k \cdot \nabla \varphi_l \, dx \, dy, \quad \text{and}$$

$$R_l(t) = \iint_{\Omega_h} \varphi_l \eta_R(x,t) \, dx \, dy. \tag{11.4.17}$$

Variation of (11.4.16) yields

$$\delta \eta_l : \quad \left(M_{kl} + \frac{\mu}{2} S_{kl} \right) \dot{\phi}_k = -\frac{\epsilon}{2} \phi_k \phi_m \iint_{\Omega_h} \varphi_l \nabla \varphi_k \cdot \nabla \varphi_m \, dx \, dy - M_{kl} \eta_k + R_l,$$

$$\tag{11.4.18a}$$

$$\delta \phi_k : \quad \left(M_{kl} + \frac{\mu}{2} S_{kl} \right) \dot{\eta}_l = S_{kl} \phi_l + \epsilon \eta_l \phi_m \iint_{\Omega_h} \varphi_l \nabla \varphi_k \cdot \nabla \varphi_m \, dx \, dy + \mu S_{kl} q_l,$$

$$\tag{11.4.18b}$$

$$\delta q_k : \quad M_{kl} q_l = \frac{2}{3} S_{kl} \phi_l. \tag{11.4.18c}$$

For $R_l = 0$, the Hamiltonian of the system is

$$H(t) = \frac{1}{2} \left(\phi_k S_{kl} \phi_l + \eta_k M_{kl} \eta_l \right) + \frac{\epsilon}{2} \eta_l \phi_k \phi_m \iint_{\Omega_h} \varphi_l \nabla \varphi_k \cdot \nabla \varphi_m \, dx \, dy$$

$$+ \frac{\mu}{3} \phi_k S_{kl} M_{ln}^{-1} S_{nm} \phi_m, \tag{11.4.19}$$

in which $\frac{\mu}{3} \phi_k S_{kl} M_{ln}^{-1} S_{nm} \phi_m$ serves as discretisation of $\frac{\mu}{3} (\Delta \Phi)^2$ in (11.4.10b).

Several limits are enclosed in (11.4.10b) or (11.4.12a) and its spatial discretisation (11.4.16): nonlinear, potential flow shallow water equations emerge for $\mu = 0$; linear Benney–Luke type equations for $\epsilon = 0$ and $\mu > 0$; and, linear, potential flow shallow water equations for $\epsilon = \mu = 0$. By defining $P_k = (M_{kl} + \frac{\mu}{2} S_{kl}) \eta_l$ and $Q_k = \phi_k$, (11.4.16) obtains the standard form (11.2.1), which discretisation in time is considered next.

11.5 Discontinuous Galerkin Time Discretisation

11.5.1 Defining the Time Derivative Across Time Nodes

Time interval $[0, T]$ is partitioned into N time intervals $I_n = [t_n, t_{n+1} = t_n + \Delta t_n]$ for $n = 0, \ldots, N-1$. The variables in the variational principle (11.2.1) or (11.2.8) are approximated using a discontinuous Galerkin finite element method in time. These approximations Q_δ, P_δ are continuous in each time interval but discontinuous across. Across each time node $t_n = \sum_{j=0}^{n-1} \Delta t_j$ for $n > 0$ with $t_0 = 0$ and $t_N = T$, the contribution of the term $P \cdot dQ/dt$ in the variational principle needs to be determined. It leads to a δ–function at each node in the integrand, but since the variational principle is a weak form, a finite sum of contributions emerges. Since the expansions are discontinuous, limiting values on either side of each time node t_n are discontinuous, with $Q^{n,-} = \lim_{\epsilon \downarrow 0} Q_\delta(t_n - \epsilon)$ and $Q^{n,+} = \lim_{\epsilon \downarrow 0} Q_\delta(t_n + \epsilon)$, and likewise for $P^{n,-}$ and $P^{n,+}$. The jump in a quantity across the node will be denoted as, e.g., $[Q] = Q^{n,+} - Q^{n,-}$ and the weighted average as, e.g., $\{P\} = \alpha P^{n,+} + (1 - \alpha) P^{n,-}$ with $\alpha \in [0, 1]$.

Limit continuous Galerkin to discontinuous Galerkin:
Consider all time nodes $n > 0$ and split each time node into a narrow element $t \in [t_{n+1} - \Delta\tau, t_{n+1} + \Delta\tau]$ of width $2\Delta\tau$ such that none of these narrow elements overlap another. Hence, the original elements shrink slightly in size. Instead of a discontinuous Galerkin finite element method in time, we have created a continuous finite element in time by using expansions within these narrow elements such that the overall expansion becomes continuous. Consider the mapping $t = t_{n+1} - \Delta\tau + (2\Delta\tau)\tilde{\tau}$ within this narrow element to a reference coordinate $\tilde{\tau} \in [0, 1]$. To enforce continuity, we have values on the left side of this narrow element $P_L = P(\tilde{\tau} = 0) = P^{n+1,-}$ and $Q_L = Q(\tilde{\tau} = 0) = Q^{n+1,-}$ and on the right $P_R = P(\tilde{\tau} = 1) = P^{n+1,+}$ and $Q_R = Q(\tilde{\tau} = 1) = Q^{n+1,+}$. The goal is now to calculate the contribution of the term $P \cdot dQ/dt$ in the variational principle at a node for a discontinuous Galerkin finite element method, in which the variables are multivalued, as a limit $\Delta\tau \to 0$ of the continuous Galerkin finite element method defined above. The result will, of course, depend on these continuous finite element expansions. Within these narrow elements, we choose a linear expansion in Q and a quadratic one for P across node t_{n+1}, but other choices can be explored as well,

$$Q \approx Q^{n+1,-} + (Q^{n+1,+} - Q^{n+1,-})\tilde{\tau}, \quad \text{and}$$
$$P \approx P^{n+1,-} + (P^{n+1,+} - P^{n+1,-})\big((4 - 6\alpha)\tilde{\tau} + (6\alpha - 3)\tilde{\tau}^2\big), \quad (11.5.1)$$

with $\alpha \in [0, 1]$ a free parameter. Hence, using (11.5.1) we obtain that

$$\int_0^1 P \cdot \frac{dQ}{d\tilde{\tau}} d\tilde{\tau} = (Q^{n+1,+} - Q^{n+1,-}) \cdot \big(\alpha P^{n+1,-} + (1-\alpha)P^{n+1,+}\big). \quad (11.5.2)$$

The result is the jump in the coordinate Q times a weighted mean of the momentum P of the limit values across the time node at t_{n+1}, and it is independent of $\Delta\tau$. The other terms in the variational principle within these narrow elements contribute to zero in this limit $\Delta\tau \to 0$ due to the absence of further time derivatives.

Consequently, for an arbitrary discontinuous Galerkin finite element expansion within each element, the (intermediate) discrete variational principle is

$$0 = \delta \sum_{n=0}^{N-1} \int_{t_n}^{t_{n+1}} \boldsymbol{P}_\delta \cdot \frac{\mathrm{d}\boldsymbol{Q}_\delta}{\mathrm{d}t} - H(\boldsymbol{P}_\delta, \boldsymbol{Q}_\delta, t)\,\mathrm{d}t$$

$$+ \sum_{n=-1}^{N-1} (\boldsymbol{Q}^{n+1,+} - \boldsymbol{Q}^{n+1,-}) \cdot \left(\alpha\boldsymbol{P}^{n+1,-} + (1-\alpha)\boldsymbol{P}^{n+1,+}\right). \qquad (11.5.3)$$

Similarly, the Kamiltonian principle becomes

$$0 = \delta \sum_{n=0}^{N-1} \int_{t_n}^{t_{n+1}} \boldsymbol{P}_\delta \cdot \frac{\mathrm{d}\boldsymbol{Q}_\delta}{\mathrm{d}t} + p_\delta \frac{\mathrm{d}\tau_\delta}{\mathrm{d}t} - K(\boldsymbol{P}_\delta, \boldsymbol{Q}_\delta, p_\delta, \tau_\delta, t)\,\mathrm{d}t$$

$$+ \sum_{n=-1}^{N-1} (\boldsymbol{Q}^{n+1,+} - \boldsymbol{Q}^{n+1,-}) \cdot \left(\alpha\boldsymbol{P}^{n+1,-} + (1-\alpha)\boldsymbol{P}^{n+1,+}\right)$$

$$+ (\tau^{n+1,+} - \tau^{n+1,-})\left(\alpha p^{n+1,-} + (1-\alpha)p^{n+1,+}\right). \qquad (11.5.4)$$

To derive definite time integration schemes, we still need to specify the expansions for $\boldsymbol{P}_\delta, \boldsymbol{Q}_\delta, p_\delta$ and τ_δ as well as the quadrature rules to evaluate the integral of the Hamiltonian.

11.5.2 Optimization Criteria

The strategy is to optimise the properties of the time discretisation scheme (11.5.3) or (11.5.4) for various values of α, choices of the approximating polynomials and different quadratures used to approximate the time slab integrals. Formally, a polynomial of order n_p will yield a time discretisation scheme that is $O(n_p + 1)$. So far [9], the following results have been obtained for *autonomous Hamiltonian systems*:

- Using piecewise constant polynomials and exact integration yields first-order accurate stable and symplectic schemes for $\alpha = 0, 1/2, 1$. The variable in a time element is represented by its mean. Classic symplectic Euler (SE) schemes (see, e.g., Hairer et al. [12]) emerge for $\alpha = 0$ and $\alpha = 1$ in the autonomous case with $H = H(\boldsymbol{Q}, \boldsymbol{P})$.
- For a second-order scheme at least piecewise linear polynomials are required. Consequently, two coefficients or degrees of freedom represent

each variable in a time element. Classic stable Störmer–Verlet (SV) schemes emerge. The four degrees of freedom per time slab in the scheme reduce to three because the variations yield that one of the variables becomes continuous. The scheme is a mixed (dis)continuous Galerkin scheme with in general two implicit stages and one explicit stage.

- Third-order schemes require at least quadratic polynomials. A symmetric quadrature (Simpson's rule) using three quadrature points for each variable leads to a six-stage scheme, only stable for $\alpha = 1/2$. Dispersion analysis by Gagarina et al. [9] based on the discrete harmonic oscillator reveals that there are two parasitic modes, the influence of which is neutralised by initialising the scheme continuously, thus removing the initial jumps. The resulting scheme remains discontinuous with in general two coupled implicit and four explicit stages.

All these schemes will be extended to the *non-autonomous case* next via a Kamiltonian treatment of time in (11.2.8). This Kamiltonian can be used to monitor that the discrete Kamiltonian energy shows no drift and has bounded fluctuations.

11.5.3 Störmer–Verlet Scheme

To derive a second-order scheme, a piecewise linear approximation is used in the time slab (t_n, t_{n+1}) with

$$P_\delta = P^{n,+}\frac{(t_{n+1}-t)}{\Delta t} + P^{n+1,-}\frac{(t-t_n)}{\Delta t}, \tag{11.5.5a}$$

$$Q_\delta = Q^{n,+}\frac{(t_n+t_{n+1}-2t)}{\Delta t} + Q^{n+1/2}\frac{2(t-t_n)}{\Delta t}, \tag{11.5.5b}$$

$$p_\delta = p^{n,+}\frac{(t_{n+1}-t)}{\Delta t} + p^{n+1,-}\frac{(t-t_n)}{\Delta t}, \quad \text{and} \tag{11.5.5c}$$

$$\tau_\delta = \tau^{n,+}\frac{(t_n+t_{n+1}-2t)}{\Delta t} + \tau^{n+1/2}\frac{2(t-t_n)}{\Delta t}. \tag{11.5.5d}$$

We use this in (11.5.4) for $\alpha = 1$. Hence,

$$\int_{t_n}^{t_{n+1}} P_\delta \cdot \frac{dQ_\delta}{dt}\, dt = (P^{n+1,-}+P^{n,+})\cdot(Q^{n+1/2}-Q^{n,+}), \tag{11.5.6}$$

while the jump term becomes $(Q^{n+1,+} - 2Q^{n+1/2} + Q^{n,+})\cdot P^{n+1,-}$, and the integral of the Hamiltonian is approximated as follows

$$\int_{t_n}^{t_{n+1}} H(Q,P,\tau) + p\, dt \approx \frac{1}{2}\Delta t\Big(H(Q^{n+1/2},P^{n,+},\tau^{n+1/2}) + p^{n,+}$$

$$+ H(Q^{n+1/2},P^{n+1,-},\tau^{n+1/2}) + p^{n+1,-}\Big). \tag{11.5.7}$$

Consequently, the discrete variational principle becomes

$$0 = \delta \sum_{n=0}^{N-1} (\boldsymbol{P}^{n+1,-} + \boldsymbol{P}^{n,+}) \cdot (\boldsymbol{Q}^{n+1/2} - \boldsymbol{Q}^{n,+}) + (p^{n+1,-} + p^{n,+})(\tau^{n+1/2} - \tau^{n,+})$$

$$-\frac{1}{2}\Delta t \Big(H(\boldsymbol{Q}^{n+1/2}, \boldsymbol{P}^{n,+}, \tau^{n+1/2}) + H(\boldsymbol{Q}^{n+1/2}, \boldsymbol{P}^{n+1,-}, \tau^{n+1/2}) \Big)$$

$$-\frac{1}{2}\Delta t (p^{n,+} + p^{n+1,-})$$

$$+\delta \sum_{n=-1}^{N-1} \boldsymbol{P}^{n+1,-} \cdot (\boldsymbol{Q}^{n+1,+} - 2\boldsymbol{Q}^{n+1/2} + \boldsymbol{Q}^{n,+})$$

$$+ p^{n+1,-}(\tau^{n+1,+} - 2\tau^{n+1/2} + \tau^{n,+}). \qquad (11.5.8)$$

Variations of (11.5.8) yield the following scheme

$$\delta \boldsymbol{Q}^{n+1/2}: \quad \boldsymbol{P}^{n+1,-} = \boldsymbol{P}^{n,+}$$

$$-\frac{1}{2}\Delta t \left(\frac{\partial H(\boldsymbol{Q}^{n+1/2}, \boldsymbol{P}^{n,+}, \tau^{n+1/2})}{\partial \boldsymbol{Q}^{n+1/2}} + \frac{\partial H(\boldsymbol{Q}^{n+1/2}, \boldsymbol{P}^{n+1,-}, \tau^{n+1/2})}{\partial \boldsymbol{Q}^{n+1/2}} \right),$$

$$(11.5.9a)$$

$$\delta \boldsymbol{Q}^{n,+}: \quad \boldsymbol{P}^{n,+} = \boldsymbol{P}^{n,-}, \qquad (11.5.9b)$$

$$\delta \boldsymbol{P}^{n,+}: \quad \boldsymbol{Q}^{n+1/2} = \boldsymbol{Q}^{n,+} + \frac{1}{2}\Delta t \frac{\partial H(\boldsymbol{Q}^{n+1/2}, \boldsymbol{P}^{n,+}, \tau^{n+1/2})}{\partial \boldsymbol{P}^{n,+}}, \qquad (11.5.9c)$$

$$\delta \boldsymbol{P}^{n+1,-}: \quad \boldsymbol{Q}^{n+1,+} = \boldsymbol{Q}^{n+1/2} + \frac{1}{2}\Delta t \frac{\partial H(\boldsymbol{Q}^{n+1/2}, \boldsymbol{P}^{n+1,-}, \tau^{n+1/2})}{\partial \boldsymbol{P}^{n+1,-}}, \qquad (11.5.9d)$$

$$\delta p^{n,+}: \quad \tau^{n+1/2} = \tau^n + \frac{1}{2}\Delta t, \qquad (11.5.9e)$$

$$\delta p^{n+1,-}: \quad \tau^{n+1,+} = \tau^{n+1/2} + \frac{1}{2}\Delta t, \qquad (11.5.9f)$$

$$\delta \tau^{n+1/2}: \quad p^{n+1,-} = p^{n,+}$$

$$-\frac{1}{2}\Delta t \left(\frac{\partial H(\boldsymbol{Q}^{n+1/2}, \boldsymbol{P}^{n,+}, \tau^{n+1/2})}{\partial \tau^{n+1/2}} + \frac{\partial H(\boldsymbol{Q}^{n+1/2}, \boldsymbol{P}^{n+1,-}, \tau^{n+1/2})}{\partial \tau^{n+1/2}} \right),$$

$$(11.5.9g)$$

$$\delta \tau^{n,+}: \quad p^{n,+} = p^{n,-}. \qquad (11.5.9h)$$

We note that $\boldsymbol{P}^{n,+} = \boldsymbol{P}^{n,-} \equiv \boldsymbol{P}^n$ and $p^{n,+} = p^{n,-} \equiv p^n$ have become continuous, while $\boldsymbol{Q}$ remains discontinuous. The final time discretisation can thus be viewed as a second-order mixed discontinuous and continuous Galerkin finite element approximation. When we also denote $\boldsymbol{Q}^{n,+}$ by $\boldsymbol{Q}^n$ and replace τ^n by t_n and $\tau^{n+1/2}$ by $t_{n+1/2}$, then the final, compactly written non-autonomous SV scheme reads

$$Q^{n+1/2} = Q^n + \frac{1}{2}\Delta t \frac{\partial H(Q^{n+1/2}, P^n, t_{n+1/2})}{\partial P^n}, \tag{11.5.10a}$$

$$P^{n+1} = P^n - \frac{1}{2}\Delta t \left(\frac{\partial H(Q^{n+1/2}, P^n, t_{n+1/2})}{\partial Q^{n+1/2}} + \frac{\partial H(Q^{n+1/2}, P^{n+1}, t_{n+1/2})}{\partial Q^{n+1/2}} \right), \tag{11.5.10b}$$

$$Q^{n+1} = Q^{n+1/2} + \frac{1}{2}\Delta t \frac{\partial H(Q^{n+1/2}, P^{n+1}, t_{n+1/2})}{\partial P^{n+1}}. \tag{11.5.10c}$$

The first two steps are, generally, implicit, while the last one is explicit, and the evaluation of time is at $t_{n+1/2}$, by construction.

Alternatively, we can reverse the approximation of Q and P (as well as replacing τ and p by q and τ), such that

$$Q_\delta = Q^{n,+} \frac{(t_{n+1} - t)}{\Delta t} + Q^{n+1,-} \frac{(t - t_n)}{\Delta t}, \tag{11.5.11a}$$

$$P_\delta = P^{n,+} \frac{(t_n + t_{n+1} - 2t)}{\Delta t} + P^{n+1/2} \frac{2(t - t_n)}{\Delta t}, \tag{11.5.11b}$$

$$q_\delta = q^{n,+} \frac{(t_{n+1} - t)}{\Delta t} + q^{n+1,-} \frac{(t - t_n)}{\Delta t}, \quad \text{and} \tag{11.5.11c}$$

$$\tau_\delta = \tau^{n,+} \frac{(t_n + t_{n+1} - 2t)}{\Delta t} + \tau^{n+1/2} \frac{2(t - t_n)}{\Delta t}, \tag{11.5.11d}$$

while taking $\alpha = 0$. The discrete variational principle then becomes

$$0 = \delta \sum_{n=0}^{N-1} P^{n+1/2} \cdot (Q^{n+1,-} - Q^{n,+}) + \tau^{n+1/2}(q^{n+1,-} - q^{n,+})$$

$$+ \frac{1}{2}\Delta t \Big(-H(Q^{n,+}, P^{n+1/2}, \tau^{n+1/2}) + q^{n,+}$$

$$- H(Q^{n+1,-}, P^{n+1/2}, \tau^{n+1/2}) + q^{n+1,-} \Big)$$

$$+ \delta \sum_{n=-1}^{N-1} P^{n+1,+} \cdot (Q^{n+1,+} - Q^{n+1,-}) + \tau^{n+1,+}(q^{n+1,+} - q^{n+1,-}). \tag{11.5.12}$$

A similar evaluation but now giving $Q^{n+1,+} = Q^{n+1,-} \equiv Q^{n+1}$ and $q^{n+1,+} = q^{n+1,-} \equiv q^{n+1}$, then yields the following second SV scheme for a non-autonomous Hamiltonian system

$$P^{n+1/2} = P^n - \frac{1}{2}\Delta t \frac{\partial H(Q^n, P^{n+1/2}, t_{n+1/2})}{\partial Q^n}, \tag{11.5.13a}$$

$$Q^{n+1} = Q^n + \frac{1}{2}\Delta t \left(\frac{\partial H(Q^n, P^{n+1/2}, t_{n+1/2})}{\partial P^{n+1/2}} + \frac{\partial H(Q^{n+1}, P^{n+1/2}, t_{n+1/2})}{\partial P^{n+1/2}} \right),$$

$$\tag{11.5.13b}$$

$$P^{n+1} = P^{n+1/2} - \frac{1}{2}\Delta t \frac{\partial H(Q^{n+1}, P^{n+1/2}, t_{n+1/2})}{\partial Q^{n+1}}, \tag{11.5.13c}$$

after denoting the discontinuous variables as $P^{n+1,+} = P^{n+1}$, while one can derive that $\tau^{n+1,+} = \tau^{n+1,-} = \tau^{n+1} = t_{n+1}$ is continuous when $\tau^{0,+} = \tau^{0,-} = 0$ is continuous initially. Again the first two steps are implicit and the last one is explicit. Depending on the factual Hamiltonian system, one can choose one or the other SV version. For the spatially discrete Miles' variational principle for potential flow water waves, we used the second version in Gagarina et al. [6]. It makes the free surface elevation continuous in time, while the velocity potential remains discontinuous in time. In addition, it is easy to see that variations on the above Kamiltonian approaches include the following modified mid-point approximations of the Hamiltonian as well

$$\frac{1}{2}\Delta t \left(H(Q^{n+1/2}, P^{n,+}, t_n) + H(Q^{n+1/2}, P^{n+1,-}, t_{n+1}) \right), \tag{11.5.14}$$

and

$$\frac{1}{2}\Delta t \left(H(Q^{n,+}, P^{n+1/2}, t_n) + H(Q^{n+1,-}, P^{n+1/2}, t_{n+1}) \right), \tag{11.5.15}$$

respectively, with corresponding changes in the discrete dynamics.

11.5.4 Third-Order Scheme

To derive a six-stage third-order scheme [9], we start with quadratic expansions

$$Q_\delta = Q^{n,+} \frac{(t_n + t_{n+1} - 2t)(t_{n+1} - t)}{\Delta t^2} + Q^{n+1/2} \frac{4(t - t_n)(t_{n+1} - t)}{\Delta t^2}$$
$$+ Q^{n+1,-} \frac{(t_n + t_{n+1} - 2t)(t_n - t)}{\Delta t^2}, \tag{11.5.16}$$

and so forth for P_δ, p_δ and τ_δ. The Kamiltonian is approximated in symmetric fashion using Simpson's rule as

$$\int_{t_n}^{t_{n+1}} H(Q, P, \tau) + p \, dt \approx \frac{1}{6}\Delta t \Big(H(Q^{n,+}, P^{n,+}, \tau^{n,+}) + p^{n,+}$$

$$+ 4H(Q^{n+1/2}, P^{n+1/2}, \tau^{n+1/2}) + 4p^{n+1/2} + H(Q^{n+1,-}, P^{n+1,-}, \tau^{n+1,-}) + p^{n+1,-} \Big).$$

$$\tag{11.5.17}$$

After using the above approximations in (11.5.4) with $\alpha = 1/2$, the discrete variational principle becomes

$$0 = \delta \sum_{n=0}^{N-1} \int_{t_n}^{t_{n+1}} \boldsymbol{P}_\delta \cdot \frac{\mathrm{d}\boldsymbol{Q}_\delta}{\mathrm{d}t} + p_\delta \frac{\mathrm{d}\tau_\delta}{\mathrm{d}t} \,\mathrm{d}t - \frac{1}{6}\Delta t \Big(H(\boldsymbol{Q}^{n,+}, \boldsymbol{P}^{n,+}, \tau^{n,+}) + p^{n,+}$$

$$+ 4H(\boldsymbol{Q}^{n+1/2}, \boldsymbol{P}^{n+1/2}, \tau^{n+1/2}) + 4p^{n+1/2}$$

$$+ H(\boldsymbol{Q}^{n+1,-}, \boldsymbol{P}^{n+1,-}, \tau^{n+1,-}) + p^{n+1,-}\Big)$$

$$+ \delta \sum_{n=-1}^{N-1} \frac{1}{2}(\boldsymbol{P}^{n+1,-} + \boldsymbol{P}^{n+1,+}) \cdot (\boldsymbol{Q}^{n+1,+} - \boldsymbol{Q}^{n+1,-})$$

$$+ \frac{1}{2}(p^{n+1,-} + p^{n+1,+})(\tau^{n+1,+} - \tau^{n+1,-}), \tag{11.5.18}$$

with

$$\int_{t_n}^{t_{n+1}} \boldsymbol{P}_\delta \cdot \frac{\mathrm{d}\boldsymbol{Q}_\delta}{\mathrm{d}t} + p_\delta \frac{\mathrm{d}\tau_\delta}{\mathrm{d}t} \,\mathrm{d}t = \boldsymbol{P}^{n+1,-} \cdot \Big(\frac{1}{6}\boldsymbol{Q}^{n,+} - \frac{2}{3}\boldsymbol{Q}^{n+1/2} + \frac{1}{2}\boldsymbol{Q}^{n+1,-} \Big)$$

$$+ \frac{2}{3}\boldsymbol{P}^{n+1/2} \cdot \big(\boldsymbol{Q}^{n+1,-} - \boldsymbol{Q}^{n,+} \big) + \boldsymbol{P}^{n,+} \cdot \Big(-\frac{1}{2}\boldsymbol{Q}^{n,+} + \frac{2}{3}\boldsymbol{Q}^{n+1/2} - \frac{1}{6}\boldsymbol{Q}^{n+1,-} \Big)$$

$$+ p^{n+1,-}\Big(\frac{1}{6}\tau^{n,+} - \frac{2}{3}\tau^{n+1/2} + \frac{1}{2}\tau^{n+1,-} \Big) + \frac{2}{3}p^{n+1/2}\big(\tau^{n+1,-} - \tau^{n,+} \big)$$

$$+ p^{n,+}\Big(-\frac{1}{2}\tau^{n,+} + \frac{2}{3}\tau^{n+1/2} - \frac{1}{6}\tau^{n+1,-} \Big). \tag{11.5.19}$$

Variations thereof yield

$$\delta \boldsymbol{P}^{n,+}: \quad \boldsymbol{Q}^{n+1/2} = \frac{1}{4}\boldsymbol{Q}^{n,+} + \frac{3}{4}\boldsymbol{Q}^{n,-} +$$

$$+ \frac{1}{4}\Delta t \Big(\frac{\partial H(\boldsymbol{Q}^{n,+}, \boldsymbol{P}^{n,+}, \tau^{n,+})}{\partial \boldsymbol{P}^{n,+}} + \frac{\partial H(\boldsymbol{Q}^{n+1/2}, \boldsymbol{P}^{n+1/2}, \tau^{n+1/2})}{\partial \boldsymbol{P}^{n+1/2}} \Big), \tag{11.5.20a}$$

$$\delta \boldsymbol{Q}^{n,+}: \quad \boldsymbol{P}^{n+1/2} = \frac{1}{4}\boldsymbol{P}^{n,+} + \frac{3}{4}\boldsymbol{P}^{n,-} +$$

$$- \frac{1}{4}\Delta t \Big(\frac{\partial H(\boldsymbol{Q}^{n,+}, \boldsymbol{P}^{n,+}, \tau^{n,+})}{\partial \boldsymbol{Q}^{n,+}} + \frac{\partial H(\boldsymbol{Q}^{n+1/2}, \boldsymbol{P}^{n+1/2}, \tau^{n+1/2})}{\partial \boldsymbol{Q}^{n+1/2}} \Big), \tag{11.5.20b}$$

$$\delta \boldsymbol{Q}^{n+1/2}: \quad \boldsymbol{P}^{n+1,-} = \boldsymbol{P}^{n,+} - \Delta t \frac{\partial H(\boldsymbol{Q}^{n+1/2}, \boldsymbol{P}^{n+1/2}, \tau^{n+1/2})}{\partial \boldsymbol{Q}^{n+1/2}}, \tag{11.5.20c}$$

$$\delta \boldsymbol{P}^{n+1/2}: \quad \boldsymbol{Q}^{n+1,-} = \boldsymbol{Q}^{n,+} + \Delta t \frac{\partial H(\boldsymbol{Q}^{n+1/2}, \boldsymbol{P}^{n+1/2}, \tau^{n+1/2})}{\partial \boldsymbol{P}^{n+1/2}}, \tag{11.5.20d}$$

$$\delta \boldsymbol{P}^{n+1,-}: \quad \boldsymbol{Q}^{n+1,+} = \frac{4}{3}\boldsymbol{Q}^{n+1/2} - \frac{1}{3}\boldsymbol{Q}^{n,+} + \frac{1}{3}\Delta t \frac{\partial H(\boldsymbol{Q}^{n+1,-}, \boldsymbol{P}^{n+1,-}, \tau^{n+1,-})}{\partial \boldsymbol{P}^{n+1,-}}, \tag{11.5.20e}$$

$$\delta \boldsymbol{Q}^{n+1,-}: \quad \boldsymbol{P}^{n+1,+} = \frac{4}{3}\boldsymbol{P}^{n+1/2} - \frac{1}{3}\boldsymbol{P}^{n,+} - \frac{1}{3}\Delta t \frac{\partial H(\boldsymbol{Q}^{n+1,-}, \boldsymbol{P}^{n+1,-}, \tau^{n+1,-})}{\partial \boldsymbol{Q}^{n+1,-}}, \tag{11.5.20f}$$

and

$$\delta p^{n+1/2}: \quad \tau^{n+1,-} = \tau^{n,+} + \Delta t, \tag{11.5.21a}$$

$$\delta p^{n,+}: \quad \tau^{n+1/2} = \frac{1}{4}\tau^{n,+} + \frac{3}{4}\tau^{n,-} + \frac{1}{2}\Delta t, \tag{11.5.21b}$$

$$\delta p^{n+1,-}: \quad \tau^{n+1,+} = \frac{4}{3}\tau^{n+1/2} - \frac{1}{3}\tau^{n,+} + \frac{1}{3}\Delta t, \tag{11.5.21c}$$

$$\delta \tau^{n+1/2}: \quad p^{n+1,-} = p^{n,+} - \Delta t \frac{\partial H(\boldsymbol{Q}^{n+1/2}, \boldsymbol{P}^{n+1/2}, \tau^{n+1/2})}{\partial \tau^{n+1/2}}, \tag{11.5.21d}$$

$$\delta \tau^{n,+}: \quad p^{n+1/2} = \frac{1}{4}p^{n,+} + \frac{3}{4}p^{n,-}$$
$$- \frac{1}{4}\Delta t \left(\frac{\partial H(\boldsymbol{Q}^{n,+}, \boldsymbol{P}^{n,+}, \tau^{n,+})}{\partial \tau^{n,+}} + \frac{\partial H(\boldsymbol{Q}^{n+1/2}, \boldsymbol{P}^{n+1/2}, \tau^{n+1/2})}{\partial \tau^{n+1/2}} \right), \tag{11.5.21e}$$

$$\delta \tau^{n+1,-}: \quad p^{n+1,+} = \frac{4}{3}p^{n+1/2} - \frac{1}{3}p^{n,+} + \frac{1}{3}\Delta t \frac{\partial H(\boldsymbol{Q}^{n+1,-}, \boldsymbol{P}^{n+1,-}, \tau^{n+1,-})}{\partial \tau^{n+1,-}}. \tag{11.5.21f}$$

Provided $\tau^{0,-} = \tau^{0,+} = 0$ is initialised continuously, we see that $\tau^{n,+} = \tau^{n,-} = \tau^n = t_n$. Hence, time can be evaluated using Simpson's rule in the approximation of the Hamiltonian in line with the other variables. The corresponding, succinct variational principle

$$0 = \delta \sum_{n=0}^{N-1} \int_{t_n}^{t_{n+1}} \boldsymbol{P}_\delta \cdot \frac{\mathrm{d}\boldsymbol{Q}_\delta}{\mathrm{d}t} \mathrm{d}t - \frac{1}{6}\Delta t \Big(H(\boldsymbol{Q}^{n,+}, \boldsymbol{P}^{n,+}, t_{n,+})$$

$$+ 4H(\boldsymbol{Q}^{n+1/2}, \boldsymbol{P}^{n+1/2}, t_{n+1/2}) + H(\boldsymbol{Q}^{n+1,-}, \boldsymbol{P}^{n+1,-}, t_{n+1}) \Big)$$

$$+ \delta \sum_{n=-1}^{N-1} \frac{1}{2}(\boldsymbol{P}^{n+1,-} + \boldsymbol{P}^{n+1,+}) \cdot (\boldsymbol{Q}^{n+1,+} - \boldsymbol{Q}^{n+1,-}), \tag{11.5.22}$$

with

$$\int_{t_n}^{t_{n+1}} \boldsymbol{P}_\delta \cdot \frac{\mathrm{d}\boldsymbol{Q}_\delta}{\mathrm{d}t}\mathrm{d}t = \boldsymbol{P}^{n+1,-} \cdot \left(\frac{1}{6}\boldsymbol{Q}^{n,+} - \frac{2}{3}\boldsymbol{Q}^{n+1/2} + \frac{1}{2}\boldsymbol{Q}^{n,+}\right)$$
$$+ \frac{2}{3}\boldsymbol{P}^{n+1/2} \cdot \left(\boldsymbol{Q}^{n+1,-} - \boldsymbol{Q}^{n,+}\right) + \boldsymbol{P}^{n,+} \cdot \left(-\frac{1}{2}\boldsymbol{Q}^{n,+} + \frac{2}{3}\boldsymbol{Q}^{n+1/2} - \frac{1}{6}\boldsymbol{Q}^{n,+}\right),$$

$$(11.5.23)$$

then yields the dynamics (11.5.20) with τ therein replaced by t, while p is superfluous. The coupled, first two steps are implicit while the last four steps are explicit.

11.6 Numerical Verification and Validation

11.6.1 Forced-Dissipative Linear Oscillator

To verify the first-order SE, second-order SV, and symmetric third-order (3S) time discretisations of the Kamiltonian dynamics, we consider the linear forced-dissipative oscillator and its exact solution (11.3.8). The first transformation (11.3.3) to the Kamiltonian form turns out to be only first-order accurate numerically for a scheme designed to be a third-order accurate autonomous scheme, while proper third-order accuracy was obtained numerically using the second transformation (11.3.6) for the non-autonomous case. The reason why the second transformation turns out to be better than the first one may be that we know that the second transformation can be brought into the archetypical form (11.2.11), while we have not been able to do that for the first transformation. From Table 11.1, we discern that the numerical solutions have the expected first-, second- and third-order accuracy, respectively.

11.6.2 Variational Benney–Luke Water Wave Model

All simulations of (11.4.13) were done within the finite element modelling environment Firedrake [13] on a quadrilateral mesh [14], to third order in space using quadratic Lagrange polynomials, and second or third order in time using symplectic integrators (11.5.10) or (11.5.20), respectively[a]

To verify the numerical implementation, we compared numerical solutions with a standing wave solution of the linearised equations (i.e., (11.4.13) with $\epsilon = 0$). We also compared small-amplitude nonlinear numerical solutions with an asymptotic KdV–soliton (i.e., (11.4.13) with both $\epsilon > 0$ and $\mu > 0$). A final validation case consists of comparisons between numerical solutions and wave

---

[a] Störmer–Verlet and third-order integrators for (11.4.18), as well as details about the numerical implementation using Firedrake can be found in Appendix A.

Table 11.1. L^∞–error $\max_{t\in[0,30]} |q_\delta(t) - q_{exact}(t)|$. *The parameters values used are* $\Omega = 0.9$, $\omega = 1$, $a = 1$, $q_0 = 1$, $p_0 = 1$, $\gamma = 0.2$.

Δt	0.05	0.025	0.0125	Order
1st SE ($\times 100$)	7.0389	3.4597	1.7148	1.02
2nd SV ($\times 1000$)	0.2795	0.0699	0.0175	2.00
3rd 3S ($\times 10^6$)	8.214	1.011	0.125	3.00

phenomena observed during a soliton splash experiment[b]. Unless otherwise stated, regular rectangular elements are used with time step $\Delta t = 0.0028$, quadratic Lagrange basis functions and SV in time.

Exact standing wave solutions
An exact standing wave solution of the linearised counterpart of (11.4.11) (with $\epsilon = 0$ and $\eta_R = 0$) in a rectangular domain of size $L_x \times L_y$ is given by

$$\eta(x,y,t) = A\cos(\omega t)\cos\left(\frac{2\pi m_1 x}{L_x}\right)\cos\left(\frac{2\pi m_2 y}{L_y}\right), \qquad (11.6.1a)$$

$$\Phi(x,y,t) = B\sin(\omega t)\cos\left(\frac{2\pi m_1 x}{L_x}\right)\cos\left(\frac{2\pi m_2 y}{L_y}\right), \qquad (11.6.1b)$$

where A is the wave amplitude, $m_1 > 0$, $m_2 > 0$ are integers and

$$\omega = \pm\kappa\frac{\sqrt{1+\frac{2\mu}{3}\kappa^2}}{1+\frac{\mu}{2}\kappa^2}, \quad B = \mp\frac{A}{\kappa\sqrt{1+\frac{2\mu}{3}\kappa^2}}, \quad \kappa^2 = \left(\frac{2\pi m_1}{L_x}\right)^2 + \left(\frac{2\pi m_2}{L_y}\right)^2.$$
$$(11.6.1c)$$

The above solution is used in a comparison with the numerical solution of the linearised Benney–Luke system with $\epsilon = 0$. The visual comparison after two wave periods can be seen in Fig. 11.2. Energy fluctuations are small and bounded, as expected.

Small amplitude soliton solutions
Leading order asymptotic solutions in ϵ are derived from (11.4.11) (for $\eta_R = 0$) by applying the following transformations in one spatial dimension

$$\xi = \sqrt{\frac{\epsilon}{\mu}}(x-t), \quad \tau = \epsilon\sqrt{\frac{\epsilon}{\mu}}t, \quad \mathcal{F} = \sqrt{\frac{\epsilon}{\mu}}\Phi, \quad \eta = \eta, \qquad (11.6.2)$$

[b] The *Python* code for all the simulations presented in this paper is available at https://github.com/kalogirou/Benney-Luke.

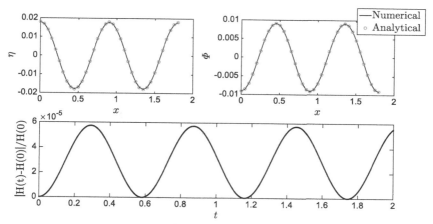

Figure 11.2. Comparison between numerical solution (solid lines) and exact standing wave solution (circles) of the linearised Benney–Luke model. The parameter values are $L_x = 1.8$, $L_y = 5$, $A = 0.1$, $m_1 = 4$, $m_2 = 2$, $T = 2$, $\mu = 0.04$ and of course $\epsilon = 0$. Cross–sections of $\eta(x, y, T)$ (top left panel) and $\Phi(x, y, T)$ (top right panel) at the mid lines are given, as well as the relative energy compared to the initial energy $H(t = 0)$ (bottom panel), see (11.4.19). 36×200 elements are used.

yielding

$$\eta = \mathcal{F}_\xi - \epsilon \mathcal{F}_\tau + \frac{\epsilon^2}{2} \mathcal{F}_{\tau\xi\xi} - \frac{\epsilon}{2} \mathcal{F}_{\xi\xi\xi} - \frac{\epsilon}{2}(\mathcal{F}_\xi)^2. \tag{11.6.3}$$

Substitution of the above scalings into the variational principle (11.4.10b) on the real line with $\lim_{|\xi| \to \infty} \mathcal{F}, \eta \to 0$, subsequent use of (11.6.3) to eliminate η and truncation to $O(\epsilon^2)$, yields the principle for the KdV–equation and its variation

$$0 = \delta \int_0^T \int_0^L \eta(\epsilon \mathcal{F}_\tau - \mathcal{F}_\xi) + \frac{\epsilon}{2} \eta_\xi(\epsilon \mathcal{F}_{\tau\xi} - \mathcal{F}_{\xi\xi})$$

$$+ \frac{1}{2}(1 + \epsilon\eta)(\mathcal{F}_\xi)^2 + \frac{\epsilon}{3}(\mathcal{F}_{\xi\xi})^2 + \frac{1}{2}\eta^2 \, d\xi \, d\tau \tag{11.6.4a}$$

$$= \epsilon \delta \int_0^T \int_0^L \mathcal{F}_\xi \mathcal{F}_\tau + \frac{1}{2}(\mathcal{F}_\xi)^3 - \frac{1}{6}(\mathcal{F}_{\xi\xi})^2 \, d\xi \, d\tau + O(\epsilon^2) \tag{11.6.4b}$$

$$= -\epsilon \int_0^T \int_0^L \left(2\mathcal{F}_{\xi\tau} + 3\mathcal{F}_\xi \mathcal{F}_{\xi\xi} + \frac{1}{3}\mathcal{F}_{\xi\xi\xi\xi} \right) \delta \mathcal{F} \, d\xi \, d\tau + O(\epsilon^2). \tag{11.6.4c}$$

Using that $\eta = \mathcal{F}_\xi$ at leading order in ϵ from (11.6.3) into the equation resulting from (11.6.4c), gives the following form of the KdV–equation

$$\eta_\tau + \frac{3}{2}\eta\eta_\xi + \frac{1}{6}\eta_{\xi\xi\xi} = 0. \tag{11.6.5}$$

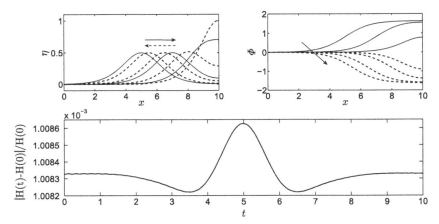

Figure 11.3.   Nonlinear numerical solutions of the Benney–Luke equations with $sech^2$–soliton initial conditions. The parameter values are $L_x = 10$, $L_y = 1$, $x_0 = L_x/2$, $c = 1.5$ and $\mu = \epsilon = 0.01$. Mid–channel cross–sections of $\eta(x, L_y/2, t)$ (top left panel) and $\Phi(x, L_y/2, t)$ (top right panel) at times $t = 0, 2, 4, 6, 7, 8.5, 9.5$ are given, as the solution is uniform in the y–direction, as well as the relative energy (bottom panel). 50×1 elements are used.

Employing a travelling wave solution ansatz $\eta = \eta(\chi = \xi - c\tau)$ with $c > 0$ into (11.6.5) then leads to the solutions (cf. Drazin & Johnson [15])

$$\eta(x,t) = \frac{c}{3} \operatorname{sech}^2\left(\frac{1}{2}\sqrt{\frac{c\epsilon}{\mu}}\left(x - x_0 - t - \frac{\epsilon c}{6}t\right)\right), \qquad (11.6.6a)$$

$$\Phi(x,t) = \frac{2}{3}\sqrt{\frac{c\mu}{\epsilon}}\left[\tanh\left(\frac{1}{2}\sqrt{\frac{c\epsilon}{\mu}}\left(x - x_0 - t - \frac{\epsilon c}{6}t\right)\right) + 1\right], \qquad (11.6.6b)$$

of the Korteweg–de–Vries (KdV) equation, which we used at $t = 0$ as initial conditions in the numerical computations. A simulation of a soliton reflecting against a solid wall using the nonlinear discrete Benney–Luke equations appears to be good, see Fig. 11.3 (solid lines correspond to the initial wave propagation phase, while dashed lines correspond to the after-reflection phase). The maximum wave amplitude during the reflection is twice the initial amplitude of the soliton, confirming the results of Power and Chwang [16]. The initial propagation before reflection matches the exact solution (11.6.6) well. In addition, the energy fluctuations are small and bounded[c].

---

[c] A tutorial on the discretisation of the Benney–Luke equations and a free *Python* code including the comparison against the $sech^2$–soliton solution can be found on the Firedrake site http://www.firedrakeproject.org/demos/benney_luke.py.html

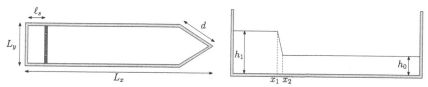

Figure 11.4. Sketch of the wave channel setup: top (left panel) and side views (right panel).

Soliton splash experiment

A soliton splash experiment with wave generation, propagation and reflection phenomena is used as a validation of the Benney–Luke model. The wave channel is sketched in Fig. 11.4. At one end of the channel, there is a lock with a sluice gate and at the other end, there is a linear symmetric V-shaped contraction. At time $t = 0$ the water is at rest with water level h_1 in the sluice compartment and $h_0 < h_1$ in the main channel. Several water levels $h_{0,1}$ have been considered but we focus on the event with smooth unbroken waves. Specific details about the soliton splash experiment, including the wavetank dimensions and location of sluice gate can be found in Table 11.2.

We model the sluice gate release in an approximate fashion by adding a horizontal gravitational component, effectively yielding an initial rest level $\eta_R(x,0)$. This free surface rest level coincides with a field line of a modified gravitational potential, such that (in dimensional units)

$$\eta_R(x,t) = (h_0 - H_0) + \begin{cases} h_1 - h_s(t) & \text{if } x < x_1 \\ (h_1 - h_s(t))\left(1 - \frac{x-x_1}{x_2-x_1}\right) & \text{if } x_1 \leq x \leq x_2 \\ 0 & \text{if } x > x_2, \end{cases} \quad (11.6.7a)$$

with

$$h_s(t) = \begin{cases} h_1 + (h_0 - h_1)\frac{(T_s-t)}{T_s} & \text{for } t < T_s \\ h_1 & \text{for } t \geq T_s, \end{cases} \quad (11.6.7b)$$

$0 < x_1 < x_2 < L_x$, a sluice gate release time $T_s > 0$, and initial water level difference $h_1 - h_0 > 0$ at $t = 0$. Initial conditions are therefore $\eta(x,y,0) = \eta_R(x,0)$ and $\phi(x,y,0) = 0$. Given the validation case, we took $x_1 = \ell_s$, $x_2 = x_1 + 0.1$ m (a sluice gate of thickness 10 cm) and $T_s \approx h_1/V_g$. The average still water level H_0 is calculated as

$$H_0 = \frac{x_1 h_1 + \frac{1}{2}(x_2 - x_1)(h_1 + h_0) + (L_x - x_2)h_0}{L_x} \approx 0.46 \text{ m.} \quad (11.6.8)$$

In the numerical simulations, we took a wave amplitude of $a_0 = 0.25$ m and wavelength of $\ell_0 = 2.25$ m, both based on observations from the soliton splash experiment. These length scales were used to obtain the dimensionless parameters μ and ϵ that appear in the Benney–Luke equations, resulting in

Table 11.2. *Details about the soliton splash experiment,*
including wavetank dimensions.

Wavetank length	$L_x = 43.63 \pm 0.1$ m
Wavetank width	$L_y = 2$ m
Wavetank height	$L_z = 1.2$ m
Contraction length	$d = 2.7$ m
Location of sluice gate	$\ell_s = 2.63$ m
Rest water level (hight)	$h_1 = 0.9$ m
Rest water level (low)	$h_0 = 0.43$ m
Sluice gate release speed	$V_g \approx 2.5$ m/s
Sluice gate removal time	$T_s = h_1/V_g \approx 0.36$ s

the values $\mu = 0.04$ and $\epsilon = 0.55$. A comparison between simulations of
the Benney–Luke model and actual snapshots of the event reveal a striking
similarity in terms of timing, wave structure and amplification heights in
the contraction, see Fig. 11.5. The maximum wave amplitude attains the
value of 1.39 m, which is about 4 times the initial amplitude of 0.34 m.
Strictly speaking, the in situ value of ϵ is no longer small; nonetheless, the
Benney–Luke model captures the wave uprush reasonably well.

The energy as function of time in the soliton splash simulations of the
nonlinear and linear Benney–Luke system is displayed in Fig. 11.6. In both
cases, the energy remains bounded and shows no drift. Additional figures from
the soliton splash simulation can be found in Appendix B.

11.7 Conclusion

Mathematical modelling of water waves in wave tanks with a sluice gate has
been illustrated by variational methods: by deriving a modified reduced model
with a time-dependent gravitational potential mimicking a removable "sluice
gate" and by discretising this model in space and time using finite element
methods, both within a variational framework. As validation, we simulated
a soliton splash in a wave channel with a contraction. Future work will
explore these methods for wave-energy devices and ships in modest to heavy
seas.

Acknowledgments

This work was mostly undertaken during the program "Water Waves" at
the Isaac Newton Institute of Mathematical Sciences (Cambridge, 2014). We
thank the organisers Tom Bridges, Mark Groves, Paul Milewski and David
Nicholls, as well as all participants for creating a stimulating environment.

Figure 11.5. Snapshots of a soliton splash event. Left column: observations. Right column: numerical solution with $\mu = 0.04$, $\epsilon = 0.55$. When taking $h_0 = 0.43$ m instead of $h_0 = 0.41$ m with $h_1 = 0.9$ m the same, no wave breaking occurs [17]. Photo courtesy: http://www.woutzweers.nl. Photo times $t = 8, 14, 15, 15 \pm 0.5$ s (relative) and simulation times $t = 8.16, 14.44, 15.08, 15.27$ s (scaled $7.56, 13.38, 13.97, 14.15$). Values displayed are in meters. 400×20 elements are used, stretched into the V-shaped end and regular without.

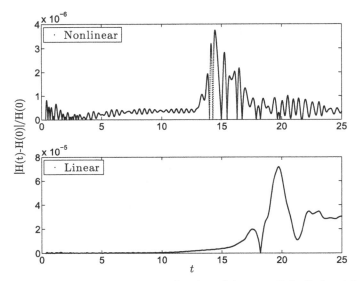

Figure 11.6. Normalised energy as function of time $t > T_s$ is displayed for simulations of the nonlinear (top panel) and linear (bottom panel) Benney–Luke equations. Even from these graphs it is discernible that the main signal in the nonlinear case arrives earlier at the contraction at (dimensionless) $t \approx 14$, whereas in the linear case the wave enters the contraction at around $t \approx 17$. The former coincides better with the observations. Note that the energy $H(0) = \lim_{\epsilon_0 \downarrow 0} H(T_s + \epsilon_0)$, see (11.4.19).

Special regards go to the Firedrake project team, in particular Colin Cotter, David Ham and Lawrence Mitchell, for assistance in using Firedrake. We acknowledge funding from EPSRC grant no. EP/L025388/1 with a link to the Dutch Technology Foundation STW for the project "FastFEM: behaviour of fast ships in waves."

References

[1] J.C. Luke 1967: A variational principle for a fluid with a free surface, *J. Fluid Mech.* **27**, 395–397.

[2] J.W. Miles 1977: On Hamilton's principle for surface waves. *J. Fluid Mech.* **83**, 153–158.

[3] C.J. Cotter, O. Bokhove 2010: Water wave model with accurate dispersion and vertical vorticity. *J. Eng. Maths.* **67**, 33–54.

[4] O. Bokhove, A.J. van der Horn, D. van der Meer, A.R. Thornton, W. Zweers 2014: On wave-driven "shingle" beach dynamics in a table-top Hele-Shaw cell. *Proc. Int. Conf. Coastal Eng.*, Seoul, 2014.

[5] A.R. Thornton, A.J. van der Horn, E. Gagarina, D. van der Meer, W. Zweers, O. Bokhove 2014: Hele-Shaw beach creation by breaking waves. *Env. Fluid Dyn.* **14**, 1123–1145.

[6] E. Gagarina, V.R. Ambati, J.J.W. van der Vegt, O. Bokhove 2014: Variational space-time discontinuous Galerkin finite element method for nonlinear free surface waves. *J. Comp. Phys.* **275**, 459–483.

[7] R.L. Pego, J.R. Quintero 1999: Two-dimensional solitary waves for a Benney–Luke equation. *Physica D* **132**, 476–496.

[8] D.J. Benney, J.C. Luke 1964: On the interactions of permanent waves of finite amplitude. *J. Math. Phys.* **43**, 309–313.

[9] E. Gagarina, V.R. Ambati, S. Nurijanyan, J.J.W. van der Vegt, O. Bokhove 2016: On variational and symplectic time integrators for Hamiltonian systems. Accepted by *J. Comp. Phys.*

[10] H. Goldstein 1980: *Classical Mechanics*, Addison-Wesley, 672 pp.

[11] P.J. Olver 2000: *Application of Lie Groups to Differential Equations*. Springer, 513 pp.

[12] E. Hairer, C. Lubich, G. Wanner 2006: *Geometric Numerical Integration*. Springer.

[13] G.R. Markall, F. Rathgeber, L. Mitchell, N. Loriant, C. Bertolli, D.A. Ham, P.H.J. Kelly 2013: Performance-Portable Finite Element Assembly Using PyOP2 and FEniCS. In Kunkel; Martin, J.; Ludwig, T.; Meuer; and Werner, H., editor(s), 28th International Supercomputing Conference, ISC, Proceedings, volume 7905, of Lecture Notes in Computer Science, pp. 279–289. Springer.

[14] A.T.T. McRae, G.-T. Bercea, L. Mitchell, D.A. Ham, C.J. Cotter 2015: Automated generation and symbolic manipulation of tensor product finite elements. Subm. *ACM TOMS*. http://arxiv.org/abs/1411.2940

[15] P.G. Drazin, R.S. Johnson 1989: *Solitons: An Introduction*. Cambridge University Press. 226 pp.

[16] H. Power, A.T. Chwang 1984: On reflection of a planar wall solitary wave at a vertical wall. *Wave Motion* **6**, 183–195.

[17] O. Bokhove, E. Gagarina, W. Zweers, A.R. Thornton 2011: Bore Soliton Splash -van spektakel tot oceaangolf? *Ned. Tijdschrift voor Natuurkunde.* **77/12**, 446–450. In Dutch.

Appendix A: Numerical Implementation

A.1 Störmer–Verlet and Symmetric Third-order Time Schemes for the Benney–Luke System

Employing (11.5.13), a second-order Störmer–Verlet time discretisation of (11.4.18) becomes

$$M_{kl}q_l^n = \frac{2}{3}S_{kl}\phi_l^n, \tag{A.1a}$$

$$\left(M_{kl} + \frac{\mu}{2}S_{kl}\right)\eta_l^{n+1/2} = \left(M_{kl} + \frac{\mu}{2}S_{kl}\right)\eta_l^n$$
$$+ \frac{1}{2}\Delta t\left(S_{kl}\phi_l^n + \epsilon\eta_m^{n+1/2}\phi_l^n \iint_{\Omega_h} \varphi_m\nabla\varphi_k\cdot\nabla\varphi_l\,\mathrm{d}x\,\mathrm{d}y + \mu S_{kl}q_l^n\right), \tag{A.1b}$$

$$\left(M_{kl} + \frac{\mu}{2}S_{kl}\right)\phi_k^{n+1} = \left(M_{kl} + \frac{\mu}{2}S_{kl}\right)\phi_k^n - \Delta t M_{kl}\eta_k^{n+1/2} + \Delta t R_l^{n+1/2}$$

$$-\frac{1}{2}\Delta t \frac{\epsilon}{2}\left(\phi_k^n\phi_m^n + \phi_k^{n+1}\phi_m^{n+1}\right)\iint_{\Omega_h}\varphi_l\nabla\varphi_k\cdot\nabla\varphi_m\,\mathrm{d}x\,\mathrm{d}y, \tag{A.1c}$$

$$M_{kl}q_l^{n+1} = \frac{2}{3}S_{kl}\phi_l^{n+1}, \tag{A.1d}$$

$$\left(M_{kl} + \frac{\mu}{2}S_{kl}\right)\eta_l^{n+1} = \left(M_{kl} + \frac{\mu}{2}S_{kl}\right)\eta_l^{n+1/2}$$

$$+\frac{1}{2}\Delta t\left(S_{kl}\phi_l^{n+1} + \epsilon\eta_m^{n+1/2}\phi_l^{n+1}\iint_{\Omega_h}\varphi_m\nabla\varphi_k\cdot\nabla\varphi_l\,\mathrm{d}x\,\mathrm{d}y + \mu S_{kl}q_l^{n+1}\right). \tag{A.1e}$$

Employing (11.5.20), a third-order symmetric time discretisation of (11.4.18) becomes

$$M_{kl}q_l^{n+1/2} = \frac{2}{3}S_{kl}\phi_l^{n+1/2}, \tag{A.2a}$$

$$M_{kl}q_l^{n,+} = \frac{2}{3}S_{kl}\phi_l^{n,+}, \tag{A.2b}$$

$$\left(M_{kl} + \frac{\mu}{2}S_{kl}\right)\eta_l^{n+1/2} = \frac{1}{4}\left(M_{kl} + \frac{\mu}{2}S_{kl}\right)\eta_l^{n,+} + \frac{3}{4}\left(M_{kl} + \frac{\mu}{2}S_{kl}\right)\eta_l^{n,-} +$$

$$+\frac{1}{4}\Delta t\left(S_{kl}\phi_l^{n,+} + S_{kl}\phi_l^{n+1/2}\right.$$

$$+\epsilon\left(\eta_m^{n,+}\phi_l^{n,+} + \eta_m^{n+1/2}\phi_l^{n+1/2}\right)\iint_{\Omega_h}\varphi_m\nabla\varphi_k\cdot\nabla\varphi_l\,\mathrm{d}x\,\mathrm{d}y + \mu S_{kl}\left(q_l^{n,+} + q_l^{n+1/2}\right)\Big), \tag{A.2c}$$

$$\left(M_{kl} + \frac{\mu}{2}S_{kl}\right)\phi_k^{n+1/2} = \frac{1}{4}\left(M_{kl} + \frac{\mu}{2}S_{kl}\right)\phi_k^{n,+} + \frac{3}{4}\left(M_{kl} + \frac{\mu}{2}S_{kl}\right)\phi_k^{n,-}$$

$$+\frac{1}{4}\Delta t\left(-\left(M_{kl}\eta_k^{n,+} + M_{kl}\eta_k^{n+1/2}\right) + \left(R_l^{n,+} + R_l^{n+1/2}\right)\right.$$

$$-\frac{\epsilon}{2}\left(\phi_k^{n,+}\phi_m^{n,+} + \phi_k^{n+1/2}\phi_m^{n+1/2}\right)\iint_{\Omega_h}\varphi_l\nabla\varphi_k\cdot\nabla\varphi_m\,\mathrm{d}x\,\mathrm{d}y\Big), \tag{A.2d}$$

$$\left(M_{kl} + \frac{\mu}{2}S_{kl}\right)\eta_l^{n+1,-} = \left(M_{kl} + \frac{\mu}{2}S_{kl}\right)\eta_l^{n,+}$$

$$+\Delta t\left(S_{kl}\phi_l^{n+1/2} + \epsilon\eta_m^{n+1/2}\phi_l^{n+1/2}\iint_{\Omega_h}\varphi_m\nabla\varphi_k\cdot\nabla\varphi_l\,\mathrm{d}x\,\mathrm{d}y + \mu S_{kl}q_l^{n+1/2}\right), \tag{A.2e}$$

$$\left(M_{kl} + \frac{\mu}{2}S_{kl}\right)\phi_k^{n+1,-} = \left(M_{kl} + \frac{\mu}{2}S_{kl}\right)\phi_k^{n,+} + \Delta t\left(-M_{kl}\eta_k^{n+1/2} + R_l^{n+1/2}\right.$$

$$-\frac{\epsilon}{2}\phi_k^{n+1/2}\phi_m^{n+1/2}\iint_{\Omega_h}\varphi_l\nabla\varphi_k\cdot\nabla\varphi_m\,\mathrm{d}x\,\mathrm{d}y\Big), \tag{A.2f}$$

$$M_{kl}q_l^{n+1,-} = \frac{2}{3}S_{kl}\phi_l^{n+1,-}, \tag{A.2g}$$

$$\left(M_{kl} + \frac{\mu}{2}S_{kl}\right)\eta_l^{n+1,+} = \frac{4}{3}\left(M_{kl} + \frac{\mu}{2}S_{kl}\right)\eta_l^{n+1/2} - \frac{1}{3}\left(M_{kl} + \frac{\mu}{2}S_{kl}\right)\eta_l^{n,+}$$
$$+ \frac{1}{3}\Delta t\left(S_{kl}\phi_l^{n+1,-} + \epsilon\eta_m^{n+1,-}\phi_l^{n+1,-}\iint_{\Omega_h}\varphi_m\nabla\varphi_k\cdot\nabla\varphi_l\,dx\,dy + \mu S_{kl}q_l^{n+1,-}\right), \tag{A.2h}$$

$$\left(M_{kl} + \frac{\mu}{2}S_{kl}\right)\phi_k^{n+1,+} = -\frac{1}{3}\left(M_{kl} + \frac{\mu}{2}S_{kl}\right)\phi_k^{n,+} + \frac{4}{3}\left(M_{kl} + \frac{\mu}{2}S_{kl}\right)\phi_k^{n+1/2}$$
$$+ \frac{1}{3}\Delta t\left(-M_{kl}\eta_k^{n+1,-} + R_l^{n+1,-} - \frac{\epsilon}{2}\phi_k^{n+1,-}\phi_m^{n+1,-}\iint_{\Omega_h}\varphi_l\nabla\varphi_k\cdot\nabla\varphi_m\,dx\,dy\right). \tag{A.2i}$$

A.2 Implementation in Finite Element Package Firedrake

The finite element package Firedrake starts from the weak formulation of the variations in (11.4.12) with the variables replaced by their respective finite element expansions (11.4.14) and (11.4.15), such that $\eta \to \eta_h$, $\Phi \to \Phi_h$, $q \to q_h$. With these substitutions, the starting point follows therefore from (11.4.12) as

$$0 = \delta\int_0^T\iint_{\Omega_h}\eta\partial_t\Phi + \frac{\mu}{2}\nabla\eta\cdot\partial_t\nabla\Phi + \frac{1}{2}(1+\epsilon\eta_h)|\nabla\Phi|^2 + \frac{1}{2}\eta^2 - \eta_R\eta$$
$$+ \mu\left(\nabla q\cdot\nabla\Phi - \frac{3}{4}q^2\right)dx\,dy\,dt \tag{A.3a}$$
$$= \int_0^T\iint_{\Omega_h}\left(\delta\eta_h\partial_t\Phi_h + \frac{\mu}{2}\nabla\delta\eta_h\cdot\partial_t\nabla\Phi_h + \delta\eta_h\left(\eta_h - \eta_R + \frac{\epsilon}{2}|\nabla\Phi_h|^2\right)\right)$$
$$-\left(\delta\Phi_h\partial_t\eta_h + \frac{\mu}{2}\nabla\delta\Phi_h\cdot\partial_t\nabla\eta_h - (1+\epsilon\eta)\nabla\delta\Phi_h\cdot\nabla\Phi_h\right.$$
$$\left.- \mu\nabla\delta\Phi_h\cdot\nabla q_h\right) + \mu\left(\nabla\delta q_h\cdot\nabla\Phi_h - \delta q_h\frac{3}{2}q_h\right)dx\,dy\,dt, \tag{A.3b}$$

in which we do not perform integration by parts, cf. our previous finite element approach. Given that $\delta\eta_h, \delta\Phi_h$ and δq_h are arbitrary variations, three weak formulations arise as a consequence by taking $\delta\eta_h = \varphi_l$, $\delta\Phi_h = 0$ and $\delta q_h = 0$, and so forth.

In mild contrast to the previous derivation in which space and time were discretised consecutively, the weak formulation implemented in Firedrake needs to be discretised in time. For the Störmer–Verlet time discretisation scheme (11.5.10) this yields the following sequence of weak formulations

$$
0 = \iint_{\Omega_h} \left(\delta\eta_h \frac{\Phi_h^{n+1/2} - \Phi_h^n}{\Delta t/2} + \frac{\mu}{2} \nabla\delta\eta_h \cdot \nabla \frac{\Phi_h^{n+1/2} - \Phi_h^n}{\Delta t/2} \right.
$$
$$
\left. + \delta\eta_h \left(\eta_h^n - \eta_R^{n+1/2} + \frac{\epsilon}{2} \left| \nabla\Phi_h^{n+1/2} \right|^2 \right) \right) \mathrm{d}x\,\mathrm{d}y,
\tag{A.4a}
$$

$$
0 = \iint_{\Omega_h} \mu \left(\nabla\delta q_h \cdot \nabla\Phi_h^{n+1/2} - \delta q_h \frac{3}{2} q_h^{n+1/2} \right) \mathrm{d}x\,\mathrm{d}y,
\tag{A.4b}
$$

$$
0 = \iint_{\Omega_h} \left(\delta\Phi_h \frac{\eta_h^{n+1} - \eta_h^n}{\Delta t} + \frac{\mu}{2} \nabla\delta\Phi_h \cdot \nabla \frac{\eta_h^{n+1} - \eta_h^n}{\Delta t} \right.
$$
$$
- \frac{1}{2} \left((1 + \epsilon\eta_h^{n+1}) + (1 + \epsilon\eta_h^n) \right) \nabla\delta\Phi_h \cdot \nabla\Phi_h^{n+1/2}
$$
$$
\left. - \mu\nabla\delta\Phi_h \cdot \nabla q_h^{n+1/2} \right) \mathrm{d}x\,\mathrm{d}y,
\tag{A.4c}
$$

$$
0 = \iint_{\Omega_h} \left(\delta\eta_h \frac{\Phi_h^{n+1} - \Phi_h^{n+1/2}}{\Delta t/2} + \frac{\mu}{2} \nabla\delta\eta_h \cdot \nabla \frac{\Phi_h^{n+1} - \Phi_h^{n+1/2}}{\Delta t/2} \right.
$$
$$
\left. + \delta\eta_h \left(\eta_h^{n+1} - \eta_R^{n+1/2} + \frac{\epsilon}{2} \left| \nabla\Phi_h^{n+1/2} \right|^2 \right) \right) \mathrm{d}x\,\mathrm{d}y,
\tag{A.4d}
$$

$$
0 = \iint_{\Omega_h} \mu \left(\nabla\delta q_h \cdot \nabla\Phi_h^{n+1} - \delta q_h \frac{3}{2} q_h^{n+1} \right) \mathrm{d}x\,\mathrm{d}y.
\tag{A.4e}
$$

These weak formulations are consecutively implemented as such in Firedrake, in the order provided. Note that the test functions $\delta\eta_h$, $\delta\Phi_h$ and δq_h are space dependent test functions only.

For the third-order time stepping scheme (11.5.20), there will be three coefficients per time slab, i.e., we have $\Phi_h^{n,\pm}, \Phi_h^{n+1/2}$ and $\eta_h^{n,\pm}, \eta_h^{n+1/2}$. The initial data need to be initialised continuously, namely $\Phi_h^{0,+} = \Phi_h^{0,-}$ and $\eta_h^{0,+} = \eta_h^{0,-}$. The first three weak formulations become

$$
0 = \iint_{\Omega_h} \left(\delta\eta_h \frac{(4\Phi_h^{n+1/2} - \Phi_h^{n,+} - 3\Phi_h^{n,-})}{\Delta t} \right.
$$
$$
+ \frac{\mu}{2} \nabla\delta\eta_h \cdot \nabla \frac{(4\Phi_h^{n+1/2} - \Phi_h^{n,+} - 3\Phi_h^{n,-})}{\Delta t}
$$
$$
+ \delta\eta_h \left(\eta_h^{n,+} - \eta_R^{n,+} + \frac{\epsilon}{2} \left| \nabla\Phi_h^{n,+} \right|^2 \right)
$$
$$
\left. + \eta_h^{n+1/2} - \eta_R^{n+1/2} + \frac{\epsilon}{2} \left| \nabla\Phi_h^{n+1/2} \right|^2 \right) \right) \mathrm{d}x\,\mathrm{d}y,
\tag{A.5a}
$$

$$0 = \iint_{\Omega_h} \mu \left(\nabla \delta q_h \cdot \nabla \Phi_h^{n+1/2} - \delta q_h \frac{3}{2} q_h^{n+1/2} \right) dx\, dy, \tag{A.5b}$$

$$0 = \iint_{\Omega_h} \left(\delta \Phi_h \frac{(4\eta_h^{n+1/2} - \eta_h^{n,+} - 3\eta_h^{n,-})}{\Delta t} \right.$$

$$+ \frac{\mu}{2} \nabla \delta \Phi_h \cdot \nabla \frac{(4\eta_h^{n+1/2} - \eta_h^{n,+} - 3\eta_h^{n,-})}{\Delta t}$$

$$- \left(1 + \epsilon \eta_h^{n,+} \right) \nabla \delta \Phi_h \cdot \nabla \Phi_h^{n,+} - \mu \nabla \delta \Phi_h \cdot \nabla q_h^{n,+}$$

$$\left. - \left(1 + \epsilon \eta_h^{n+1/2} \right) \nabla \delta \Phi_h \cdot \nabla \Phi_h^{n+1/2} - \mu \nabla \delta \Phi_h \cdot \nabla q_h^{n+1/2} \right) dx\, dy, \tag{A.5c}$$

which need to be solved in unison to find $\eta_h^{n+/12}, \Phi_h^{n+1/2}$ and $q_h^{n+1/2}$.

The remaining weak formulations are explicit once these half-time updates have been solved, for $\Phi_h^{n+1,-}, \eta_h^{n+1,-}, q_h^{n+1,-}$, as follows

$$0 = \iint_{\Omega_h} \left(\delta \eta_h \frac{(\Phi_h^{n+1,-} - \Phi_h^{n,+})}{\Delta t} + \frac{\mu}{2} \nabla \delta \eta_h \cdot \nabla \frac{(\Phi_h^{n+1,-} - \Phi_h^{n,+})}{\Delta t} \right.$$

$$\left. + \delta \eta_h \left(\eta_h^{n+1/2} - \eta_R^{n+1/2} + \frac{\epsilon}{2} |\nabla \Phi_h^{n+1/2}|^2 \right) \right) dx\, dy, \tag{A.5d}$$

$$0 = \iint_{\Omega_h} \mu \left(\nabla \delta q_h \cdot \nabla \Phi_h^{n+1,-} - \delta q_h \frac{3}{2} q_h^{n+1,-} \right) dx\, dy, \tag{A.5e}$$

$$0 = \iint_{\Omega_h} \left(\delta \Phi_h \frac{(\eta_h^{n+1,-} - \eta_h^{n,+})}{\Delta t} + \frac{\mu}{2} \nabla \delta \Phi_h \cdot \nabla \frac{(\eta_h^{n+1,-} - \eta_h^{n,+})}{\Delta t} \right.$$

$$\left. - \left(1 + \epsilon \eta_h^{n+1/2} \right) \nabla \delta \Phi_h \cdot \nabla \Phi_h^{n+1/2} - \mu \nabla \delta \Phi_h \cdot \nabla q_h^{n+1/2} \right) dx\, dy, \tag{A.5f}$$

and for $\Phi_h^{n+1,+}, \eta_h^{n+1,+}, q_h^{n+1,+}$, as follows

$$0 = \iint_{\Omega_h} \left(\delta \eta_h \frac{(3\Phi_h^{n+1,+} - 4\Phi_h^{n+1/2} + \Phi_h^{n,+})}{\Delta t} \right.$$

$$+ \frac{\mu}{2} \nabla \delta \eta_h \cdot \nabla \frac{(3\Phi_h^{n+1,+} - 4\Phi_h^{n+1/2} + \Phi_h^{n,+})}{\Delta t}$$

$$\left. + \delta \eta_h \left(\eta_h^{n+1,-} - \eta_R^{n+1,-} + \frac{\epsilon}{2} |\nabla \Phi_h^{n+1,-}|^2 \right) \right) dx\, dy, \tag{A.5g}$$

$$0 = \iint_{\Omega_h} \mu \left(\nabla \delta q_h \cdot \nabla \Phi_h^{n+1,+} - \delta q_h \frac{3}{2} q_h^{n+1,+} \right) dx\, dy, \tag{A.5h}$$

$$0 = \iint_{\Omega_h} \left(\delta \Phi_h \frac{(3\eta_h^{n+1,+} - 4\eta_h^{n+1/2} + \eta_h^{n,+})}{\Delta t} \right.$$

$$+ \frac{\mu}{2} \nabla \delta \Phi_h \cdot \nabla \frac{(3\eta_h^{n+1,+} - 4\eta_h^{n+1/2} + \eta_h^{n,+})}{\Delta t}$$

$$- \left(1 + \epsilon \eta_h^{n+1,-}\right) \nabla \delta \Phi_h \cdot \nabla \Phi_h^{n+1,-} - \mu \nabla \delta \Phi_h \cdot \nabla q_h^{n+1,-}\right) \mathrm{d}x \, \mathrm{d}y. \quad \text{(A.5i)}$$

Appendix B: Soliton Splash Simulations

A comparison between simulations of the linear Benney–Luke system (with $\epsilon = 0$) in Fig. B.1 and the nonlinear Benney–Luke system (with $\epsilon = 0.55$) in Fig. 11.5, shows that the (solitary) waves in the nonlinear simulations have more coherence, as expected, resulting in a good correspondence with the observed soliton splash event. The linear simulations produce a wave that arrives in the channel contraction later and exhibits a smaller amplitude amplification than the nonlinear one.

Mid channel and width averaged profiles of the nonlinear numerical solution for η can be found in Fig. B.2. Double resolution runs with 800×40 elements confirm that our simulations have converged.

Figure B.1. Snapshots of the soliton splash event. Left column: observations. Right column: numerical, linear solution. Photo courtesy: http://www.woutzweers.nl. Photo times: $t = 8, 14, 15, 15 \pm 0.5$ s (relative) and simulation times $t = 11.06, 19.49, 20.58, 20.82$ s (scaled $t = 11.06, 18.06, 19.07, 19.29$). Values displayed are in meters. 400×20 elements are used, stretched into the V-shaped end and regular without.

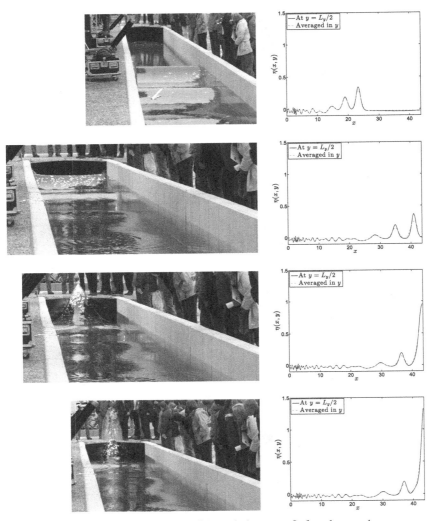

Figure B.2. Snapshots of the soliton splash event. Left column: observations. Right column: numerical solution with $\mu = 0.04$, $\epsilon = 0.55$, mid–line and width averaged profiles of η. Photo courtesy: http://www.woutzweers. nl. Photo times: $t = 8, 14, 15, 15 \pm 0.5$ s (relative) and simulation times $t = 8.16, 14.44, 15.08, 15.27$ s (scaled $7.56, 13.38, 13.97, 14.15$). Values displayed are in meters. 400×20 elements have been used, stretched into the V-shaped end and regular without.

12

Symmetry, Modulation, and Nonlinear Waves

Thomas J. Bridges

Abstract

These lecture notes provide an introduction to the theory of "modulation" and its role in the derivation of model equations, such as the KdV equation, Boussinesq equation, KP equation, and Whitham modulation equations, and their role in the theory of water waves. The classical theory of modulation, such as Whitham modulation theory, will be discussed, and a new approach will be introduced, based on modulation of background flow. Methodology that is key to the theory is symmetry and conservation laws, relative equilibria, Hamiltonian and Lagrangian structures, multiple scale perturbation theory, and elementary differential geometry. By basing the theory on modulation of relative equilibria, new settings are discovered for the emergence of KdV and other modulation equations. For example, it is shown that the KdV equation can be a valid model for deep water as well as shallow water. The lecture notes are introductory, and no prior knowledge is assumed.

12.1 Introduction

Modulation is one of the most widely used concepts in the theory of nonlinear waves. In linear theory modulation is normally the process of varying the envelope of a signal. In electronics this concept is expanded further to include digital modulation, analog modulation, pulse modulation, frequency modulation, and so on. In the theory of nonlinear waves it is used to describe "modulation equations," which typically are nonlinear equations governing the envelope of a wave, although the term is much more widely used now, with any equation on a slow space and time scale called a modulation equation. The term modulation in these notes is closest in spirit to Whitham modulation theory [1]. The idea is that given a basic state, dependent on a phase and a parameter or parameters, the phase and parameters are treated as slowly-varying functions of space and time, and governing equations are derived for these slowly-varying functions.

An abstraction of the idea of a basic state dependent on a phase is a *relative equilibrium*. A central theme of these notes is modulation of relative equilibria (RE). A definition of RE is given in §12.2. The context throughout will be conservative (Lagrangian, Hamiltonian, multisymplectic).

Given a basic state, dependent on a phase and a parameter, represented by $\widehat{Z}(\theta,k)$, with $\theta = kx + \theta_0$, satisfying a given partial differential equation (PDE), a classical multiple scales perturbation would be of the form

$$Z(x,t) = \widehat{Z}(\theta,k) + \varepsilon^d \widetilde{W}(\theta,X,T,\varepsilon), \qquad (12.1.1)$$

with slow space and time scales $T = \varepsilon^\alpha t, X = \varepsilon^\beta x$. With $d = \alpha = \beta = 1$, this is the approach that is used in the justification of Whitham modulation theory [2].

The approach taken here is different. In addition to the slow space and time scales, the phase and parameter are also modulated

$$Z(x,t) = \widehat{Z}(\theta + \varepsilon^a\phi, k + \varepsilon^b q) + \varepsilon^d W(\theta,X,T,\varepsilon), \qquad (12.1.2)$$

with slow space and time scales $T = \varepsilon^\alpha t, X = \varepsilon^\beta x$. A key feature of this approach is that the solution is shifted to a different vantage point: instead of the basic state being at θ, it is viewed at $\theta + \varepsilon^a\phi$. An RE has the property that $\widehat{Z}(\theta + s,k)$ is a solution whenever $\widehat{Z}(\theta,k)$ is a solution for any s in the family of RE, so the ansatz $\theta \mapsto \theta + \varepsilon^a\phi(X,T,\varepsilon)$ is taking a modulated path through the family of RE.

In principle the two expressions (12.1.1) and (12.1.2) are equivalent: expand the first term in (12.1.2) in a Taylor series in ε and absorb into W to form $\widetilde{W}$. However, the separation of perturbation of θ, k, and W in (12.1.2) gives more information that feeds into the equations for $\phi(X,T,\varepsilon)$ and $q(X,T,\varepsilon)$. A new derivation of the dispersionless Whitham equations can be obtained from (12.1.2) by taking $a = 0$, $b = 1$, $d = 2$, and $\alpha = \beta = 1$.

The lecture notes start with the simplest possible case: modulation of RE of conservative ordinary differential equations (ODEs). The theory is then extended to conservative partial differential equations (PDEs), deriving equations like the Korteweg-de Vries (KdV) equation and the Kadomtsev-Petviashvili (KP) equation. In §12.7 the theory is compared with Whitham modulation theory. The main examples are a reduction of a Boussinesq equation from dispersive shallow water hydrodynamics to KdV, and the reduction of defocussing NLS, which is a model for deep water wave, to KdV. In §12.9 various extensions are the theory are discussed.

12.2 Relative Equilibria of Hamiltonian ODEs

What are RE? The simplest solutions of Hamiltonian ODEs are equilibria – these are solutions for which the gradient of the Hamiltonian function vanishes.

The next simplest class of solutions is RE. RE are associated with symmetry and are steady relative to a frame moving along a group orbit. In this section RE are introduced in the context of Hamiltonian ODEs with a one-parameter symmetry group. Further detail on the theory of RE can be found in Chapter 4 of MARSDEN [3].

Consider a Hamiltonian ODE on $\mathbb{R}^4$:

$$\mathbf{J}Z_x = \nabla H(Z), \quad Z \in \mathbb{R}^4, \tag{12.2.1}$$

where the symplectic operator is in standard form

$$\mathbf{J} = \begin{pmatrix} \mathbf{0} & -\mathbf{I} \\ \mathbf{I} & \mathbf{0} \end{pmatrix}, \tag{12.2.2}$$

where $\mathbf{I}$ is the identity on $\mathbb{R}^2$, and the Hamiltonian function $H(Z)$ is a given smooth function. A four-dimensional phase space is the lowest dimension where the modulation is nontrivial. Evolution is considered in the $x-$direction as that will dovetail with the space-time nonlinear wave modulation later.

The system (12.2.1) is assumed to be equivariant with respect to a one-parameter Lie group with action $\mathbf{G}_\theta$:

$$\mathbf{G}_\theta \mathbf{J} = \mathbf{J}\mathbf{G}_\theta \quad \text{and} \quad H(\mathbf{G}_\theta Z) = H(Z).$$

The two most common cases are where $\mathbf{G}_\theta$ is affine

$$\mathbf{G}_\theta Z = Z + \theta\eta, \quad \forall \theta \in \mathbb{R}, \tag{12.2.3}$$

for some fixed vector $\eta \in \mathbb{R}^4$, and where $\mathbf{G}_\theta$ is an orthogonal action: $\mathbf{G}_\theta^{-1} = \mathbf{G}_\theta^T$, the principal example being

$$\mathbf{G}_\theta = \mathbf{R}_\theta \oplus \mathbf{R}_\theta, \quad \mathbf{R}_\theta = \begin{bmatrix} \cos\theta & -\sin\theta \\ \sin\theta & \cos\theta \end{bmatrix}, \quad \forall \theta \in S^1. \tag{12.2.4}$$

A third case of interest is periodic solutions of (12.2.1); solutions of the form $\widehat{Z}(\theta)$, with $\theta = kx + \theta_0$, which are $2\pi-$periodic in θ.

A RE associated with a one-parameter symmetry is a solution of the form

$$Z(x) := \widehat{Z}(\theta, k) = \mathbf{G}_\theta Z_0(k), \quad \theta = kx + \theta_0. \tag{12.2.5}$$

The most important property of an RE is that $\dot\theta = k$ and k is a constant. Assume throughout that $k \neq 0$, as $k = 0$ reduces the RE to the trivial case of an equilibrium. RE travel at constant speed along the group and are equilibria in a frame of reference following the group action.

The generator of the group is defined by

$$\mathfrak{g}(Z) = \frac{\partial}{\partial\theta}\mathbf{G}_\theta Z\bigg|_{\theta=0}, \quad Z \in \mathbb{R}^4. \tag{12.2.6}$$

For the three cases above, the generators are

$$\text{affine case} \quad \mathfrak{g}(\widehat{Z}) = \eta$$

$$\text{orthogonal case} \quad \mathfrak{g}(\widehat{Z}) = (\mathbf{J}_2 \oplus \mathbf{J}_2)Z_0(k)$$

$$\text{periodic orbit} \quad \mathfrak{g}(\widehat{Z}) = \widehat{Z}_\theta,$$

where where $\mathbf{J}_2$ is just a 2×2 version of (12.2.2).

By symplectic Noether's Theorem, there exists a functional $B : \mathbb{R}^4 \to \mathbb{R}$, satisfying

$$\mathbf{J}\mathfrak{g}(Z) = \nabla B(Z), \qquad (12.2.7)$$

where the gradient is defined with respect to the standard inner product on $\mathbb{R}^4$. Let $\mathscr{B}(k) = B(\widehat{Z})$, the functional B evaluated on a RE for each of the above three cases is

$$\text{affine case} \quad \mathscr{B}(k) = \langle \mathbf{J}\eta, Z_0(k) \rangle$$

$$\text{orthogonal case} \quad \mathscr{B}(k) = \tfrac{1}{2}\langle (\mathbf{J}_2 \otimes \mathbf{J}_2)Z_0(k), Z_0(k) \rangle$$

$$\text{periodic orbit} \quad \mathscr{B}(k) = \tfrac{1}{2}\langle\langle \mathbf{J}\widehat{Z}_\theta, \widehat{Z} \rangle\rangle,$$

using $\mathbf{J}(\mathbf{J}_2 \oplus \mathbf{J}_2) = \mathbf{J}_2 \otimes \mathbf{J}_2$. The inner product $\langle\langle \cdot, \cdot \rangle\rangle$ includes averaging over θ

$$\langle\langle U, V \rangle\rangle := \frac{1}{2\pi} \int_0^{2\pi} \langle U, V \rangle \, d\theta. \qquad (12.2.8)$$

The inner product with averaging is only needed for the case when $\widehat{Z}(\theta, k)$ is a periodic solution. The averaged inner product will be used throughout with the understanding that $\langle\langle \cdot, \cdot \rangle\rangle$ reduces to $\langle \cdot, \cdot \rangle$ for the cases of RE associated with the affine or orthogonal symmetry groups.

Substitution of (12.2.5) into (12.2.1) and use of (12.2.7) shows that $Z_0(k)$ satisfies

$$\nabla H(Z_0) - k\nabla B(Z_0) = 0, \qquad (12.2.9)$$

or $\nabla H(\widehat{Z}) - k\mathbf{J}\widehat{Z}_\theta$ in the case of periodic solutions. This latter equation and (12.2.9) can be characterized as the Lagrange necessary condition for the constrained variational principle: find critical points of H restricted to level sets of B with k as Lagrange multiplier. A family of RE is said to be non-degenerate if

$$\mathscr{B}_k \neq 0. \qquad (12.2.10)$$

A typical $(k, \mathscr{B}(k))$ diagram is shown in Figure 12.1. Points away from the critical points $\mathscr{B}_k = 0$ are non-degenerate. It will be shown that the critical points are also interesting, as they generate a higher-order modulation equation.

In order to be as general as possible, the family of RE will be denoted by the first representation $\widehat{Z}(\theta, k)$, and the tangent vector at any point denoted by $\widehat{Z}_\theta$. This representation includes the case where $\widehat{Z}(\theta, k)$ is a family of periodic

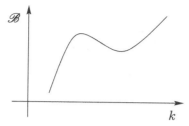

Figure 12.1. Typical $(k, \mathcal{B})$ Diagram.

solutions of (12.2.1) of period $2\pi/k$. The most general form of the invariant B evaluated on a family of RE is

$$\mathcal{B}(k) = \tfrac{1}{2} \langle\langle \mathbf{J}\widehat{Z}_\theta, \widehat{Z}\rangle\rangle \,. \tag{12.2.11}$$

The derivatives of this functional will play a prominent role in the modulation theory. The first and second derivatives with respect to k are

$$\mathcal{B}_k = \langle\langle \mathbf{J}\widehat{Z}_\theta, \widehat{Z}_k\rangle\rangle \,, \tag{12.2.12}$$

and

$$\mathcal{B}_{kk} = \langle\langle \mathbf{J}\widehat{Z}_\theta, \widehat{Z}_{kk}\rangle\rangle + \langle\langle \mathbf{J}\widehat{Z}_{k\theta}, \widehat{Z}_k\rangle\rangle \,. \tag{12.2.13}$$

12.3 Modulation of RE

If the non-degeneracy condition (12.2.10) is satisfied, the appropriate modulation ansatz is

$$Z(x) = \widehat{Z}(\theta + \phi, k + \varepsilon q) + \varepsilon^2 W(\theta, X, \varepsilon) \,, \tag{12.3.1}$$

with $\phi(X, \varepsilon)$, $q(X, \varepsilon)$, $X = \varepsilon x$. Substitution of (12.3.1) into (12.2.1), expansion of everything in powers of ε, and equating terms proportion to $\varepsilon^0, \varepsilon^1, \varepsilon^2$ to zero, shows that ϕ and q satisfy the trivial equations

$$\phi_X = q \quad \text{and} \quad \mathcal{B}_k q_X = 0 \,, \tag{12.3.2}$$

or, since $\mathcal{B}_k \neq 0$,

$$\begin{bmatrix} 0 & -1 \\ 1 & 0 \end{bmatrix} \begin{pmatrix} \phi \\ q \end{pmatrix}_X = \begin{pmatrix} 0 \\ q \end{pmatrix} \,.$$

It is just an ODE version of the Whitham modulation theory. The theory gets more interesting when the non-degeneracy condition (12.2.10) is *not* satisfied. In this case the proposed modulation ansatz is revised to

$$Z(x) = \widehat{Z}(\theta + \varepsilon\phi, k + \varepsilon^2 q) + \varepsilon^3 W(\theta, X, \varepsilon) \,, \tag{12.3.3}$$

with

$$X = \varepsilon x \quad \text{and} \quad q = \phi_X \,. \tag{12.3.4}$$

This ansatz is justified *a posteriori* by the fact that a necessary condition for this scaling is $\mathscr{B}_k = 0$. Note that both the phase and the wavenumber k are modulated. They could be treated independently, but for simplicity they are taken to be *a priori* related as in the second expression in (12.3.4).

The aim is to show that the generalisation of (12.3.2) is

$$\phi_X = q \quad \text{and} \quad \mathscr{B}_{kk}\, q q_X + \mathscr{K}\, q_{XXX} = 0, \tag{12.3.5}$$

where $\mathscr{B}_{kk}$ is the second derivative (12.2.13) and $\mathscr{K}$ is defined below as part of the derivation.

The strategy is to substitute (12.3.3) into (12.2.1), expand all terms in Taylor series in ε and then solve order by order. The Taylor expansion of $\widehat{Z}(\theta + \varepsilon\phi, k + \varepsilon^2 q)$ is

$$\widehat{Z}(\theta + \varepsilon\phi, k + \varepsilon^2 q) = \widehat{Z} + \varepsilon\phi\widehat{Z}_\theta + \tfrac{1}{2}\varepsilon^2\left(\phi^2\widehat{Z}_{\theta\theta} + 2q\widehat{Z}_k\right)$$

$$+ \frac{1}{3!}\varepsilon^3\left(\phi^3\widehat{Z}_{\theta\theta\theta} + 6q\phi\widehat{Z}_{k\theta}\right)$$

$$+ \frac{1}{4!}\varepsilon^4\left(\phi^4\widehat{Z}_{\theta\theta\theta\theta} + 12q\phi^2\widehat{Z}_{k\theta\theta} + 12q^2\widehat{Z}_{kk}\right)$$

$$+ \frac{1}{5!}\varepsilon^5\left(\phi^5\widehat{Z}_{\theta\theta\theta\theta\theta} + 20q\phi^3\widehat{Z}_{k\theta\theta\theta} + 60\phi q^2\widehat{Z}_{\theta kk}\right) + \cdots. \tag{12.3.6}$$

The substitution of all terms produces lengthy expansions, and just a summary is given here. The ε^0 equation just reproduces the governing equation for the RE

$$k\mathbf{J}\widehat{Z}_\theta = \nabla H(\widehat{Z}), \tag{12.3.7}$$

and the ε^1 equation just reproduces the tangent equation

$$\phi D^2 H(\widehat{Z})\widehat{Z}_\theta - k\phi\mathbf{J}\widehat{Z}_{\theta\theta} = 0. \tag{12.3.8}$$

Define the linear operator

$$\mathbf{L} := D^2 H(\widehat{Z}) - k\mathbf{J}\frac{d}{d\theta}, \tag{12.3.9}$$

with

$$\text{Ker}(\mathbf{L}) = \text{coKer}(\mathbf{L}) = \text{span}\{\widehat{Z}_\theta\}. \tag{12.3.10}$$

It is assumed that the kernel is no larger. Since $\mathbf{L}$ is symmetric the solvability condition for $\mathbf{L}U = V$ is $\langle\langle\widehat{Z}_\theta, V\rangle\rangle = 0$. With the operator $\mathbf{L}$, equation (12.3.8) can be written in the succinct form

$$\phi\mathbf{L}\widehat{Z}_\theta = 0, \tag{12.3.11}$$

which is identically satisfied. At second order the terms are

$$\tfrac{1}{2}\phi^2\left[\mathbf{L}\widehat{Z}_{\theta\theta} + D^3 H(\widehat{Z})(\widehat{Z}_\theta, \widehat{Z}_\theta)\right] + q\left[\mathbf{L}\widehat{Z}_k - \mathbf{J}\widehat{Z}_\theta\right] = (\phi_X - q)\mathbf{J}\widehat{Z}_\theta.$$

The first term in brackets vanishes because it is the exact equation for $\widehat{Z}_{\theta\theta}$ obtained by differentiating $\mathbf{L}\widehat{Z}_\theta = 0$, the second term in brackets is the exact equation for $\widehat{Z}_k$, obtained by differenting (12.3.7) with respect to k. The right hand side vanishes due to the hypothesis (12.3.4). Alternatively, the vanishing of the right hand side could be taken as confirming that $q = \phi_X$ ($\mathbf{J}\widehat{Z}_\theta$ can not vanish or else $\widehat{Z}$ would be an equilibrium; see (12.3.7)).

At third order the equations start to get more interesting as a solvability condition enters. Expand W in (12.3.3) in a Taylor series

$$W(\theta, X, \varepsilon) = W_1(\theta, X) + \varepsilon W_2(\theta, X) + \varepsilon^2 W_3(\theta, X) + \cdots.$$

Then the ε^3 equation, after eliminating the terms which vanish identically, is

$$\mathbf{L}W_1 = q_X \mathbf{J}\widehat{Z}_k, \qquad (12.3.12)$$

This is the same equation that arises in the expansion (12.3.2) where there solvability requires $q_X = 0$, but here the condition $\mathscr{B}_k = 0$ will make it non-trivially solvable.

The equation (12.3.12) is solvable if and only if

$$0 = \langle\!\langle \widehat{Z}_\theta, \mathbf{J}\widehat{Z}_k \rangle\!\rangle = -\langle\!\langle \mathbf{J}\widehat{Z}_\theta, \widehat{Z}_k \rangle\!\rangle = -\mathscr{B}_k,$$

using (12.2.12). The condition $\mathscr{B}_k = 0$ therefore shows up in a natural way as a solvability condition. To solve the equation (12.3.12) the *Jordan chain* is required. The required Jordan chain theory is given in §12.3.1. With this theory it is immediate that the solution for W_1 is

$$W_1 = q_X \xi_3 + \alpha \xi_1, \qquad (12.3.13)$$

where at this stage α is an arbitrary function of X, and ξ_1, ξ_3 are defined in the next subsection.

12.3.1 Jordan Chain Theory

For a linear operator $\mathbf{A}$ acting on a finite-dimensional space, a Jordan chain of length J associated with an eigenvalue λ_0 has the form

$$[\mathbf{A} - \lambda_0\mathbf{I}]\xi_1 = 0, \quad [\mathbf{A} - \lambda_0\mathbf{I}]\xi_j = \xi_{j-1}, \quad j = 2, \cdots, J.$$

In this case the geometric multiplicity (dimension of the kernel of $[\mathbf{A} - \lambda_0\mathbf{I}]$) is one, and the algebraic multiplicity is J. The chain continues as long as ξ_{j-1} is in the range of $[\mathbf{A} - \lambda_0\mathbf{I}]$; that is, it is solvable [18]. For the solvability condition the adjoint Jordan chain is needed. The theory is essentially the same for linear operators on a function space with appropriate modification [4].

The situation here is simplified by the fact that $\mathbf{A}$ is the product of an invertible skew-symmetric operator, $\mathbf{J}$, and a symmetric operator, $\mathbf{L}$:

$$\mathbf{A} = \mathbf{J}^{-1}\mathbf{L},$$

and the eigenvalue $\lambda_0 = 0$. Hence the Jordan chain is of the form

$$\mathbf{L}\xi_1 = 0, \quad \mathbf{L}\xi_j = \mathbf{J}\xi_{j-1}, \quad j = 2, \ldots, J. \qquad (12.3.14)$$

Since $\xi_1 = \widehat{Z}_\theta$ and the kernel is one-dimensional (12.3.10), the equation is solvable if and only if the right hand side is orthogonal to ξ_1. Let $\xi_2 = \widehat{Z}_k$, the the Jordan chain is generically of length two:

$$\mathbf{L}\xi_1 = 0 \quad \text{and} \quad \mathbf{L}\xi_2 = \mathbf{J}\xi_1. \qquad (12.3.15)$$

The chain has length 3 if the equation

$$\mathbf{L}\xi_3 = \mathbf{J}\xi_2, \qquad (12.3.16)$$

is solvable. This equation is solvable if $\langle\!\langle \xi_1, \mathbf{J}\xi_2 \rangle\!\rangle = 0$. But

$$0 = \langle\!\langle \xi_1, \mathbf{J}\xi_2 \rangle\!\rangle = -\langle\!\langle \mathbf{J}\widehat{Z}_\theta, \widehat{Z}_k \rangle\!\rangle = -\mathscr{B}'(k).$$

Hence with this condition the Jordan chain has length 3, and there exists a solution ξ_3 (although one may not be able to compute it explicitly). The chain has length 4 if the equation

$$\mathbf{L}\xi_4 = \mathbf{J}\xi_3, \qquad (12.3.17)$$

is solvable. This equation is solvable if $\langle\!\langle \xi_1, \mathbf{J}\xi_3 \rangle\!\rangle = 0$. A calculation shows that this equation is always satisfied (this result also follows from the fact that the Hamiltonian structure assures that the algebraic multiplicity is always even). The chain has length 5 if the equation

$$\mathbf{L}\xi_5 = \mathbf{J}\xi_4, \qquad (12.3.18)$$

is solvable, which requires $\langle\!\langle \xi_1, \mathbf{J}\xi_4 \rangle\!\rangle = 0$. Hence for the chain to terminate at 4 it is required that

$$\mathscr{K} := \langle\!\langle \mathbf{J}\xi_1, \xi_4 \rangle\!\rangle \neq 0. \qquad (12.3.19)$$

The parameter $\mathscr{K}$ will appear in the modulation theory.

12.3.2 Fourth- and Fifth-Order Terms

The fourth-order equation, after eliminating the terms, which vanish identically, is

$$\mathbf{L}W_2 = \mathbf{J}(W_1)_X + \phi q_X \mathbf{J}\widehat{Z}_{k\theta} - \phi D^3 H(\widehat{Z})(\widehat{Z}_\theta, W_1). \qquad (12.3.20)$$

Substitute for W_1

$$\mathbf{L}W_2 = q_{XX}\mathbf{J}\xi_3 + \alpha_X\mathbf{J}\xi_1 + \phi q_X\mathbf{J}\widehat{Z}_{k\theta} - \phi q_X D^3 H(\widehat{Z})(\widehat{Z}_\theta, \xi_3) - \phi\alpha D^3 H(\widehat{Z})(\widehat{Z}_\theta, \widehat{Z}_\theta).$$

Now, ξ_3 satisfies

$$D^2 H(\widehat{Z})\xi_3 - k\mathbf{J}(\xi_3)_\theta = \mathbf{J}\widehat{Z}_k.$$

Differentiate with respect to θ

$$\mathbf{L}(\xi_3)_\theta + D^3 H(\widehat{Z})(\widehat{Z}_\theta, \xi_3) = \mathbf{J}\widehat{Z}_{k\theta} \,.$$

Substitute into the W_2 equation,

$$\mathbf{L}W_2 = q_{XX}\mathbf{L}\xi_4 + \alpha_X\mathbf{L}\xi_2 + \phi q_X\mathbf{L}(\xi_3)_\theta + \phi\alpha\mathbf{L}\widehat{Z}_{\theta\theta} \,.$$

In the first term, the fourth element in the Jordan chain (§12.3.1) has been used. From this expression the solution for W_2 is immediate

$$W_2 = q_{XX}\xi_4 + \alpha_X\xi_2 + \phi q_X(\xi_3)_\theta + \phi\alpha\widehat{Z}_{\theta\theta} + \beta\xi_1 \,, \qquad (12.3.21)$$

where at this stage β is an arbitrary function of X.

At fifth order, after eliminating terms, which vanish identically, the terms are

$$\mathbf{L}W_3 = qq_X\mathbf{J}\widehat{Z}_{kk} + \mathbf{J}(W_2)_X - qD^3 H(\widehat{Z})(\widehat{Z}_k, W_1) \,.$$

Substitute for W_1 and W_2 and neglect terms that vanish identically in the solvability condition, leaving

$$-\mathbf{L}W_3 = qq_X\left(-\mathbf{J}\widehat{Z}_{kk} + D^3 H(\widehat{Z})(\widehat{Z}_k, \xi_3) - \mathbf{J}(\xi_3)_\theta\right) - q_{XXX}\mathbf{J}\xi_4 \,, \qquad (12.3.22)$$

where the minus sign has been added for convenience.

Using $\mathscr{K}$ defined in (12.3.19) in §12.3.1, solvability gives

$$\mathsf{c}\,qq_X + \mathscr{K}\,q_{XXX} = 0 \,,$$

with

$$\mathsf{c} = \langle\langle \widehat{Z}_\theta, -\mathbf{J}\widehat{Z}_{kk} + D^3 H(\widehat{Z})(\widehat{Z}_k, \xi_3) - \mathbf{J}(\xi_3)_\theta \rangle\rangle \,. \qquad (12.3.23)$$

It remains to prove that this coefficent is indeed $\mathsf{c} = \mathscr{B}_{kk}$ as in (12.3.5).

First, differentiate $\mathbf{L}\widehat{Z}_\theta = 0$ with respect to k,

$$D^3 H(\widehat{Z})(\xi_2, \xi_1) + \mathbf{L}\widehat{Z}_{\theta k} = \mathbf{J}\widehat{Z}_{\theta\theta} \,.$$

Now use this in the third term in c,

$$
\begin{aligned}
-\langle\langle \widehat{Z}_\theta, \mathbf{J}(\xi_3)_\theta \rangle\rangle &= -\langle\langle \mathbf{J}\widehat{Z}_{\theta\theta}, \xi_3 \rangle\rangle \\
&= -\langle\langle D^3 H(\widehat{Z})(\xi_2, \xi_1) + \mathbf{L}\widehat{Z}_{\theta k}, \xi_3 \rangle\rangle \\
&= -\langle\langle D^3 H(\widehat{Z})(\xi_2, \xi_1), \xi_3 \rangle\rangle - \langle\langle \mathbf{L}\widehat{Z}_{\theta k}, \xi_3 \rangle\rangle \\
&= -\langle\langle D^3 H(\widehat{Z})(\xi_2, \xi_3), \xi_1 \rangle\rangle - \langle\langle \widehat{Z}_{\theta k}, \mathbf{J}\xi_2 \rangle\rangle \,,
\end{aligned}
$$

using permutation invariance of the triple product and the Jordan chain. Substitute into c

$$
\begin{aligned}
\mathsf{c} &= \langle\!\langle \widehat{Z}_\theta, -\mathbf{J}\widehat{Z}_{kk} + D^3 H(\widehat{Z})(\widehat{Z}_k, \xi_3) - \mathbf{J}(\xi_3)_\theta \rangle\!\rangle \\
&= \langle\!\langle \mathbf{J}\widehat{Z}_\theta, \widehat{Z}_{kk} \rangle\!\rangle + \langle\!\langle \widehat{Z}_\theta, D^3 H(\widehat{Z})(\xi_2, \xi_3) \rangle\!\rangle - \langle\!\langle \widehat{Z}_\theta, \mathbf{J}(\xi_3)_\theta \rangle\!\rangle \\
&= \langle\!\langle \mathbf{J}\widehat{Z}_\theta, \widehat{Z}_{kk} \rangle\!\rangle + \langle\!\langle \mathbf{J}\widehat{Z}_{\theta k}, \widehat{Z}_k \rangle\!\rangle \\
&= \mathscr{B}_{kk},
\end{aligned}
$$

with the last equality following from (12.2.13). Hence the governing equation, to leading order, for modulation of RE is (12.3.5).

Remarkably, even though this modulation equation arises from the simple ODE (12.2.1) it is the steady KdV equation. Moreover the coefficient of the nonlinearity is determined by the curvature of the mapping $\mathscr{B}(k)$ evaluated on the branch of RE.

The modulation equation (12.3.5) is unfamiliar in the dynamical systems literature. However, it becomes more familiar by integrating (12.3.5) and writing it as a first-order system

$$
\begin{aligned}
-\mu_X &= 0 \\
-p_X &= \mu - \tfrac{1}{2}\mathscr{B}_{kk}q^2 \\
\phi_X &= q \\
q_X &= sp, \quad s = -\mathscr{K}.
\end{aligned}
\tag{12.3.24}
$$

This system is the normal form for the linearization about a quadruple zero eigenvalue with symmetry in Hamiltonian systems. It is derived using classical nonlinear normal form theory in [5]. It is a Hamiltonian system on $\mathbb{R}^4$ with Hamiltonian function

$$
H(Z) = \mu q + \tfrac{1}{2}sp^2 - \frac{1}{6}\mathscr{B}_{kk}q^3,
$$

and $Z = (\phi, q, \mu, p)$. It has a one-parameter affine symmetry: $\phi \mapsto \phi + \theta$ for all $\theta \in \mathbb{R}$.

12.3.3 Dynamical Systems Interpretation of $\mathscr{K}$

An analysis of the system (12.3.24) shows up some interesting solution properties. RE of this system are reduced versions of the RE of the original system. RE of the reduced system are of the form: $\phi(X) = cX + \phi_0$. Substitution into (12.3.24) shows that there are 4 classes depending on the sign of s and $\kappa = \mathscr{B}_{kk}$, and they are shown in Figure 12.2. The branches are labelled with h (hyperbolic) or $e^\pm$ (elliptic with Krein signature ± 1). See Appendix C in [6] for a definition of Krein signature in this case. The elliptic-hyperbolic

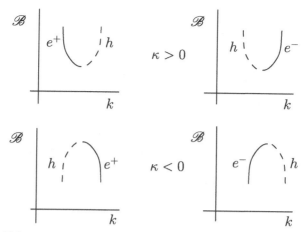

Figure 12.2. Four Classes of RE in the Normal Form (12.3.24).

classification is obtained by linearizing (12.3.24) about the RE and calculating eigenvalues. A branch is hyperbolic (elliptic) if eigenvalues are real (purely imaginary). There are two cases for $\kappa > 0$ and two cases for $\kappa < 0$.

12.4 Hamiltonian PDEs

The theory for modulation of RE extends in a natural way to PDEs. We will need a canonical form for PDEs in order to derive general results. The starting point for conservative PDEs is a Lagrangian. It is conceivable that the theory can be derived directly from a Lagrangian. For example, the Whitham theory works in the Lagrangian context. However we found from the analysis of the ODE case that the advantage of the Hamiltonian form is that the additional structure allows for a precise connection between the tangent vector to the RE and the conserved quantity (12.2.7).

Start with a *Lagrangian* formulation for some PDE,

$$\mathscr{L}(U) = \int \int L(U_t, U_x, U) \, dxdt.$$

Legendre transform $V = \delta L / \delta U_t$, giving a *Hamiltonian formulation*

$$\widehat{\mathscr{L}}(W) = \int \int \left[\tfrac{1}{2} \langle \mathbf{M} W_t, W \rangle - H(W_x, W) \right] dxdt,$$

with new coordinates $W = (U, V)$. The density is still the same Lagrangian, but the advantage is that it has been split into two parts: a Hamiltonian function, which is scalar valued, and a part defined by a symplectic (or pre-symplectic) form with operator $\mathbf{M}$.

Legendre transform again $P = \delta L/\delta W_x$, giving a *multisymplectic Hamiltonian* formulation

$$\widetilde{\mathscr{L}}(Z) = \int\int \left[\tfrac{1}{2}\langle \mathbf{M}Z_t, Z\rangle + \tfrac{1}{2}\langle \mathbf{J}Z_x, Z\rangle - S(Z)\right]\mathrm{d}x\mathrm{d}t, \qquad (12.4.1)$$

with new coordinates $Z = (U, V, P)$. The density is again the same Lagrangian in terms of new coordinates, but now it is split into three parts: a new Hamiltonian function $S(Z)$, which does not contain any derivatives with respect to t or x, and two symplectic (or pre-symplectic) structures [7, 20]. The principal advantage of the multisymplectic structure in the context of modulation is that the symplectic structures appear in the equations, and relate linearly the tangent vector to the group to the components of a conservation law.

The above sequence of Legendre transforms is schematic, as in general non-degeneracy conditions are required, and each PDE has to be treated with care. An example of the above sequence of Legendre transforms is given in the subsection 12.4.2.

12.4.1 Multisymplectic Noether Theory

It is assumed henceforth that the Lagrangian is in the canonical form (12.4.1) – a multisymplectic Hamiltonian PDE – and therefore the Euler Lagrange equation is

$$\mathbf{M}Z_t + \mathbf{J}Z_x = \nabla S(Z), \quad Z \in \mathbb{R}^4. \qquad (12.4.2)$$

The matrices $\mathbf{M}$ and $\mathbf{J}$ are constant skew-symmetric matrices, and $S(Z)$ is a given scalar-valued function of Z only. It is assumed that $\mathbf{J}$ is non-degenerate, but $\mathbf{M}$ is not necessarily non-degenerate. For simplicity, the phase space is restricted to $\mathbb{R}^4$.

The PDE (12.4.2) is assumed to be equivariant with respect to a one-parameter Lie group with action G_θ:

$$G_\theta \mathbf{M} = \mathbf{M}G_\theta, \quad G_\theta \mathbf{J} = \mathbf{J}G_\theta \quad \text{and} \quad S(G_\theta Z) = S(Z). \qquad (12.4.3)$$

By multisymplectic Noether theory (cf. Appendix of [7]), there exists functions $A : \mathbb{R}^4 \to \mathbb{R}$ and $B : \mathbb{R}^4 \to \mathbb{R}$ satisfying

$$\nabla A(Z) = \mathbf{M}\mathfrak{g}(Z) \quad \text{and} \quad \nabla B(Z) = \mathbf{J}\mathfrak{g}(Z). \qquad (12.4.4)$$

It is immediate from (12.4.4) and invariance of S in (12.4.3) that A and B are the components of a conservation law since

$$
\begin{aligned}
A_t + B_x &= \langle \nabla A, Z_t \rangle + \langle \nabla B, Z_x \rangle \\
&= \langle \mathbf{M}\mathfrak{g}(Z), Z_t \rangle + \langle \mathbf{J}\mathfrak{g}(Z), Z_x \rangle \\
&= -\langle \mathfrak{g}(Z), \mathbf{M}Z_t \rangle - \langle \mathfrak{g}(Z), \mathbf{J}Z_x \rangle \\
&= -\langle \mathfrak{g}(Z), \nabla S(Z) \rangle \\
&= -\frac{d}{d\theta} S(G_\theta Z)\bigg|_{\theta=0} = 0.
\end{aligned}
$$

12.4.2 Example: a Boussinesq Model for Shallow Water Waves

An example from the theory of water waves which illustrates the transformation from a Lagrangian to a multisymplectic Hamiltonian formulation is the Boussinesq model

$$
\begin{aligned}
h_t + (hu)_x &= 0 \\
u_t + uu_x + gh_x &= \tau h_{xxx},
\end{aligned}
\tag{12.4.5}
$$

where, for the gravity water-wave problem,

$$
\tau = -\frac{1}{3}gh_o^2.
\tag{12.4.6}
$$

$h(x,t)$ is the surface elevation, $u(x,t)$ is the vertical average of the horizontal velocity, g is the gravitational constant and h_0 is a reference depth. The system (12.4.5) is derived in Chapter 5 of DINGEMANS [8]. The system (12.4.5) with negative τ is ill-posed but it is valid for long waves, which is the case where it will be needed here.

Introduce a velocity potential $u = \psi_x$. Then the equations can be written in the form

$$
\begin{aligned}
-h_t - (h\psi_x)_x &= 0 \\
\psi_t + gh &= \tau h_{xx} + R - \tfrac{1}{2}\psi_x^2
\end{aligned}
\tag{12.4.7}
$$

where R is the Bernoulli function. They are generated by the Lagrangian functional

$$
\mathscr{L}(h, \psi) = \int_{t_1}^{t_2} \int_{x_1}^{x_2} L(h, h_x, \psi_x, \psi_t)\, dx dt,
$$

where x_1, x_2 are arbitrary values of x associated with fixed endpoint variations, and

$$
L(h, h_x, \psi_x, \psi_t) = h\psi_t + \tfrac{1}{2}h\psi_x^2 + \tfrac{1}{2}\tau h_x^2 + \tfrac{1}{2}gh^2 - Rh.
$$

The time-derivative term is already in canonical form, so a time-direction Legendre transform is not necessary. Take a Legendre transform with respect to the x–direction

$$p = \frac{\partial \mathcal{L}}{\partial \psi_x} = h\psi_x \quad \text{and} \quad w = \frac{\partial \mathcal{L}}{\partial h_x} = \tau h_x.$$

Hence the fully Legendre transformed system is

$$\mathcal{L}(Z, Z_x, Z_t) = \tfrac{1}{2}\langle \mathbf{M}Z_t, Z \rangle + \tfrac{1}{2}\langle \mathbf{J}Z_x, Z \rangle - S(Z), \qquad (12.4.8)$$

with

$$Z = \begin{pmatrix} \psi \\ h \\ p \\ w \end{pmatrix}, \quad \mathbf{M} = \begin{pmatrix} 0 & -1 & 0 & 0 \\ 1 & 0 & 0 & 0 \\ 0 & 0 & 0 & 0 \\ 0 & 0 & 0 & 0 \end{pmatrix}, \quad \mathbf{J} = \begin{pmatrix} 0 & 0 & -1 & 0 \\ 0 & 0 & 0 & -1 \\ 1 & 0 & 0 & 0 \\ 0 & 1 & 0 & 0 \end{pmatrix}, \qquad (12.4.9)$$

and

$$S(Z) = Rh - \tfrac{1}{2}gh^2 + \frac{p^2}{2h} + \frac{1}{2\tau}w^2. \qquad (12.4.10)$$

The Hamiltonian function $S(Z)$ does not depend explicitly on ψ and so there is an affine symmetry of the form (12.2.3)

$$G_\theta Z = Z + \theta\eta, \quad \text{with} \quad \eta = \begin{pmatrix} 1 \\ 0 \\ 0 \\ 0 \end{pmatrix}. \qquad (12.4.11)$$

The associated conservation law is $A_t + B_x = 0$ with

$$\nabla A(Z) = \mathbf{M}\eta \quad \text{and} \quad \nabla B(Z) = \mathbf{J}\eta, \qquad (12.4.12)$$

which can be integrated to give $A(Z) = \langle \mathbf{M}\eta, Z \rangle = h$ and $B(Z) = \langle \mathbf{J}\eta, Z \rangle = h\psi_x$, showing that the affine symmetry (12.4.11) generates the mass conservation law.

Modulation of the RE of the Boussinesq model will lead to the classical KdV equation in shallow water. Before considering this example further, the general theory of modulation leading to KdV is developed.

12.5 Emergence of KdV

The modulation theory for Hamiltonian PDEs closely follows that for ODEs. The main new result is that the modulation equation (12.3.5) now includes a time derivative term

$$2\mathscr{A}_k q_T + \mathscr{B}_{kk} qq_X + \mathscr{K} q_{XXX} = 0, \qquad (12.5.1)$$

and the coefficient of the time-derivative term comes from the density A, evaluated on the family of RE, of the conservation law for which B is the flux. In this section the argument is sketched that leads from (12.3.5) to (12.5.1). A detailed derivation is given in [9].

The starting point is again RE. It is assumed that the steady part of (12.4.2) has a RE. Since the steady problem reduces exactly to (12.2.1), the theory for RE is exactly the same as in §12.2. The basic state is

$$Z(x,t) = \widehat{Z}(\theta, k), \quad \theta = kx + \theta_0. \tag{12.5.2}$$

This RE can be generalized to be a space-time RE, taking $\theta = kx + \omega t + \theta_0$, and the implications of this generalization are discussed in §12.6.

The case $\mathscr{B}_k \neq 0$ leads to the dispersionless Whitham equations (see §12.7). In order to get the KdV equation via modulation, the assumption $\mathscr{B}_k = 0$ will be necessary. Hence, the generalization of the modulation ansatz (12.3.3) is

$$Z(x,t) = \widehat{Z}(\theta + \varepsilon\phi, k + \varepsilon^2 q) + \varepsilon^3 W(\theta, X, T, \varepsilon), \tag{12.5.3}$$

where

$$q(X,T,\varepsilon) = \phi_X(X,T,\varepsilon), \quad \text{and} \quad X = \varepsilon x, \quad T = \varepsilon^3 t.$$

Everything is expanded in a Taylor series and substituted into the Euler-Lagrange equation (12.4.2), and then solved order by order. The basic linear operator (12.3.9) is the same as in §12.3. Similarly, the solvability condition is the same, and the order $\varepsilon^j, j = 0, 1, 2, 3$ equations are also exactly the same. The first new term appears at order ε^4. The W_2 equation in (12.3.20) is replaced by

$$\mathbf{L}W_2 = \phi_T \mathbf{M}\widehat{Z}_\theta + \mathbf{J}(W_1)_X + \phi q_X \mathbf{J}\widehat{Z}_{k\theta} - \phi D^3 H(\widehat{Z})(\widehat{Z}_\theta, W_1). \tag{12.5.4}$$

The new term is solvable ($\mathbf{M}\widehat{Z}_\theta$ is in the range of $\mathbf{L}$) and so the new W_2 solution is

$$W_2 = \phi_T \zeta + q_{XX}\xi_4 + \alpha_X \xi_2 + \phi q_X(\xi_3)_\theta + \phi a \widehat{Z}_{\theta\theta} + \beta \xi_1, \tag{12.5.5}$$

where ζ is the solution of $\mathbf{L}\zeta = \mathbf{M}\widehat{Z}_\theta$. The coefficients α and β and the eigenfunctions $\xi_1, \ldots, \xi_4$ are the same as in §12.2.

At fifth order the equation (12.3.22) is replaced by

$$-\mathbf{L}W_3 = -q_T(\mathbf{M}\widehat{Z}_k + \mathbf{J}\zeta) + qq_X\left(-\mathbf{J}\widehat{Z}_{kk} + D^3 H(\widehat{Z})(\widehat{Z}_k, \xi_3) - \mathbf{J}(\xi_3)_\theta\right) - q_{XXX}\mathbf{J}\xi_4, \tag{12.5.6}$$

Application of the solvability condition then produces (12.3.5) with an additional term

$$a q_T + \mathscr{B}_{kk} qq_X + \mathscr{K} q_{XXX} = 0, \tag{12.5.7}$$

where

$$a = -\langle\!\langle \widehat{Z}_\theta, \mathbf{M}\widehat{Z}_k + \mathbf{J}\zeta \rangle\!\rangle.$$

Now,

$$\langle\!\langle \widehat{Z}_\theta, \mathbf{J}\zeta \rangle\!\rangle = -\langle\!\langle \mathbf{J}\widehat{Z}_\theta, \zeta \rangle\!\rangle = -\langle\!\langle \mathbf{L}\widehat{Z}_k, \zeta \rangle\!\rangle = -\langle\!\langle \widehat{Z}_k, \mathbf{L}\zeta \rangle\!\rangle$$
$$= -\langle\!\langle \widehat{Z}_k, \mathbf{M}\widehat{Z}_\theta \rangle\!\rangle = \langle\!\langle \mathbf{M}\widehat{Z}_k, \widehat{Z}_\theta \rangle\!\rangle,$$

and so

$$a = -\langle\langle \widehat{Z}_\theta, \mathbf{M}\widehat{Z}_k + \mathbf{J}\zeta \rangle\rangle = 2\langle\langle \mathbf{M}\widehat{Z}_\theta, \widehat{Z}_k \rangle\rangle = 2\langle\langle \nabla A(\widehat{Z}), \widehat{Z}_k \rangle\rangle = 2\mathscr{A}_k.$$

This completes the derivation of (12.5.1). The function $q(X, T, \varepsilon)$ depends on ε, and so the q appearing here should be considered as the first term in a Taylor expansion; that is, $q := q(X, T, 0)$.

12.5.1 Example: Application to Shallow Water Waves

In this subsection the theory for emergence of KdV is applied to the shallow water Boussinesq equation in §12.4.2.

The family of RE associated with the symmetry (12.4.11) is

$$\widehat{Z}(\theta, k) = \widehat{z}(k) + \theta\eta, \quad \widehat{z}(k) = \begin{pmatrix} 0 \\ h_0 \\ p_0 \\ 0 \end{pmatrix}, \quad \text{with} \quad \theta = kx + \theta_0,$$

and k represents the background velocity of the uniform flow. This family of RE satisfies

$$\nabla S(\widehat{z}) = k\nabla B(\widehat{z}). \tag{12.5.8}$$

Writing out (12.5.8) gives $w = 0$, $p = h_0 k$ and $h_0 = \frac{1}{g}\left(R - \frac{1}{2}k^2\right)$, where R (the Bernoulli constant) is the specified value of total head.

The components of the conservation law evaluated on the family of RE are

$$\mathscr{A}(k) = h_0(k) = \frac{1}{g}\left(R - \frac{1}{2}k^2\right) \quad \text{and} \quad \mathscr{B}(k) = kh_0(k) = \frac{k}{g}\left(R - \frac{1}{2}k^2\right).$$

The necessary condition for the emergence of KdV is

$$0 = \mathscr{B}'(k) \quad \Rightarrow \quad R = \frac{3}{2}k^2 \quad \Rightarrow \quad gh_0 + \frac{1}{2}k^2 = \frac{3}{2}k^2 \quad \Rightarrow \quad gh_0 = k^2,$$

which is the usual condition for criticality. By defining the Froude number $F = |k|/\sqrt{gh_0}$, the "Froude number unity" condition is recovered.

Differentiating the conservation laws further

$$\mathscr{A}'(k) = -\frac{k}{g} \quad \text{and} \quad \mathscr{B}''(k) = -\frac{3}{g}k.$$

Hence the KdV equation should be of the form

$$-2\frac{k}{g}q_T - \frac{3}{g}kqq_X + \mathscr{K}q_{XXX} = 0.$$

To determine $\mathcal{K}$ linearize (12.5.8) and solve for the Jordan chain $\{\xi_1, \xi_2, \xi_3, \xi_4\}$ in §12.3.1. The result is

$$\xi_1 = \begin{pmatrix} 1 \\ 0 \\ 0 \\ 0 \end{pmatrix}, \quad \xi_2 = -\frac{h_0}{k}\begin{pmatrix} 0 \\ 1 \\ 0 \\ 0 \end{pmatrix}, \quad \xi_3 = -\frac{\tau h_0}{k}\begin{pmatrix} 0 \\ 0 \\ 0 \\ 1 \end{pmatrix}, \quad \xi_4 = -\frac{\tau h_0^2}{k^2}\begin{pmatrix} 0 \\ 1/k \\ 1 \\ 0 \end{pmatrix}.$$

Hence

$$\mathcal{K} = \langle\langle \mathbf{J}\xi_1, \xi_4 \rangle\rangle = -\tau\frac{h_0^2}{k^2} = \frac{1}{3}h_0^3,$$

and so the emergent form for KdV is

$$-2\frac{k}{g}q_T - \frac{3}{2}kqq_X + \frac{1}{3}h_0^3 q_{XXX} = 0.$$

or

$$q_T + \frac{3}{2}qq_X - \frac{1}{6}kh_0^2 q_{XXX}, \tag{12.5.9}$$

which is precisely the KdV equation in shallow water (e.g., equation (6.9c) on page 693 of DINGEMANS [8] with $k \mapsto -k$ noting that both signs of k are admissible). In summary, the classic KdV in shallow water arises due to the modulation of degenerate RE, and the RE in this case are uniform flows in shallow water at criticality.

12.5.2 Example: KdV for Deep Water Waves

An example showing that shallow water is neither necessary nor sufficient for the emergence of the KdV equation as a model in the theory of water waves is to look at the case of infinite depth when the surface tension coefficient is nonzero. In this case, the nonlinear Schrödinger (NLS) equation is defocussing if the surface tension coefficient is large enough. The form of the 1+1 NLS in deep water is

$$ic_0 A_\tau + c_1 A_{\xi\xi} = c_2|A|^2 A,$$

where τ and ξ are slow time and space variables, the coefficients c_0, c_1, c_2 are all real, and $A(\xi, \tau)$ is a complex amplitude function. Precise expressions for the coefficients are given in equation (2.20) of [10]. When the surface tension coefficient is large enough,

$$\tilde{\sigma} := \frac{\sigma a^2}{g} > -1 + \frac{2}{\sqrt{3}} \approx 0.155, \tag{12.5.10}$$

the coefficients c_1 and c_2 are positive so the 1+1 NLS equation is defocussing. In (12.5.10) σ is the dimensional coefficient of surface tension and a is the wavenumber about which the NLS equation is derived. The value of surface tension in (12.5.10) is not large: for comparison, Wilton ripples, which have been experimentally observed, arise near the larger value of $\widetilde{\sigma} = 0.5$.

There is a well-known reduction from NLS to KdV in the defocussing case (see [6] and references therein). Here it is shown that it emerges via the modulation of degenerate RE.

Scaling τ, ξ, and A, the defocussing 1+1 NLS can be put in the canonical form

$$i\Psi_t + \Psi_{xx} + \Psi - |\Psi|^2\Psi = 0, \tag{12.5.11}$$

where x, y, and t are used for the scaled independent variables so as not to confuse with the slow time and space variables associated with the KP reduction.

The PDE (12.5.11) is the Euler-Lagrange equation for an associated Lagrangian functional; the precise form for the Lagrangian can be written down but is not needed. The system has a natural S^1 symmetry of the form (12.2.4); that is, $e^{i\theta}\Psi$ is a solution whenever Ψ is a solution, for any θ. The conservation law associated with this symmetry is of the form $A_t + B_x = 0$ with

$$A = \tfrac{1}{2}|\Psi|^2 \quad \text{and} \quad B = \mathrm{Im}\big(\overline{\Psi}\Psi_x\big). \tag{12.5.12}$$

The background state, which is an RE associated with the S^1 symmetry, is $\widehat{\Psi}(\theta,k) = \Psi_0(k)e^{i\theta}$, with $\theta = kx + \theta_0$. Substitution into (12.5.11) shows that

$$k^2 + |\Psi_0|^2 = 1.$$

The components of the conservation law (12.5.12) evaluated on the background state are

$$\mathscr{A}(k) = \tfrac{1}{2}(1 - k^2) \quad \text{and} \quad \mathscr{B}(k) = k(1 - k^2).$$

The necessary condition for emergence of KdV is

$$0 = \mathscr{B}_k = 1 - 3k \quad \Rightarrow \quad k^2 = \frac{1}{3}.$$

The condition $\mathscr{B}_k = 0$ is a generalization of criticality to the finite-amplitude periodic state.

Differentiating further

$$\mathscr{A}_k = -k \quad \text{and} \quad \mathscr{B}_{kk} = -6k.$$

The value of $\mathscr{K} = -\tfrac{1}{2}$ is calculated in the 1+1 NLS to KdV reduction in [9]. Substituting the coefficients into the form (12.5.1), gives the emergent KdV equation

$$-2kq_T - 6kqq_X - \tfrac{1}{2}q_{XXX} = 0, \tag{12.5.13}$$

noting that $k = \pm\frac{\sqrt{3}}{3}$ at criticality. Further detail on this example can be found in [6].

This example serves two purposes: it shows that KdV is not restricted to a model for shallow water waves, and secondly it shows how the KdV equation can emerge from a finite-amplitude state.

12.6 Generalizing Relative Equilibria

In extending from the ODE setting in §12.2 to the PDE setting in §12.4 the form of the RE stayed the same. What happens when the RE is generalized to be a space-time RE?

The extension of the RE to be space time simply amounts to replacing θ by

$$\theta = kx + \omega t + \theta_0, \tag{12.6.1}$$

and the basic state is modified to $\widehat{Z}(\theta, k, \omega)$. The modulation ansatz (12.5.3) is modified to

$$Z(x,t) = \widehat{Z}(\theta + \varepsilon\phi, k + \varepsilon^2 q, \omega + \varepsilon^4\Omega) + \varepsilon^3 W(\theta, X, T, \varepsilon). \tag{12.6.2}$$

The ε^4 term for modulation of ω is needed in order to achieve balance in the conservation of waves: $q_T = \Omega_X$.

One then proceeds with the modulation, and Taylor expansions, as before. The key change is that the emergent KdV equation has the new form

$$(\mathscr{A}_k + \mathscr{B}_\omega)q_T + \mathscr{B}_{kk}qq_X + \mathscr{K} q_{XXX} = 0. \tag{12.6.3}$$

However, because both $\mathscr{A}$ and $\mathscr{B}$ come from a Lagrangian ($\mathscr{A} = \mathscr{L}_\omega$ and $\mathscr{B} = \mathscr{L}_k$), it follows that $\mathscr{B}_\omega = \mathscr{A}_k$. Hence (12.6.3) reduces exactly to (12.5.1). This argument also explains where the 2 comes from in (12.5.1). It also shows a closer connection with Whitham Modulation Theory.

12.7 Connection with Whitham Modulation Theory

Although Whitham theory uses the classical method of multiple scales [1, 2] it can be reconfigured using the modulation ansatz (12.5.3) with the Whitham scaling.

Given an RE of the form

$$Z(x,t) = \widehat{Z}(\theta, \omega, k), \quad \theta = kx + \omega t + \theta_0,$$

which is $2\pi-$periodic in θ, associated with a conservation law $A_t + B_x = 0$, introduce the modulation

$$Z(x,t) = \widehat{Z}(\theta + \phi, \omega + \varepsilon\Omega, k + \varepsilon q) + \varepsilon^2 W(\theta, X, T, \varepsilon),$$

with ϕ, q, Ω functions of X, T, ε and scaling $T = \varepsilon t$ and $X = \varepsilon x$.

Expand everything in a Taylor series and substitute into the standard form (12.4.2). Zeroth order generates the equation for the RE, first order confirms the identites $q = \phi_X$ and $\Omega = \phi_T$. At second order the dispersionless Whitham modulation equations are generated,

$$\mathscr{A}_\omega \Omega_T + (\mathscr{A}_k + \mathscr{B}_\omega) q_T + \mathscr{B}_k q_X = 0. \tag{12.7.1}$$

When $\mathscr{B}_k = 0$ this equation degenerates, and can be immediately integrated. At degeneracy, change the scaling and use the ansatz

$$Z(x,t) = \widehat{Z}(\theta + \varepsilon \phi, k + \varepsilon^2 q, \omega + \varepsilon^4 \Omega) + \varepsilon^3 W(\theta, X, T, \varepsilon), \tag{12.7.2}$$

with $T = \varepsilon^3 t$ and $X = \varepsilon x$ and $\mathscr{B}_k = 0$ gives

$$(\mathscr{A}_k + \mathscr{B}_\omega) q_T + \mathscr{B}_{kk} q q_X + \mathscr{K} q_{XXX} = 0. \tag{12.7.3}$$

The comparison with (12.7.1) is striking. In going from (12.7.1) to (12.7.3) the higher-order scaling in (12.7.2) forces the Ω_T term to drop out, the condition $\mathscr{B}_k = 0$ leads to the addition of the $\mathscr{B}_{kk} q q_X$ term, but the surprising result is the generation of dispersion with the new q_{XXX} term. Details of the morphing of the generic conservation of wave action (12.7.1) into the KdV equation (12.7.3) are given in [11].

12.7.1 Dual KdV when $\mathscr{A}_\omega = 0$

There is a dual version of the above theory when space and time are reversed and $\mathscr{A}_\omega = 0$ resulting in a KdV equation in Ω. Replace the condition $\mathscr{B}_k = 0$ with

$$\mathscr{A}_\omega = 0 \quad \text{but} \quad \mathscr{B}_k \neq 0.$$

Switch the scalings to

$$X = \varepsilon^3 x \quad \text{and} \quad T = \varepsilon t,$$

and switch the scalings on Ω and q in order to balance conservation of waves. Then re-doing the above argument in §12.5 shows that the conservation of wave action morphs into the Ω−KdV equation

$$(\mathscr{A}_k + \mathscr{B}_\omega) \Omega_X + \mathscr{A}_{\omega\omega} \Omega \Omega_T + \mathscr{K} \Omega_{TTT} = 0, \tag{12.7.4}$$

where $\mathscr{K}$ is determined by a Jordan chain argument associated with the degeneracy $\mathscr{A}_\omega = 0$. This version of the KdV equation is less interesting as the initial-value problem is unstable. Nevertheless it appears in applications.

The most well known example of the singularity $\mathscr{A}_\omega = 0$ is the super-harmonic instability of water waves (e.g. SAFFMAN [12], JANSSEN [13], KATAOKA [14] and references therein). A graph of wave action $\mathscr{A}$ versus ω (or energy versus ω) for a Stokes wave in infinite depth (in finite depth it is shown by KATAOKA [14] that mean flow enters so the argument is slightly different) shows that there is a maximum at $\omega = \omega^{crit}$ and an exchange from stability to instability when $\omega > \omega^{crit}$. JANSSEN [13] has carried Whitham modulation theory to higher order in this case to predict the exchange of stability. In [11] it is shown how the linearization of (12.7.4) about a basic state representing a finite amplitude Stokes wave captures the instability argument for superharmonic (SH) instability in [13].

12.8 Generalization to 2+1: the KP Equation

The generalization to two space dimensions and time follows the same strategy as in §12.4. An additional Legendre transform is introduced in the second space direction generating a third symplectic structure. The Euler-Lagrange equation (12.4.2) is extended to

$$\mathbf{M}Z_t + \mathbf{J}Z_x + \mathbf{K}Z_y = \nabla S(Z), \quad Z \in \mathbb{R}^n, \tag{12.8.1}$$

with $\mathbf{K}$ a third constant skew-symmetric matrix. In $2 + 1$ dimensions, the dimension n of the phase space will be greater than 4. The PDE is again assumed to be equivariant with respect to a one-parameter Lie group as in (12.4.3) with additionally $G_\theta \mathbf{K} = \mathbf{K}G_\theta$, and with the conservation law associated with the symmetry now having two fluxes

$$A_t + B_x + C_y = 0, \tag{12.8.2}$$

with $C(Z)$ related to the group generator by $\mathbf{K}\mathfrak{g}(Z) = \nabla C(Z)$. The RE needs to be extended to a two-parameter RE $\widehat{Z}(\theta, k, \ell)$ with $\theta = kx + \ell y + \theta_0$. Keep the same scaling in x, t: $X = \varepsilon x$ and $T = \varepsilon^3 t$. There are various options for the y scaling, but *a posteriori* it is found that a scaling $Y = \varepsilon^2 y$ leads to the KP equation. Hence the proposed modulation ansatz is

$$Z(x, y, t) = \widehat{Z}(\theta + \varepsilon\phi, k + \varepsilon^2 q, \ell) + \varepsilon^3 W(\theta, X, Y, T, \varepsilon), \tag{12.8.3}$$

with ϕ, q now dependent on X, Y, T, and ε and $q = \phi_X$.

Expand everything in Taylor series and substitute. The expansions and order by order equations are very similar to the $1 + 1$ case with minor changes showing up at fourth and fifth order. For example, there are now two necessary conditions,

$$\mathscr{B}_k = 0 \quad \text{and} \quad \mathscr{C}_k = 0,$$

where $\mathscr{C}(k, \ell)$ is $C(Z)$ evaluated on the family of RE. At fifth order the solvability condition gives the KP equation

$$2\mathscr{A}_k q_T + \mathscr{B}_{kk} q q_X + \mathscr{K} q_{XXX} + \mathscr{C}_\ell p_Y = 0, \quad p_X = q_Y. \tag{12.8.4}$$

Applying this theory to uniform flows in shallow water hydrodynamics shows that it recovers the familiar KP equation in the theory of water waves [19].

12.9 Concluding Remarks

The emphasis in these notes has been on conservative systems. However, the modulation strategy works equally well in the non-conservative setting. Indeed, one of the motivations for the theory was the modulation of periodic travelling waves of reaction-diffusion equations leading to a reduced Burgers equation (DOELMAN ET AL [15]).

Another area of interest is mixed systems, where the time-dependence is nonconservative, but the steady equation is conservative; examples are the real Ginzburg-Landau equation and the Swift-Hohenberg equation. For these models, roll solutions can be modulated leading to a gradient-Boussinesq model [16] and KdV planforms [17]. KdV planforms are solutions of the KdV equation where the evolution is in a spatial direction, and so the solution leads to localized and multi-pulse patterns in the plane.

Two interesting directions for further study are modulation in higher space dimension and modulation of multidimensional RE. The KP modulation in §12.8 is a step in the direction of higher space dimension. However, the modulation leading to KP only modulated the k direction, and not the ℓ direction. Higher dimensional RE are of great practical interest and so modulation would be of interest. Multidimensional RE have multiple phases and so are of the form

$$\widehat{Z}(\theta_1, \ldots, \theta_n, k_1, \ldots, k_n, \ell_1, \ldots, \ell_n, \cdots),$$

and so modulation $k_j \mapsto k_j + \varepsilon^p q_j$, with potentially additional modulation of the ℓ_j, will lead to more complex PDEs for the wavenumber modulation.

Acknowledgments

Many thanks to Claudia Wulff and Daniel Ratliff for reading and commenting on the manuscript. This work was partially supported by EPSRC grant EP/K008188/1.

References

[1] G.B. WHITHAM. *Linear and Nonlinear Waves*, Wiley-Interscience: New York (1974).

[2] G.B. WHITHAM. *Two-timing, variational principles and waves*, J. Fluid Mech. **44** 373–395 (1970).

[3] J.E. MARSDEN. *Lectures on Mechanics*, London Mathematical Society Lecture Notes **174**, Cambridge University Press.

[4] A.E. TAYLOR & D.C. LAY. *Introduction to Functional Analysis*, Krieger Publishers (1986).

[5] T.J. BRIDGES & N.M. DONALDSON. *Degenerate periodic orbits and homoclinic torus bifurcation*, Phys. Rev. Lett. **95** 104301 (2005).

[6] T.J. BRIDGES. *Emergence of unsteady dark solitary waves from coalescing spatially-periodic patterns*, Proc. Roy. Soc. Lond. A **468** 3784–3803 (2012).

[7] T.J. BRIDGES. *Multi-symplectic structures and wave propagation*, Math. Proc. Camb. Phil. Soc. **121** 147–190 (1997).

[8] M.W. DINGEMANS. *Water wave propagation over uneven bottoms. Part 2 – Non-linear wave propagation*, World Scientific Publisher: Singapore (1997).

[9] T.J. BRIDGES. *A universal form for the emergence of the Korteweg-de Vries equation*, Proc. Roy. Soc. Lond. A **469** 20120707 (2013).

[10] V.D. DJORDJEVIC & L.G. REDEKOPP. *On two-dimensional packets of capillary-gravity waves*, J. Fluid Mech. **79** 703–714 (1977).

[11] T.J. BRIDGES. *Breakdown of the Whitham modulation theory and the emergence of dispersion*, Stud. Appl. Math. doi: 10.1111/sapm.12086 (2015).

[12] P.G. SAFFMAN. *The superharmonic instablility of finite-amplitude water waves*, J. Fluid Mech. **159** 169–174 (1985).

[13] P.A.E.M. JANSSEN. *Stability of steep gravity waves and the average Lagrangian method*, KNMI Preprint (1989).

[14] T. KATAOKA. *On the superharmonic instability of surface gravity waves on fluid of finite depth*, J. Fluid Mech. **547** 175–184 (2006).

[15] A. DOELMAN, B. SANDSTEDE, A. SCHEEL & G. SCHNEIDER. *The dynamics of modulated wave trains*, AMS Memoirs **934**, American Mathematical Society: Providence (2009).

[16] T.J. BRIDGES. *Bifurcation from rolls to multi-pulse planforms via reduction to a parabolic Boussinesq model*, Physica D **275** 8–18 (2014).

[17] T.J. BRIDGES. *Dimension breaking from spatially-periodic patterns to KdV planforms*, J. Dyn. Diff. Eqns. DOI: 10.1007/s10884-014-9405-y (2014).

[18] J. WILKENING. *An algorithm for computing Jordan chains and inverting analytic matrix functions*, Lin. Alg. Appl. **427** 6–25 (2007).

[19] T.J. BRIDGES. *Emergence of dispersion in shallow water hydrodynamics via modulation of uniform flow*, J. Fluid Mech. **761** R1-R9 (2014).

[20] T.J. BRIDGES, P.E. HYDON, & J.K. LAWSON. *Multisymplectic structures and the variational bicomplex*, Math. Proc. Camb. Phil. Soc. **148**, 159–178 (2010).